# Damage, Fracture, and Fatigue of Ceramic-Matrix Composites

Longbiao Li

# Damage, Fracture, and Fatigue of Ceramic-Matrix Composites

Springer

Longbiao Li
Nanjing University of Aeronautics
and Astronautics
Nanjing, Jiangsu, China

ISBN 978-981-13-1782-8    ISBN 978-981-13-1783-5 (eBook)
https://doi.org/10.1007/978-981-13-1783-5

Library of Congress Control Number: 2018952884



This Springer imprint is published by the registered company Springer Nature Singapore Pte Ltd.
The registered company address is: 152 Beach Road, #21-01/04 Gateway East, Singapore 189721, Singapore

*To Shengning*

# Contents

# Chapter 1
# Tensile Behavior of Ceramic-Matrix Composites

## 1.1 Introduction

Continuous fiber-reinforced ceramic-matrix composites (CMCs) possess high specific strength and specific modulus, and high toughness at elevated temperatures [1]. The non-oxide and oxide CMCs are recently being incorporated in gas turbine engines for high-pressure and high-temperature section components and exhaust nozzles. However, the complexity and variability of aerospace ceramic processing methods, compositions and microstructures, the relatively low fracture toughness of the ceramic materials, still remain the challenging factors for CMCs component design, validation, life prediction, and thus broader applications [2].

Under tensile loading of fiber-reinforced CMCs, the failure strain of the matrix is less than that of the fibers. When the stress in the matrix approaches to its local strength, matrix cracking occurs. With increasing of applied stress, matrix crackings will deflect along the fiber/matrix interface, leading to the interface debonding and sliding [3]. The fiber/matrix interface frictional shear stress will transfer loads between the fibers and the matrix, and the interface properties (i.e., the interface shear stress and the interface debonded energy) affect the nonlinear behavior of fiber-reinforced CMCs [4, 5]. When matrix cracking approaches to saturation, the fiber/matrix interface debonding may continue with increasing applied stress, and after completely interface debonding, the matrix stress will remain constant [6]. The additional load will be carried by intact fibers, and when the broken fibers approach to the critical value, the composite fracture occurs [7].

In this chapter, the micromechanical approach to predict the tensile stress–strain curves of fiber-reinforced CMCs is developed. When matrix cracking, fiber/matrix interface debonding and fibers failure occur, the shear-lag model is adopted to analyze the microstress field of the damaged fiber-reinforced CMCs, i.e., the fiber and matrix axial stress distributions. Combining the shear-lag model with damage models of matrix statistical cracking, fracture mechanics fiber/matrix interface debonding criterion and Global Load Sharing (GLS) fibers failure criterion, the



L. Li, *Damage, Fracture, and Fatigue of Ceramic-Matrix Composites*,
https://doi.org/10.1007/978-981-13-1783-5_1

matrix cracking spacing, fiber/matrix interface debonding length and fibers broken fraction are determined. The tensile stress–strain curves of fiber-reinforced CMCs corresponding to different damage stages are modeled. The tensile stress–strain curves of unidirectional, cross-ply, 2D, and 2.5D woven CMCs are predicted.

## 1.2 Unidirectional Ceramic-Matrix Composites

In this section, the tensile stress–strain behavior of unidirectional fiber-reinforced CMCs at room temperature is investigated. An approach to model the tensile stress–strain curve of unidirectional fiber-reinforced CMCs considering different damage mechanisms of matrix multicracking, fiber/matrix interface debonding, and fibers failure is developed. The tensile stress–strain curves of unidirectional SiC/Calcium Aluminosilicate (SiC/CAS), SiC/CAS-II (CAS-II is the Corning designation for the calcium aluminosilicate matrix in this composite system) and SiC/Borosilicate composites corresponding to different damage stages are predicted.

### *1.2.1 Stress Analysis*

Under tensile loading, matrix cracking and fiber/matrix interface debonding occur, leading to the nonlinear behavior of fiber-reinforced CMCs. When matrix cracking and fiber/matrix interface debonding occur, the shear-lag model is used to analyze the microstress field of the damaged composite.

Cox [8] first introduced the concept of shear-lag in the micromechanical analysis. The purpose of the introduction of the shear-lag concept is that the stress and strain fields of the composite material are not solved concretely, and the main characteristics of the material structure are considered, and the response of the material structure to the load is calculated by constructing a mathematical model. The model simplifies the mechanical analysis of composite materials and provides directional theoretical guidance for interpreting experimental data and designing material structures with better damage tolerance. However, the shear-lag model developed by Cox [8] only considers the stress distribution after single fiber fracture in the elastic matrix and does not consider the stress distribution of other adjacent fibers, so the stress concentration problem cannot be analyzed.

Hedgepeth and Dyke [9] expanded the shear-lag model developed by Cox [8], investigated the stress distribution of an unidirectional fiber-reinforced composite with multiple fibers failure, and predicted the stress concentration caused by multiple fibers fracture. The main assumption of the shear-lag model is that the fiber is one-dimensional axial stress transfer entity; the fiber is only subjected to tensile stress; the displacement is only along the longitudinal direction; the fiber is arranged at the same distance; the interface between the fiber and matrix is strong bonding; the matrix cannot transfer axial stress, and only the shear stress is transferred; the interfacial shear strength is constant. The shear-lag model can better describe the stress concentration around fibers broken in unidirectional composite with lower matrix tensile modulus and higher fibers volume fraction.

Dyke and Hedgepeth [10] investigated the effect of plastic deformation and the fiber/matrix interface debonding on the stress concentration of single fiber broken. Zweben [11] proposed an approximate analysis method to analyze the stress distribution in the unidirectional fiber-reinforced composite with a long cracking perpendicular to the fiber axial under tensile loading, considering the matrix inelastic effect before the cracking.

The shear-lag models mentioned above predict the stress field distribution of fiber-reinforced soft matrix composites (i.e., the fiber elastic modulus is far greater than that of the matrix). With the appearance of high temperature-resistant composite, i.e., metal and ceramic matrix composites, the difference of elastic modulus between the matrix and the fiber is small. The classical shear-lag theory cannot accurately describe the microstress field of the damaged composite. Budiansky et al. [12] considered the effect of matrix shear deformation on the matrix axial stress distribution, modified the classic shear-lag theory, and analyzed the microstress field of fiber-reinforced CMCs after multiple damage.

In the present analysis, the Budiansky–Hutchinson–Evans shear-lag model [12] is adopted to analyze the microstress field inside of damaged fiber-reinforced CMCs. The unit cell is extracted from the fiber-reinforced CMCs, as shown in Fig. 1.1. The unit cell contains a single fiber surrounded by a hollow cylinder of matrix. The fiber radius is $r_f$ and the matrix radius is $R$ ($R = r_f/V_f^{1/2}$). The length of the unit cell is half of the matrix crack spacing, i.e., $l_c/2$. The fiber/matrix interface debonded length is $l_d$. The unit cell can be divided into two regions, i.e., the interface debonded region ($x \in [0, l_d]$) and interface bonded region ($x \in [l_d, l_c/2]$). On the matrix crack plane, the fibers carry all the stress ($\sigma/V_f$), where $\sigma$ denotes far-field applied stress and $V_f$ denotes the fiber volume fraction. In the Budiansky–Hutchinson–Evans shear-lag model, it is assumed that the matrix axial stress is

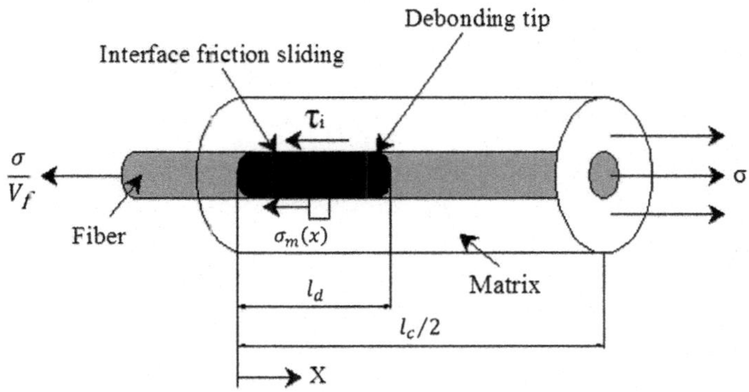


**Fig. 1.1** Unit cell of Budiansky–Hutchinson–Evans shear-lag model. Reprinted with permission from Ref. [5]. Copyright 2009, Springer Science Business Media New York

concentrated at $\overline{R}$ and the region between $r_f$ and $\overline{R}$ carries only the shear stress. The expression of $\overline{R}$ is given by the following equation [12].

$$\ln\left(\frac{\overline{R}}{r_f}\right) = -\frac{2\ \ln\ V_f + V_m(3 - V_f)}{4V_m^2} \tag{1.1}$$

where $V_m$ denotes the volume fraction of the matrix ($V_m = 1 - V_f$).

In the fiber/matrix interface debonded region ($x \in [0,\ l_d]$), the fiber/matrix interface shear stress transfers the stress between the fibers and the matrix. Consider the equilibrium of the axial force acting on the element of length d$x$ in the debonded fiber, leads to the following differential equation.

$$\frac{d\sigma_f}{dx} = -\frac{2\tau_i(x)}{r_f} \tag{1.2}$$

The axial stresses of fiber and matrix satisfy the following equation.

$$V_f\sigma_f(x) + V_m\sigma_m(x) = \sigma \tag{1.3}$$

By assuming the fiber/matrix interface shear stress in the interface debonded region is constant, the fiber and matrix axial stress can be solved by Eqs. (1.2) and (1.3) with the boundary conditions of $\sigma_f(x = 0) = \sigma/V_f$ and $\sigma_m(x = 0) = 0$. The fiber, matrix, and interface shear stress in the fiber/matrix interface debonded region are determined by the following equations.

$$\sigma_f(x) = \frac{\sigma}{V_f} - \frac{2\tau_i}{r_f}x, \quad x \in (0, l_d) \tag{1.4}$$

$$\sigma_m(x) = \frac{V_f}{V_m}\frac{2\tau_i}{r_f}x, \quad x \in (0, l_d) \tag{1.5}$$

$$\tau_i(x) = \tau_i, \quad x \in (0, l_d) \tag{1.6}$$

In the fiber/matrix interface bonded region ($x \in [l_d,\ l_c/2]$), considering the equilibrium of the radial force acting on the differential element d$x$(d$r$)$r$ dθ in the domain $r_f < r < \overline{R}$, it leads to the following differential equation.

$$\frac{\partial\tau_{rx}}{\partial r} + \frac{\tau_{rx}}{r} = 0 \tag{1.7}$$

As the matrix in the domain $r_f < r < \overline{R}$ carries only the shear stress, the stress–strain relation is determined by the following equation.

$$\tau_{rx} = G_m\frac{\partial w}{\partial r} \tag{1.8}$$

where $G_m$ denotes the matrix shear modulus and $w$ is the axial displacement.

Solving Eq. (1.7), the shear stress $\tau_{rx}$ is described by the following equation.

$$\tau_{rx}(r,x) = \frac{r_f \tau_i(x)}{r} \tag{1.9}$$

Substituting Eq. (1.9) into Eq. (1.8), the fiber/matrix interface shear stress in the fiber/matrix interface bonded region is described using the following equation.

$$\tau_i(x) = \frac{G_m(w_m - w_f)}{r_f \ln(\bar{R}/r_f)} \tag{1.10}$$

where $w_f = w(r_f, x), w_m = w(\overline{R}, x)$ denote the fiber and matrix axial displacement, respectively, and satisfy the following differential equations.

$$\frac{dw_f}{dx} = \frac{\sigma_f(x) - \sigma_{ft}}{E_f} \tag{1.11}$$

$$\frac{dw_m}{dx} = \frac{\sigma_m(x) - \sigma_{mt}}{E_m} \tag{1.12}$$

where $E_f$ and $E_m$ denote the elastic modulus of the fiber and matrix; and $\sigma_{ft}$ and $\sigma_{mt}$ denote the thermal residual stress of the fiber and the matrix, respectively.

$$\sigma_{ft} = E_f(\alpha_c - \alpha_f)\Delta T \tag{1.13}$$

$$\sigma_{mt} = E_m(\alpha_c - \alpha_m)\Delta T \tag{1.14}$$

where $\alpha_f$, $\alpha_m$ and $\alpha_c$ denote the fiber, matrix, and composite thermal expansion coefficient; and $\Delta T$ denotes the temperature difference between the fabricated temperature $T_0$ and testing temperature $T_1 (\Delta T = T_1 - T_0)$.

Substituting Eqs. (1.10)–(1.14) into Eq. (1.2) and requiring the fiber axial stress continuity at $x = l_d$, it leads to the fiber and matrix axial stresses in the fiber/matrix interface bonded region. The fiber, matrix, and interface shear stress in the fiber/matrix interface bonded region can be described using the following equations.

$$\sigma_f(x) = \sigma_{fo} + \left(\frac{V_m}{V_f}\sigma_{mo} - 2\frac{l_d}{r_f}\tau_i\right)\exp\left(-\rho\frac{x - l_d}{r_f}\right), \quad x \in (l_d, l_c/2) \tag{1.15}$$

$$\sigma_m(x) = \sigma_{mo} - \left(\sigma_{mo} - 2\tau_i\frac{V_f}{V_m}\frac{l_d}{r_f}\right)\exp\left[-\frac{\rho(x - l_d)}{r_f}\right], \quad x \in (l_d, l_c/2) \tag{1.16}$$

$$\tau_i(x) = \frac{\rho}{2}\left(\frac{V_m}{V_f}\sigma_{mo} - 2\tau_i\frac{l_d}{r_f}\right)\exp\left[-\frac{\rho(x - l_d)}{r_f}\right], \quad x \in (l_d, l_c/2) \tag{1.17}$$

where $\rho$ denotes the shear-lag parameter [12].

$$\rho^2 = \frac{4E_c G_m}{V_m E_m E_f \varphi} \tag{1.18}$$

where

$$\varphi = -\frac{2\ln V_f + V_m(3 - V_f)}{2V_m^2} \tag{1.19}$$

$\sigma_{fo}$ and $\sigma_{mo}$ denote the far-field fiber and matrix axial stress in the fiber/matrix interface bonded region, respectively, which are also the fiber and matrix axial stress when there exists no damage in the composite.

$$\sigma_{fo} = \frac{E_f}{E_c}\sigma + E_f(\alpha_c - \alpha_f)\Delta T \tag{1.20}$$

$$\sigma_{mo} = \frac{E_m}{E_c}\sigma + E_m(\alpha_c - \alpha_m)\Delta T \tag{1.21}$$

where $E_c$ denotes the composite elastic modulus. The axial elastic modulus of the composite is approximated by the rule of mixture.

$$E_c = V_f E_f + V_m E_m \tag{1.22}$$

When the fiber fails, the intact fiber axial stress distributions in the fiber/matrix interface debonded region and bonded region are determined using the following equation.

$$\sigma_f(x) = \begin{cases} T - \frac{2\tau_i}{r_f}x\,, & x \in (0, l_d) \\ \sigma_{fo} + \left(T - \sigma_{fo} - 2\frac{l_d}{r_f}\tau_i\right)\exp\left(-\rho\frac{x - l_d}{r_f}\right), & x \in (l_d, l_c/2) \end{cases} \tag{1.23}$$

where $T$ denotes the intact fiber axial stress at the matrix crack plane.

The axial stress distributions of fiber, matrix, and fiber/matrix interface shear stress of fiber-reinforced CMCs in the interface debonded region and interface bonded region are shown in Fig. 1.2.

## *1.2.2 Damage Models*

### 1.2.2.1 Matrix First Cracking

According to the internal flaw size of the matrix, the matrix crackings can be divided into the long matrix cracking and short matrix cracking. When the matrix

**Fig. 1.2** Axial stress distributions of the fiber, matrix, and interface shear stress in the interface debonded region and interface bonded region

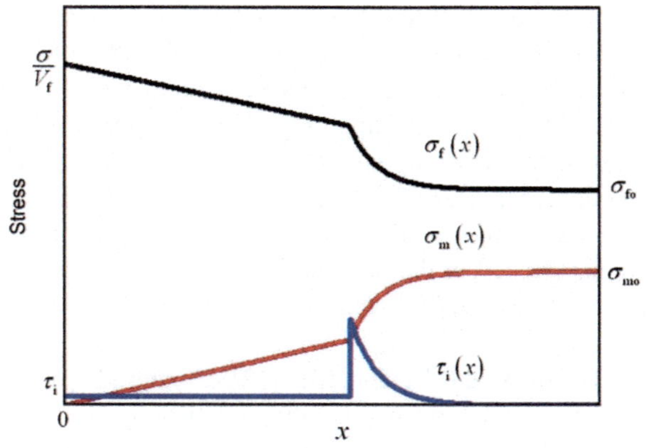


internal flaw is greater than the characteristic value of $c_o$, the first matrix cracking stress of $\sigma_{mc}$ is independent of the internal flaw inside of the matrix. During the propagation of matrix cracking, the stress field remains unchanged before and after the crack tip, and the matrix cracking propagates steadily. When the flaw size inside of matrix is less than the characteristic value of $c_o$, the first matrix cracking stress is affected by the size of matrix flaw and the cracking propagation process is unsteady. The characteristic matrix flaw size of $c_o$ can be described using the following equation [13].

$$c_{\mathrm{o}} = \frac{1}{2}(\zeta_{\mathrm{m}})^{1/3}\left(\frac{V_{\mathrm{m}} r_{\mathrm{f}} E_{\mathrm{c}}\sqrt{E_{\mathrm{m}}}}{V_{\mathrm{f}}^2 \tau_{\mathrm{i}} E_{\mathrm{f}}}\right)^{2/3} \tag{1.24}$$

Aveston et al. [14] investigated the first matrix cracking problem based on the assumption of fiber/matrix interface frictional bonding. The fiber/matrix interface debonding occurs during the propagation of matrix steady-state cracking, and the fiber/matrix interface shear stress is assumed to be constant. The energy balance approach is used to determine the steady-state first matrix cracking stress, which can be described using the following equation.

$$\sigma_{\mathrm{cr}}^{\mathrm{ACK}} = \left(\frac{6V_{\mathrm{f}}^2 E_{\mathrm{f}} E_{\mathrm{c}}^2 \tau_{\mathrm{i}} \zeta_{\mathrm{m}}}{r_{\mathrm{f}} V_{\mathrm{m}} E_{\mathrm{m}}^2}\right)^{1/3} - E_{\mathrm{c}}(\alpha_{\mathrm{c}} - \alpha_{\mathrm{m}})\Delta T \tag{1.25}$$

where $\zeta_{\mathrm{m}}$ denotes the matrix fracture energy.

Aveston and Kelly [15] investigated the first matrix cracking stress of fiber-reinforced CMCs with the bonding fiber/matrix interface.

Budiansky et al. [12] investigated the first matrix cracking stress in fiber-reinforced CMCs for different fiber/matrix interface debonding conditions using the energy balance approach, i.e., (1) the fiber/matrix interface bonding with sliding; (2) the fiber/matrix interface frictional bonding with constant interface shear stress; and (3) the fiber/matrix interface initially bonding and debonding at high applied stress. When the fiber/matrix interface frictional bonding and the fiber/matrix interface shear stress is assumed to be constant, the first matrix crack stress predicted by BHE model the same with that of ACK model [14], i.e., $\sigma_{cr}^{BHE} = \sigma_{cr}^{ACK}$; when the matrix cracking propagation without fiber/matrix interface debonding, the first matrix cracking stress predicted by BHE model is the same with that of AK model [15], i.e., $\sigma_{cr}^{BHE} = \sigma_{cr}^{AK}$; and when the fiber/matrix interface debonding occurs during the propagation of first matrix cracking and the fiber/matrix interface frictional existed in the debonded region, the first matrix cracking stress predicted by BHE model is between the result of ACK model [14] and AK model [15], i.e., $\sigma_{cr}^{ACK} < \sigma_{cr}^{BHE} < \sigma_{cr}^{AK}$.

Based on the BHE model [12], Chiang [16] introduced the fracture mechanics fiber/matrix interface debonding criterion for investigating the first matrix cracking problem. It was found that the fiber/matrix interface debonding affects the first matrix cracking stress. When the fiber/matrix interface is the non-frictional interface, i.e., $\tau_i = 0$, the first matrix cracking stress is the same with the first fiber/matrix interface debonding stress; when the fiber/matrix interface is the frictional interface without chemical bonding, i.e., $\tau_i > 0$, at low fiber/matrix interface shear stress, the first matrix cracking stress predicted by Chiang is the same with the result of ACK model [14], i.e., $\sigma_{cr}^{Chiang} = \sigma_{cr}^{ACK}$; however, with increasing fiber/matrix interface shear stress, the first matrix cracking stress predicted by Chiang is lower than the result of ACK model [14], i.e., $\sigma_{cr}^{Chiang} < \sigma_{cr}^{ACK}$, as the ACK model did not consider the effect of matrix shear deformation on first matrix cracking stress of fiber-reinforced CMCs; and when the fiber/matrix interface is the frictional-bonding interface, the first matrix cracking stress increases with fiber/matrix interface shear stress and interface debonded energy, and when the fiber/matrix interface without debonding during matrix cracking propagation, the first matrix cracking stress predicted by Chiang is the same with the result of AK model [15], i.e., $\sigma_{cr}^{Chiang} = \sigma_{cr}^{AK}$.

Chiang [17] investigated the effect of fiber Poisson contraction on first matrix cracking stress of fiber-reinforced CMCs. Coulomb's frictional law is used to describe the fiber/matrix interface shear stress in the debonded region. With increasing fiber/matrix interface frictional coefficient, the first matrix cracking stress predicted by ACK model [14] and BHE model [12] are both higher than the result of Chiang model. In the analysis of ACK model [14] and BHE model [12], the fiber/matrix interface shear stress in the debonded region is assumed to be constant; however, in the analysis of Chiang model, the fiber/matrix interface shear stress changes along the interface debonded region considering fiber Poisson contraction effect.

Li [18–21] investigated the effect of fiber fracture and interface oxidation on the first matrix cracking stress of fiber-reinforced CMCs at elevated temperature. The relationships between the first matrix cracking stress, fiber/matrix interface debonding and sliding, fibers fracture, oxidation temperature and oxidation time have been established. The effects of fiber volume fraction, fiber/matrix interface properties, fibers strength, and oxidation temperatures on the evolution of first matrix cracking stress versus oxidation time have been analyzed.

Marshall et al. [13] investigated the short matrix first cracking problem in fiber-reinforced CMCs using the stress intensity factor method. The fiber/matrix interface shear stress in the debonded region is assumed to be constant, and the relationship between the fiber bridging traction stress of $p(x)$ and the matrix cracking opening displacement of $u(x)$ is obtained using the shear-lag analysis and combined with stress intensity factor method; the first matrix non-steady cracking stress can be described using the following equation.

$$\frac{\sigma_{\mathrm{cr}}^{\mathrm{MCE}}}{\sigma_{\mathrm{cr}}^{\mathrm{ACK}}} = 1 + \frac{1}{8}\left(\frac{a}{c_{\mathrm{o}}}\right)^{-3/4} \tag{1.26}$$

where $a$ denotes the initial matrix flaw length.

The relationship between the first matrix non-steady cracking stress predicted by MCE model and the matrix flaw length is shown in Fig. 1.3. With increasing of matrix flaw length, the first matrix cracking stress predicted by MCE model [13] approaches to the result of ACK model [14].

Based on the MCE model [13], Marshall and Cox considered the effect of fibers failure [22], residual stress [23], and specimen geometry [24, 25] on the first matrix non-steady cracking stress in fiber-reinforced CMCs.

Chiang et al. [26] obtained the relationship between the fiber bridging traction stress of $p(x)$ and matrix cracking opening displacement of $u(x)$ using the BHE shear-lag model. The first matrix non-steady cracking stress predicted using the stress intensity factor method is higher than the result of MCE model, and the characteristic matrix flaw length of $c_{\mathrm{o}}$ is also larger than that of MCE model.

Chiang [27] investigated the effect of fibers failure on short matrix first cracking problem by assuming the single fibers strength distribution and failing at the matrix cracking plane. The effect of fiber failure location on the fiber bridging traction

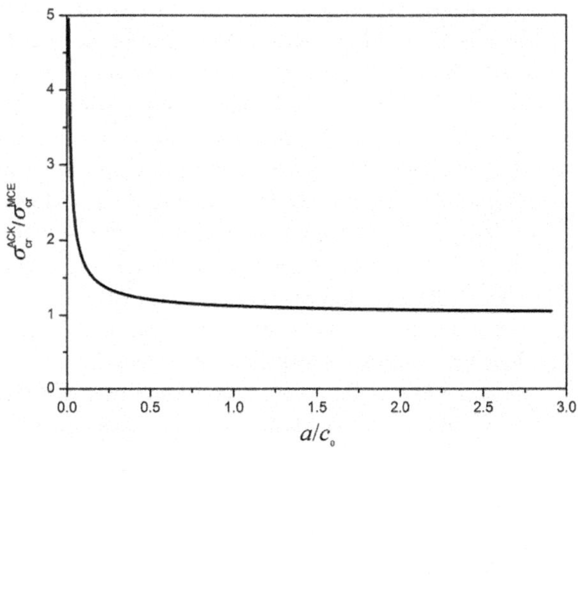


**Fig. 1.3** Relationship between the first matrix non-steady cracking stress predicted by MCE model and the matrix flaw length

stress and then the short matrix first cracking stress was not considered in the Chiang's model.

Thouless and Evans [28] investigated the fibers failure on the short matrix first cracking problem considering the effect of intact fibers, pullout fibers on matrix opening displacement, and the effect of fiber radius, fibers strength and fiber/matrix interface bonding strength on short matrix first cracking stress has been discussed.

Cao et al. [29] performed the monotonic tensile experiments on SiC/LAS composite, and it was found that the first matrix cracking stress agreed with the predicted result of ACK model [14], which is mainly due to the large flaw inside of matrix [30]. Kim and Pagano [31] monitored the first matrix cracking behavior of unidirectional SiC/CAS, SiC/1723, SiC/BMAS, and SiC/LAS-III composites using the acoustic emission (AE) and optical microscope observation. It was found that the first matrix cracking stress is much lower than the result predicted by ACK model [14]. The first matrix cracking occurs in the matrix-rich region with low fibers volume fraction. Barsoum et al. [32] found that the first matrix cracking depends on the fiber distribution inside of matrix. The microcrackings do not affect the modulus of the composite; however, with increasing applied stress, these microcrackings will propagate and evolve into the short cracking defined by MCE model [13] and then the long cracking defined by ACK model [14].

#### 1.2.2.2 Matrix Multicracking

When the applied stress is higher than the first matrix cracking stress, with continually increasing stress level, new matrix cracking will appear between the existed crackings. There are mainly four different approaches to predict the matrix multicracking in unidirectional fiber-reinforced CMCs, i.e., the maximum stress approach, energy balance approach, critical matrix strain energy approach and statistical approach.

(1) **The maximum stress approach**. Aveston et al. [14] first predicted the matrix multicracking evolution using the maximum stress theory. It is assumed that the matrix has a single strength, and the matrix cracking occurs when the matrix stress approaches to the matrix strength. The matrix cracking predicted using this method is saturated under a single stress, which cannot describe the gradual evolution of matrix cracking with increasing applied stress and the predicted tensile stress–strain curve appears step at the stress of matrix cracking. Lee and Daniel [33] combined the maximum stress theory and the modified shear-lag model, and considered the effect of fiber/matrix interface debonding and thermal residual stress on matrix multicracking evolution; however, the model still cannot describe the matrix cracking evolution with increasing applied stress and the predicted tensile stress–strain curve still existed the step behavior.
(2) **Energy balance approach**. Zok and Spearing [34] considered the effect of matrix internal flaw and interference between adjacent matrix crackings on the evolution of matrix multicracking and obtained the strain energy release rate for

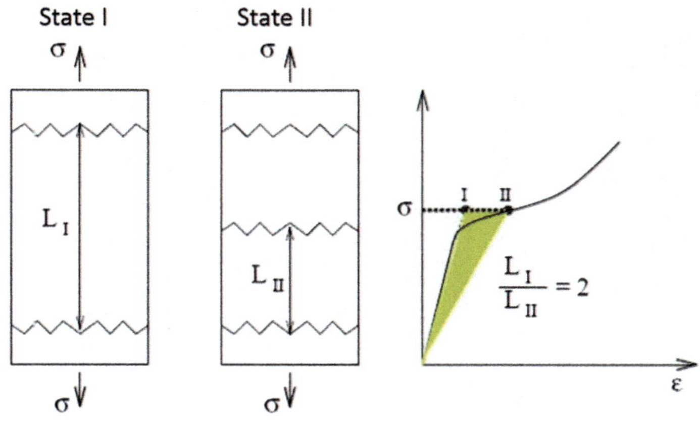


**Fig. 1.4** Matrix multicracking evolution

matrix statistical cracking using the energy balance approach. When the strain energy release rate at the matrix cracking tip approaches to the critical value, the matrix cracking propagation occurs. Weitsman and Zhu [35] analyzed the energy balance relationship during matrix multicracking for two different matrix cracking space, as shown in Fig. 1.4. The matrix crack spacing in state I is $L_{\mathrm{I}}$, and the matrix cracking space in state II is $L_{\mathrm{II}}$. It is assumed that the matrix cracking space in state I is twice of that at state II, i.e., $L_{\mathrm{I}} = 2L_{\mathrm{II}}$. When new matrix cracking appears between existed crackings, the energy balance relationship can be described using the following equation [35].

$$(W - U)_{\mathrm{I}} = (W - U)_{\mathrm{II}} + \Delta A_{\mathrm{m}} g_{\mathrm{m}}^{\mathrm{c}} + \Delta A_{\mathrm{db}} g_{\mathrm{db}}^{\mathrm{c}} \tag{1.27}$$

where $(W - U)_{\mathrm{I}}$ denotes the matrix strain release energy at state I; $(W - U)_{\mathrm{II}}$ denotes the matrix strain release energy at state II; $\Delta A_{\mathrm{m}}$ denotes the new cracking area; $\Delta A_{\mathrm{db}}$ denotes the new fiber/matrix interface debonding area; $g_{\mathrm{m}}^{\mathrm{c}}$ denotes the matrix strain energy release rate; and $g_{\mathrm{db}}^{\mathrm{c}}$ denotes the strain energy release rate when fiber/matrix interface debonding.

(3) **Critical matrix strain energy (CMSE) criterion**. Solti et al. [36] developed the critical matrix strain energy criterion to predict the matrix multicracking evolution in fiber-reinforced CMCs. The concept of a CMSE criterion presupposes the existence of a critical matrix strain energy limit beyond which the matrix fractures. Beyond the critical matrix strain energy, as more energy is placed into the composite, the matrix, unable to support the additional stress, continues to fail. As more energy is placed into the composite, the matrix fails such that all the additional energy is transferred to the fibers. Failure may consist of the formation of matrix crackings, the propagation of existing crackings or fiber/matrix interface debonding. Based on the CMSE criterion, the Coulomb frictional law is used to describe the fiber/matrix interface shear

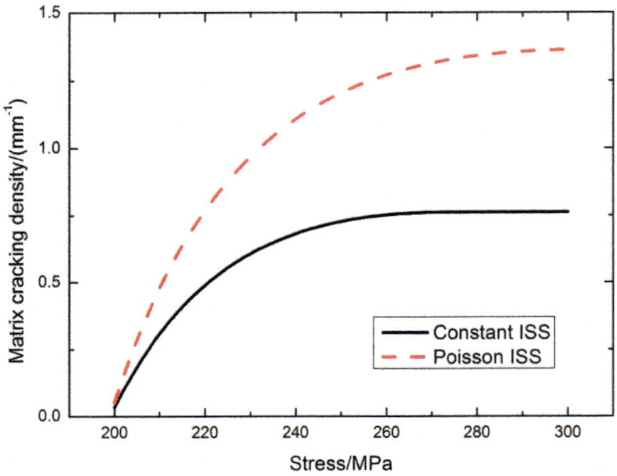


**Fig. 1.5** Comparison of matrix multicracking evolution for constant interface shear stress and the changing interface shear stress along the debonded region considering fiber Poisson effect

stress in the debonded region. The comparisons of matrix multicracking evolution in fiber-reinforced CMCs for constant interface shear stress and the changing interface shear stress along the debonded region considering fiber Poisson effect are shown in Fig. 1.5. When considering fiber Poisson effect, the matrix cracking density decreases at the same applied stress. The effects of fiber/matrix interface frictional coefficient, fiber Poisson ratio, fiber/matrix interface debonding energy, and fiber volume fraction on matrix multicracking evolution are discussed, as shown in Fig. 1.6. When the fiber/matrix interface frictional coefficient increases, the stress transfer between the fiber and the matrix increases, and the matrix carries more stress, and the matrix cracking space decreases when the strain energy reaches the critical value, leading to the increase of the matrix cracking density, as shown in Fig. 1.6a. When the fiber Poisson ratio increases, the fiber/matrix interface shear stress decreases in the debonded region, the stress transfer between the fiber and the matrix decreases, and the matrix cracking space increases when the strain energy reaches the critical value, leading to the decrease of the matrix cracking density, as shown in Fig. 1.6b. When the fiber/matrix interface debonded energy increases, the fiber/matrix interface debonding length decreases, the matrix carries more stress, and the matrix cracking space decreases when the strain energy reaches the critical value, leading to the increase of matrix cracking density, as shown in Fig. 1.6c. When the fiber volume fraction increases, the stress transfer between the fiber and the matrix increases, the matrix carries more stress, and the matrix cracking space decreases when the strain energy reaches the critical value, leading to the increase of matrix cracking density, as shown in Fig. 1.6d.

(4) **Statistical matrix cracking approach**. The statistical failure approach assumes that the matrix cracking is governed by the statistical relations, which relate the size and spatial distribution of matrix flaws to their relative

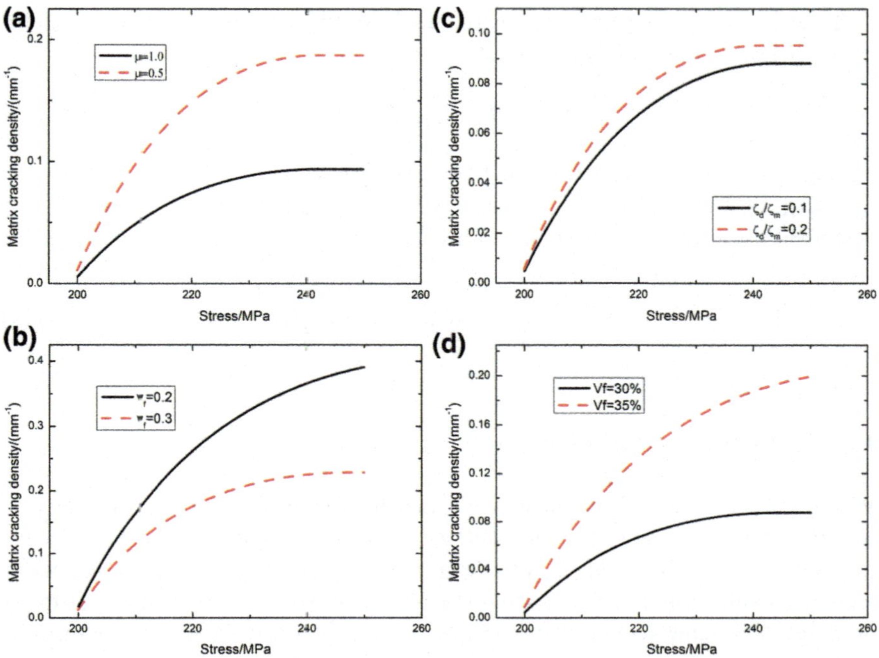


**Fig. 1.6** Effect of **a** fiber/matrix interface frictional coefficient; **b** fiber Poisson ratio; **c** fiber/matrix interface debonded energy; and **d** fiber volume fraction on matrix multicracking evolution in fiber-reinforced CMCs

propagation stress [37]. Curtin [38] investigated the thermal residual stress, fiber/matrix interface debonding on matrix statistical cracking in fiber-reinforced CMCs.

The brittle nature of the matrix material and the possible formation of initial cracks distribution throughout the microstructure suggest that a statistical approach to matrix multicracking evolution is warranted in fiber-reinforced CMCs. The matrix strength is assumed to be described using the two-parameter Weibull distribution. The matrix failure probability of $P_{\mathrm{m}}$ is determined using the following equation [37].

$$P_{\mathrm{m}} = 1 - \exp\left[-\left(\frac{\sigma_{\mathrm{m}}}{\sigma_{\mathrm{R}}}\right)^{m}\right] \tag{1.28}$$

where $\sigma_{\mathrm{m}}$ denotes the matrix stress; and $\sigma_{\mathrm{R}}$ and $m\sigma_{\mathrm{o}}, m$ denote the matrix characteristic strength and matrix Weibull modulus, respectively.

To estimate the instantaneous matrix crack spacing of $l_c$ with increasing applied stress, it leads to the following equation.

$$P_m = l_{sat}/l_c \tag{1.29}$$

where $l_{sat}$ denotes the saturation matrix crack spacing.

Using Eqs. (1.28) and (1.29), the instantaneous matrix crack spacing can be described using the following equation [37].

$$l_c = l_{sat}\left\{1 - \exp\left[-\left(\frac{\sigma_m}{\sigma_R}\right)^m\right]\right\}^{-1} \tag{1.30}$$

### 1.2.2.3 Interface Debonding

When matrix cracking propagates to the fiber/matrix interface, the interface debonding occurs. The maximum interface shear strength approach and the fracture mechanics approach can be used to determine the fiber/matrix interface debonding length. The maximum interface shear strength approach determines the fiber/matrix interface debonding length when the fiber/matrix interface shear stress approaches to the interface shear strength [39]. The fracture mechanics approach determines the fiber/matrix interface debonding length when the strain energy release rate at the fiber/matrix interface tip achieves the fiber/matrix interface debonding toughness [40]. For the case of weak bonding between the fiber and the matrix, the fracture mechanics approach is suitable for determining the fiber/matrix interface debonding [41]. The fiber/matrix interface debonding criterion is determined using the following equation [40].

$$\zeta_d = \frac{F}{4\pi r_f}\frac{\partial w_f(0)}{\partial l_d} - \frac{1}{2}\int_0^{l_d} \tau_i \frac{\partial v(x)}{\partial l_d}\mathrm{d}x \tag{1.31}$$

where $F(=\pi r_f^2\sigma/V_f)P = \pi r_f^2\sigma/V_f$ denotes the fiber load at the matrix crack plane; $w_f(0)$ denotes the fiber axial displacement at the matrix crack plane; $v(x)$ denotes the relative displacement between fiber and matrix.

The fiber and matrix axial displacement, i.e., $w_f(x)$ and $w_m(x)$, are determined using the following equations.

$$\begin{aligned} w_f(x) &= \int_x^{l_c/2} \frac{\sigma_f}{E_f}\mathrm{d}x \\ &= \frac{\sigma}{V_f E_f}(l_d - x) - \frac{\tau_i}{r_f E_f}\left(l_d^2 - x^2\right) - \frac{2\tau_i}{\rho E_f} l_d + \frac{r_f V_m E_m}{\rho V_f E_f E_c}\sigma + \frac{\sigma}{E_c}(l_c/2 - l_d) \end{aligned} \tag{1.32}$$

$$\begin{aligned} w_{\mathrm{m}}(x) &= \int_{x}^{l_{\mathrm{c}}/2} \frac{\sigma_{\mathrm{m}}}{E_{\mathrm{m}}} \mathrm{d}x \\ &= \frac{V_{\mathrm{f}}\tau_{i}}{V_{\mathrm{m}}E_{\mathrm{m}}r_{\mathrm{f}}}\left(l_{\mathrm{d}}^{2} - x^{2}\right) + \frac{2V_{\mathrm{f}}\tau_{\mathrm{i}}}{\rho V_{\mathrm{m}}E_{\mathrm{m}}} l_{\mathrm{d}} - \frac{r_{\mathrm{f}}}{\rho E_{\mathrm{c}}}\sigma + \frac{\sigma}{E_{\mathrm{c}}}(l_{\mathrm{c}}/2 - l_{\mathrm{d}}) \end{aligned} \quad (1.33)$$

Using Eqs. (1.32) and (1.33), the relative displacement between the fiber and matrix, i.e., $v(x)$, can be described using the following equation.

$$\begin{aligned} v(x) &= |w_{\mathrm{f}}(x) - w_{\mathrm{m}}(x)| \\ &= \frac{\sigma}{V_{\mathrm{f}}E_{\mathrm{f}}}(l_{\mathrm{d}} - x) - \frac{\tau_{\mathrm{i}}E_{\mathrm{c}}}{V_{\mathrm{m}}E_{\mathrm{m}}E_{\mathrm{f}}r_{\mathrm{f}}}\left(l_{\mathrm{d}}^{2} - x^{2}\right) - \frac{2\tau_{\mathrm{i}}E_{\mathrm{c}}l_{\mathrm{d}}}{\rho V_{\mathrm{m}}E_{\mathrm{m}}E_{\mathrm{f}}} + \frac{r_{\mathrm{f}}}{\rho V_{\mathrm{f}}E_{\mathrm{f}}}\sigma \end{aligned} \quad (1.34)$$

Substituting $w_{\mathrm{f}}(x = 0)$ and $v(x)w_{\mathrm{f}}(z = 0)$ and $v(z)$ into Eq. (1.31), it leads to the form of the following equation.

$$\frac{E_{\mathrm{c}}\tau_{\mathrm{i}}^{2}}{V_{\mathrm{m}}E_{\mathrm{m}}E_{\mathrm{f}}r_{\mathrm{f}}} l_{\mathrm{d}}^{2} + \left(\frac{E_{\mathrm{c}}\tau_{\mathrm{i}}^{2}}{\rho V_{\mathrm{m}}E_{\mathrm{m}}E_{\mathrm{f}}} - \frac{\tau_{\mathrm{i}}\sigma}{V_{\mathrm{f}}E_{\mathrm{f}}}\right) l_{\mathrm{d}} + \frac{r_{\mathrm{f}}V_{\mathrm{m}}E_{\mathrm{m}}\sigma^{2}}{4V_{\mathrm{f}}^{2}E_{\mathrm{f}}E_{\mathrm{c}}} - \frac{r_{\mathrm{f}}\tau_{\mathrm{i}}}{2\rho V_{\mathrm{f}}E_{\mathrm{f}}}\sigma - \zeta_{\mathrm{d}} = 0 \quad (1.35)$$

Solving Eq. (1.35), the fiber/matrix interface debonded length of $l_{\mathrm{d}}$ can be described using the following equation.

$$l_{\mathrm{d}} = \frac{r_{\mathrm{f}}}{2}\left(\frac{V_{\mathrm{m}}E_{\mathrm{m}}\sigma}{V_{\mathrm{f}}E_{\mathrm{c}}\tau_{\mathrm{i}}} - \frac{1}{\rho}\right) - \sqrt{\left(\frac{r_{\mathrm{f}}}{2\rho}\right)^{2} + \frac{r_{\mathrm{f}}V_{\mathrm{m}}E_{\mathrm{m}}E_{\mathrm{f}}}{E_{\mathrm{c}}\tau_{\mathrm{i}}^{2}}\zeta_{\mathrm{d}}} \quad (1.36)$$

#### 1.2.2.4 Fibers Failure

When matrix cracking density approaches to saturation at the stress of $\sigma_{\mathrm{sat}}$, the stress carried by intact fibers increases gradually, leading to fibers failure and composite fracture. Compared with damage models of matrix cracking and fiber/matrix interface debonding, the damage models for fibers failure of fiber-reinforced CMCs are few. Based on the analysis of load sharing between the intact and broken fibers, the Global Load Sharing (GLS) criterion and Local Load Sharing (LLS) criterion can be used to determine the fibers failure. The GLS criterion assumes that the load will be equally assigned to the intact fibers after the fibers broken, and the LLS criterion assumes that the load will be carried by the neighboring fibers after fibers fracture. The GLS criterion neglects the stress concentration caused by fibers failure and is appropriate for the condition of weak fiber/matrix interface bonding and low fiber/matrix interface shear stress. Thouless and Evans [28] determined the fibers failure probability and fibers pullout length using the weakest chain statistical theory and investigated the effect of fiber failure on the steady-state matrix first cracking stress. Cao and Thouless [42] predicted the tensile strength of SiC/LAS composite by assuming multiple matrix cracking occurred at the initial loading and neglecting the loading carrying of broken fibers. The predicted tensile strength is lower than the experimental value. Sutcu [43] analyzed the fibers pullout length, the work of fracture, and the tensile strength for single matrix cracking and multiple matrix cracking condition. Schwietert and Steif [44] determined the multiple fracture probability of a single fiber and found that the second fracture of fiber has little effect on the tensile strength of fiber-reinforced CMCs; however, the influence of interface shear stress on the tensile strength is greater. Curtin [45] assumed that the fiber fracture is independent of each other and obtained the fibers pullout length distribution, fibers pullout work and the ultimate tensile strength of the composite using the GLS criterion to determine the load sharing between intact and broken fibers. Solti et al. [46] developed the critical fiber strain energy criterion to predict the fibers failure in fiber-reinforced CMCs. Liao and Reifsnider [47] predicted the fibers failure at the matrix cracking plane, the fiber/matrix interface debonded region and interface bonded region, obtained the tensile strength and fracture toughness of fiber-reinforced CMCs using the Coulomb friction law to describe the fiber/matrix interface shear stress in the debonded region. For the LLS criterion, the models include Zhou and Curtin [48], Ibnabdeljali and Curtin [49], Beyerlein and Phoenix [50], Dutton et al. [51], Xia and Curtin [52] and are suitable for the strong fiber/matrix interface bonding condition.

The GLS criterion is adopted to determine the load sharing between the intact and broken fibers, and the two-parameter Weibull model is used to describe fiber strength distribution [45].

$$\frac{\sigma}{V_{\mathrm{f}}} = T(1 - P(T)) + \frac{2\tau_{\mathrm{i}}}{r_{\mathrm{f}}}\langle L\rangle P(T) \tag{1.37}$$

where $\langle L\rangle$ denotes the average fiber pullout length; and $P(T)$ denotes broken fibers fraction.

$$P(T) = 1 - \exp\left[-\left(\frac{T}{\sigma_{\mathrm{fc}}}\right)^{m_{\mathrm{f}}+1}\right] \tag{1.38}$$

where $m_{\mathrm{f}}$ is the fiber Weibull modulus; $\sigma_{\mathrm{fc}}$ is the fiber characteristic strength of a length $\delta$ of fiber.

Then $\sigma$ is given by the following equation [45].

$$\frac{\sigma}{V_{\mathrm{f}}} = T\left\{1 - \frac{1}{2}\left(\frac{T}{\sigma_{\mathrm{fc}}}\right)^{m_{\mathrm{f}}+1}\right\} \tag{1.39}$$

Calculating the maximum value of intact fiber stress of $T$, the composite ultimate tensile strength of $\sigma_{\mathrm{UTS}}$ can be described using the following equation [45].

$$\sigma_{\mathrm{UTS}} = V_{\mathrm{f}}\sigma_{\mathrm{fc}}\left(\frac{2}{m_{\mathrm{f}}+2}\right)^{\frac{1}{m_{\mathrm{f}}+1}}\left(\frac{m_{\mathrm{f}}+1}{m_{\mathrm{f}}+2}\right) \tag{1.40}$$

#### 1.2.2.5 Stress–Strain Relationship

When the fiber-reinforced CMCs are without damage, the composite stress–strain relationship can be determined using the following equation.

$$\varepsilon_{\mathrm{c}} = \sigma/E_{\mathrm{c}} \tag{1.41}$$

When damages of matrix multicracking, fiber/matrix interface debonding and fibers failure occurred inside of the fiber-reinforced CMCs, the composite strain can be determined using the following equation.

$$\varepsilon_{\mathrm{c}} = \frac{2}{E_{\mathrm{f}}l_{\mathrm{c}}}\int_{l_{\mathrm{c}}/2}\sigma_{\mathrm{f}}(x)\mathrm{d}x - (\alpha_{\mathrm{c}}-\alpha_{\mathrm{f}})\Delta T \tag{1.42}$$

Substituting Eqs. (1.4) and (1.15) into Eq. (1.42), the tensile stress–strain relationship for the fiber-reinforced CMCs at the damage stage of matrix multicracking and fiber/matrix interface debonding can be described using the following equation.

$$\begin{aligned}\varepsilon_{\mathrm{c}} = {} & \frac{\sigma}{V_{\mathrm{f}}E_{\mathrm{f}}}\frac{2l_{\mathrm{d}}}{l_{\mathrm{c}}} - \frac{\tau_{\mathrm{i}}}{E_{\mathrm{f}}}\frac{2l_{\mathrm{d}}^2}{r_{\mathrm{f}}l_{\mathrm{c}}} + 2\frac{\sigma_{\mathrm{fo}}}{E_{\mathrm{f}}l_{\mathrm{c}}}\left(\frac{l_{\mathrm{c}}}{2}-l_{\mathrm{d}}\right) - \frac{1}{\rho E_{\mathrm{f}}}\frac{2r_{\mathrm{f}}}{l_{\mathrm{c}}}\left(\frac{V_{\mathrm{m}}}{V_{\mathrm{f}}}\sigma_{\mathrm{mo}} - 2\frac{l_{\mathrm{d}}}{r_{\mathrm{f}}}\tau_{\mathrm{i}}\right)\\ & \times\left[\exp\left(-\rho\frac{l_{\mathrm{c}}/2-l_{\mathrm{d}}}{r_{\mathrm{f}}}\right)-1\right] - (\alpha_{\mathrm{c}}-\alpha_{\mathrm{f}})\Delta T\end{aligned} \tag{1.43}$$

Substituting Eq. (1.23) into Eq. (1.42), the tensile stress–strain relationship for the fiber-reinforced CMCs at the damage stage of fibers failure can be described using the following equation.

$$\begin{aligned}\varepsilon_{\mathrm{c}} = {} & \frac{T}{E_{\mathrm{f}}}\frac{2l_{\mathrm{d}}}{l_{\mathrm{c}}} - \frac{\tau_{\mathrm{i}}}{E_{\mathrm{f}}}\frac{2l_{\mathrm{d}}^2}{r_{\mathrm{f}}l_{\mathrm{c}}} + \frac{2\sigma_{\mathrm{fo}}}{E_{\mathrm{f}}l_{\mathrm{c}}}\left(\frac{l_{\mathrm{c}}}{2}-l_{\mathrm{d}}\right) - \frac{1}{\rho E_{\mathrm{f}}}\frac{2r_{\mathrm{f}}}{l_{\mathrm{c}}}\left(T - \sigma_{\mathrm{fo}} - 2\frac{l_{\mathrm{d}}}{r_{\mathrm{f}}}\tau_{\mathrm{i}}\right)\\ & \times\left[\exp\left(-\rho\frac{l_{\mathrm{c}}/2-l_{\mathrm{d}}}{r_{\mathrm{f}}}\right)-1\right] - (\alpha_{\mathrm{c}}-\alpha_{\mathrm{f}})\Delta T\end{aligned} \tag{1.44}$$

where the matrix crack spacing $l_{\mathrm{c}}$ is determined by Eq. (1.30), the fiber/matrix interface debonded length is determined by Eq. (1.36), and the intact fiber stress at the matrix crack plane is determined by Eq. (1.39). When the fiber/matrix interface completely debonds, i.e., $l_{\mathrm{d}} = l_{\mathrm{c}}/2$, the stress–strain relationship for the interface completely debonding is by setting $l_{\mathrm{d}} = l_{\mathrm{c}}/2$ in Eqs. (1.43) and (1.44).

## 1.2.3 Results and Discussions

When damage occurred inside of the composite, the fiber, matrix, and interface properties affect the damage extent and the nonlinear behavior of fiber-reinforced CMCs. The effects of matrix Weibull modulus, matrix cracking characteristic stress, fiber/matrix interface shear stress, fiber/matrix interface debonded energy, fiber Weibull modulus, and fiber characteristic strength on the damage evolution and then the tensile stress–strain curve of unidirectional SiC/CAS composite are discussed. The material properties of unidirectional SiC/CAS composite are given by $E_f = 190$ GPa, $E_m = 90$ GPa, $V_f = 34\%$, $r_f = 7.5$ μm, $\alpha_f = 3.3\times10^{-6}$/°C, $\alpha_m = 4.6\times10^{-6}$/°C, and $\Delta T = -1000$°C.

### 1.2.3.1 Effect of Matrix Weibull Modulus

The effect of matrix Weibull modulus (i.e., $m$ = 3 and 5) on matrix multicracking evolution and tensile stress–strain curves is shown in Fig. 1.7. The matrix Weibull modulus affects the shape of matrix multicracking evolution. At low applied stress, the matrix cracking density of low matrix Weibull modulus is higher than that of high matrix Weibull modulus; and at high applied stress, the matrix cracking density of low matrix Weibull modulus is lower than that of high matrix Weibull modulus. When the matrix Weibull modulus is high, the matrix strength distribution is more concentrated, and the matrix cracking approaches quickly to saturation, as shown in Fig. 1.7a. For the tensile stress–strain curve, at low applied stress, the matrix crack spacing of low matrix Weibull modulus is less than that of high matrix Weibull modulus, leading to the increase of strain at the same applied stress; and at high applied stress, the matrix crack spacing of low matrix Weibull modulus is larger than that of high matrix Weibull modulus, leading to the decrease of strain at the same applied stress, as shown in Fig. 1.7b.

### 1.2.3.2 Effect of Matrix Cracking Characteristic Stress

The effect of matrix cracking characteristic stress (i.e., $\sigma_R$ = 150 and 200 MPa) on the matrix multicracking evolution and tensile stress–strain curve is shown in Fig. 1.8. When the matrix cracking characteristic stress is low, the matrix cracking approaches to the saturation at low applied stress, as shown in Fig. 1.8a. For the tensile stress–strain curve, at low applied stress, the matrix cracking spacing of low matrix cracking characteristic stress is less than that of high matrix characteristic cracking stress, leading to the increase of composite strain, and at high applied stress, as the matrix cracking gradually approaches to the saturation, the composite strain for the low and high matrix cracking characteristic stress approaches to the same value, as shown in Fig. 1.8b.

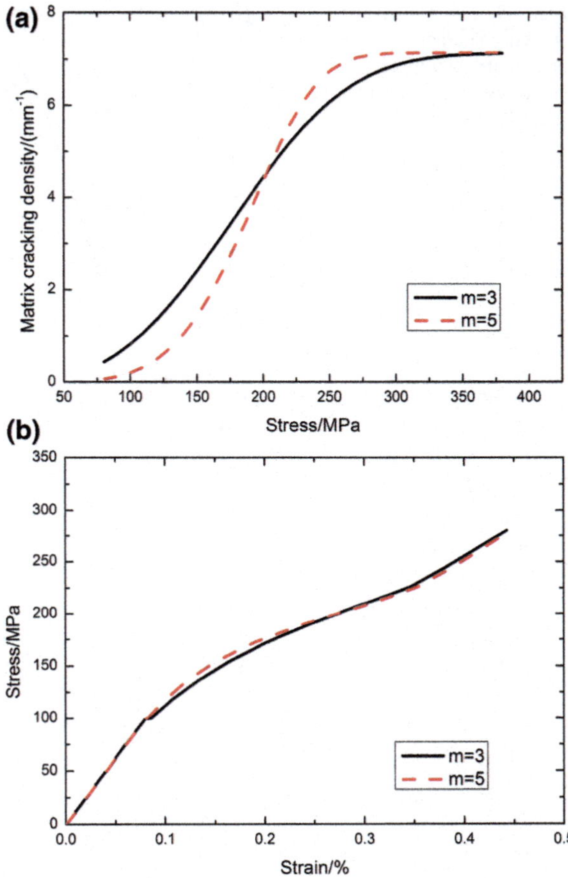


**Fig. 1.7** Effect of matrix Weibull modulus on **a** matrix multicracking evolution; and **b** tensile stress–strain curve

#### 1.2.3.3 Effect of Fiber/Matrix Interface Shear Stress

The effect of fiber/matrix interface shear stress (i.e., $\tau_i$ = 10 and 15 MPa) on the fiber/matrix interface debonding ratio and tensile stress–strain curve is shown in Fig. 1.9. When the fiber/matrix interface partially debonds, the fiber/matrix interface debonding ratio ($2l_d/l_c$) decreases with increasing interface shear stress, and the fiber/matrix interface completely debonding stress increases with increasing interface shear stress, as shown in Fig. 1.9a. For the tensile stress–strain curve, the composite strain decreases with increasing interface shear stress, as shown in Fig. 1.9b.

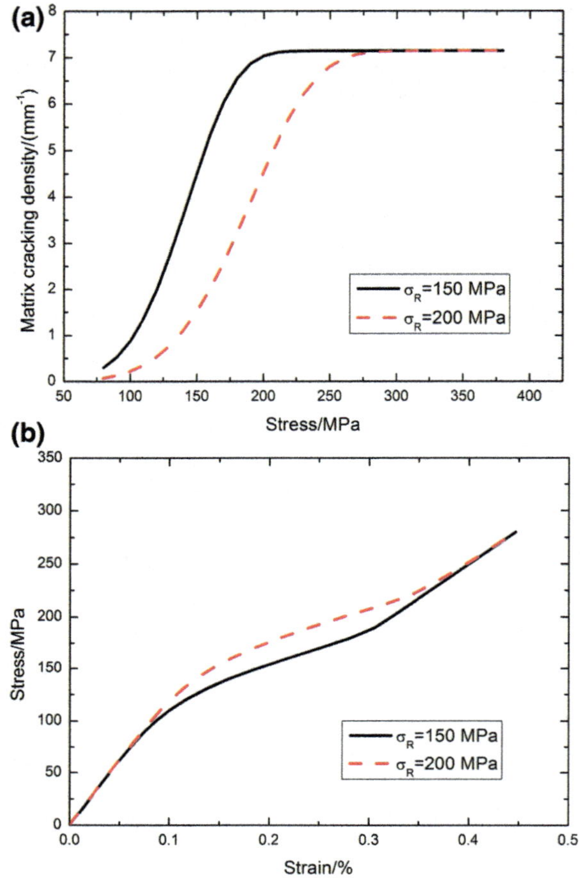


**Fig. 1.8** Effect of matrix cracking characteristic stress on **a** matrix multicracking evolution; and **b** tensile stress–strain curve

#### 1.2.3.4 Effect of Fiber/Matrix Interface Debonded Energy

The effect of fiber/matrix interface debonded energy (i.e., $\zeta_d$ = 0.1 and 0.5 J/m$^2$) on the fiber/matrix interface debonding ratio and tensile stress–strain curve is shown in Fig. 1.10. When the fiber/matrix interface debonded energy increases, the fiber/matrix interface debonding ratio decreases at the same applied stress, and the fiber/matrix interface completely debonding stress increases, as shown in Fig. 1.10a. For the tensile stress–strain curve, the composite strain decreases with increasing of fiber/matrix interface debonded energy when the interface partially debonds; however, when the fiber/matrix interface completely debonds, the composite strain with low or high interface debonded energy approaches to the same value, as shown in Fig. 1.10b.

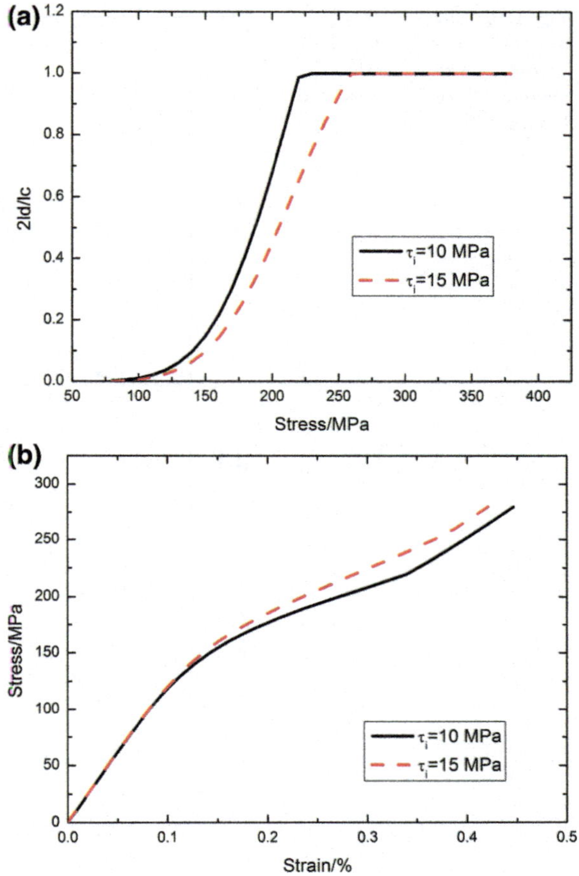


**Fig. 1.9** Effect of fiber/matrix interface shear stress on **a** fiber/matrix interface debonding ratio; and **b** tensile stress–strain curve

#### 1.2.3.5 Effect of Fiber Weibull Modulus

The effect of fiber Weibull modulus (i.e., $m_f$ = 2 and 3) on the broken fibers fraction and the tensile stress–strain curve is shown in Fig. 1.11. When the fiber Weibull modulus increases, the broken fibers fraction decreases, leading to the decrease of composite strain.

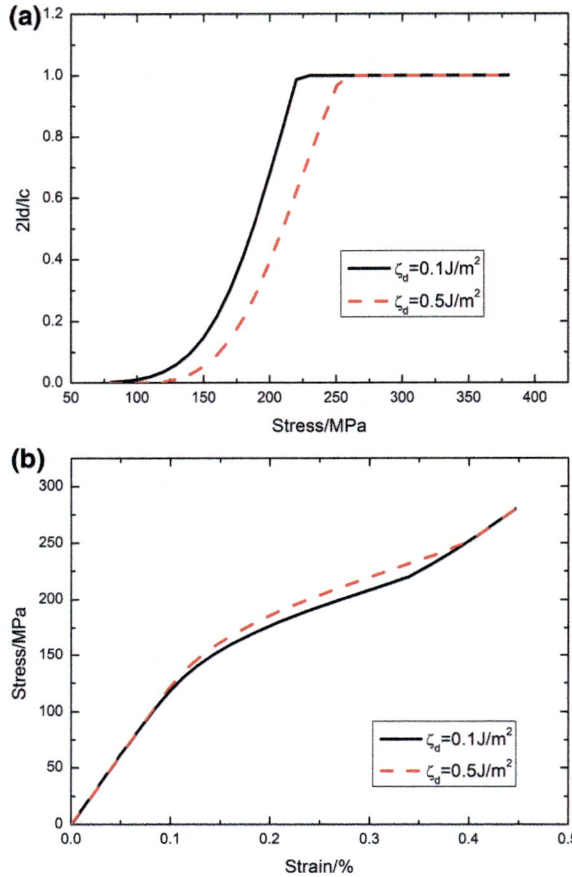


**Fig. 1.10** Effect of fiber/matrix interface debonded energy on **a** fiber/matrix interface debonding ratio; and **b** tensile stress–strain curve

#### 1.2.3.6 Effect of Fiber Characteristic Strength

The effect of fiber strength (i.e., $\sigma_c$ = 1.8 and 2.0 GPa) on the broken fibers fraction and the tensile stress–strain curve is shown in Fig. 1.12. When the fiber characteristic strength increases, the composite tensile strength increases, and the broken fibers fraction decreases at the same applied stress, leading to the decrease of the composite strain.

### *1.2.4 Experimental Comparisons*

The tensile stress–strain curves of unidirectional SiC/CAS [53–55], SiC/CAS-II [56], SiC/Borosilicate [57], and SiC/1723 [58] composites are predicted using the present analysis.

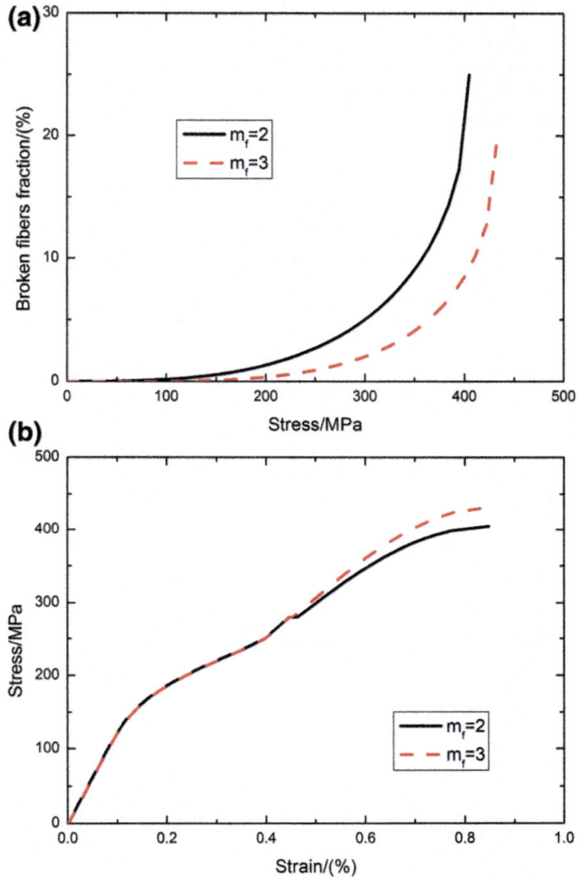


**Fig. 1.11** Effect of fiber Weibull modulus on **a** broken fibers fraction versus applied stress curve; and **b** tensile stress–strain curve

#### 1.2.4.1 SiC/CAS Composites

Pryce and Smith [53] investigated the tensile behavior of unidirectional SiC/CAS composite at room temperature. The experimental and theoretical matrix cracking density and tensile stress–strain curves of unidirectional SiC/CAS composite at room temperature are shown in Fig. 1.13. Under tensile loading, the matrix multicracking evolution is observed using the optical microscope, and the experimental matrix cracking density versus the applied stress curve is shown in Fig. 1.13a. The first matrix cracking stress of $\sigma_{mc}$ is about 100 MPa and the saturation matrix cracking stress of $\sigma_{sat}$ is about 270 MPa. The experimental tensile stress–strain curve of undirectional SiC/CAS composite is shown in Fig. 1.13b. The tensile stress–strain curve of unidirectional SiC/CAS composite can be divided into three stages, i.e., the initial linear stage (from initial loading to the first matrix cracking), matrix cracking and fiber/matrix interface debonding nonlinear stage (from first matrix cracking to the saturation matrix cracking), and fibers failure stage (from the

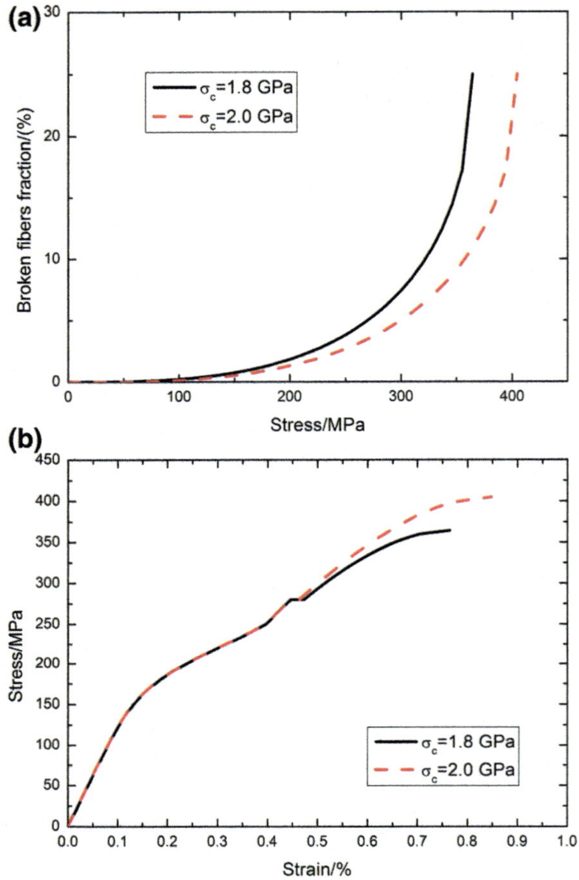


**Fig. 1.12** Effect of fiber strength on **a** broken fibers fraction versus applied stress curve; and **b** tensile stress–strain curve

saturation matrix cracking to the final fracture of the composite). The composite tensile strength is about 400 MPa, and the failure strain is about 0.8%. The material properties of SiC/CAS composite are listed in Table 1.1. The predicted tensile stress–strain curve of unidirectional SiC/CAS composite at each damage stage, the composite tensile strength and failure strain agree with experimental data.

Beyerle et al. [54] investigated the tensile behavior of unidirectional SiC/CAS composite at room temperature. The experimental and theoretical matrix cracking density and tensile stress–strain curves of unidirectional SiC/CAS composite at room temperature are shown in Fig. 1.14. Under tensile loading, the optical microscope was used to observe the specimen surface of 6–9 $mm^2$ region, and the matrix cracking density versus applied stress curve under tensile loading was shown in Fig. 1.14a. The first matrix cracking stress is about 120 MPa, and the saturation matrix cracking stress is about 290 MPa. The experimental tensile stress–strain

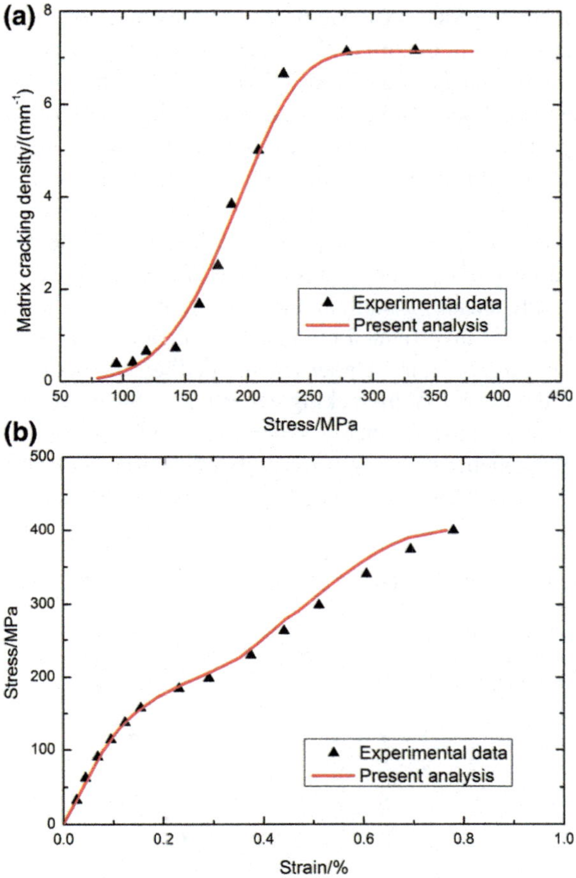


**Fig. 1.13** Experimental and the theoretical **a** matrix crack density versus applied stress curves; and **b** tensile stress–strain curves of unidirectional SiC/CAS composite. Reprinted with permission from Ref. [71]. Copyright 2013, Springer Science Business Media New York

**Table 1.1** Material properties of unidirectional SiC/CAS, SiC/CAS-II, SiC/borosilicate, and SiC/1723 composites

| Items | SiC/CAS [53] | SiC/CAS [54] | SiC/CAS [55] | SiC/CAS-II [56] | SiC/borosilicate [57] | SiC/1723 [58] |
|---|---|---|---|---|---|---|
| $E_f$ (GPa) | 190 | 200 | 210 | 200 | 230 | 190 |
| $E_m$ (GPa) | 90 | 97 | 95.5 | 98 | 60 | 88 |
| $V_f$ | 0.34 | 0.37 | 0.35 | 0.30 | 0.31 | 0.45 |
| $r_f$ (μm) | 7.5 | 7.5 | 8 | 7.5 | 8 | 6.25 |
| $\nu_f$ | 0.2 | 0.2 | 0.2 | 0.2 | 0.2 | 0.2 |
| $\nu_m$ | 0.2 | 0.2 | 0.2 | 0.2 | 0.2 | 0.2 |
| $\alpha_f$ ($10^{-6} \times$ °C$^{-1}$) | 3.3 | 4 | 4 | 4 | 3.1 | 3.1 |
| $\alpha_m$ ($10^{-6} \times$ °C$^{-1}$) | 4.6 | 5 | 5 | 5 | 3.25 | 4.36 |
| $\Delta T$ (°C) | −1000 | −1000 | −1000 | −1000 | −1000 | −1000 |
| $\tau_i$ (MPa) | 10 | 15 | 20 | 20 | 7.6 | 35 |

curve of undirectional SiC/CAS composite is shown in Fig. 1.14b. The tensile stress–strain curve of unidirectional SiC/CAS composite can be divided into three stages, i.e., the initial linear stage (from initial loading to the first matrix cracking), matrix cracking and fiber/matrix interface debonding nonlinear stage (from first matrix cracking to the saturation matrix cracking), and fibers failure stage (from the saturation matrix cracking to the final fracture of the composite). The composite tensile strength is about 450 MPa, and the failure strain is about 0.82%. The material properties of SiC/CAS composite are listed in Table 1.1. The predicted tensile stress–strain curve of unidirectional SiC/CAS composite at each damage stage, the composite tensile strength and failure strain agree with experimental data.

Kuo and Chou [55] investigated the tensile behavior of unidirectional SiC/CAS composite at room temperature. Under tensile loading, the optical microscope was used to observe the matrix multicracking evolution of SiC/CAS composite, and the experimental matrix cracking density versus the applied stress curve is shown in Fig. 1.15a. The first matrix cracking stress is about 90 MPa, and the saturation matrix cracking stress is about 390 MPa. The experimental tensile stress–strain curve of

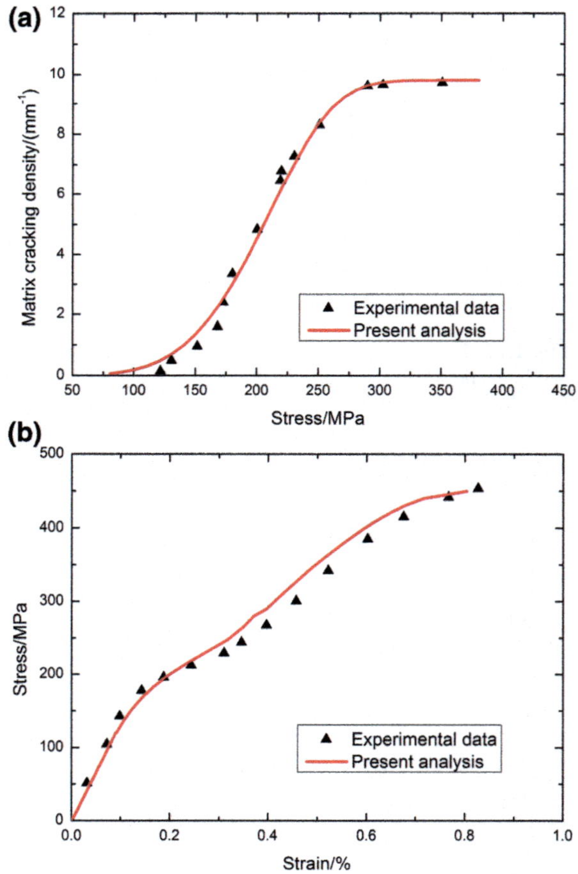


**Fig. 1.14** Experimental and the theoretical **a** matrix crack density versus applied stress curves; and **b** tensile stress–strain curves of unidirectional SiC/CAS composite

unidirectional SiC/CAS composite is shown in Fig. 1.15b. The tensile stress–strain curve of unidirectional SiC/CAS composite can be divided into three stages, i.e., the initial linear stage (from initial loading to the first matrix cracking), matrix cracking and fiber/matrix interface debonding nonlinear stage (from first matrix cracking to the saturation matrix cracking), and fibers failure stage (from the saturation matrix cracking to the final fracture of the composite). The composite tensile strength is about 410 MPa, and the failure strain is about 0.67%. The material properties of SiC/CAS composite are listed in Table 1.1. The predicted tensile stress–strain curve of unidirectional SiC/CAS composite at each damage stage, the composite tensile strength and failure strain agree with experimental data.

#### 1.2.4.2 SiC/CAS-II Composite

Sorensen and Holmes [56] investigated the tensile behavior of unidirectional SiC/CAS-II composite at room temperature. The experimental and theoretical matrix

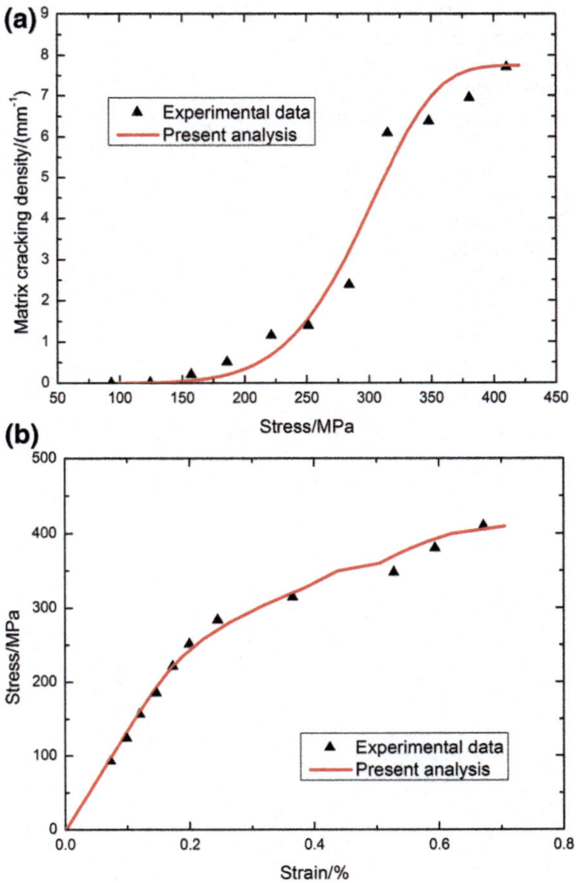


**Fig. 1.15** Experimental and the theoretical **a** matrix crack density versus applied stress curves; and **b** tensile stress–strain curves of unidirectional SiC/CAS composite

cracking density and tensile stress–strain curves of unidirectional SiC/CAS-II composite at room temperature are shown in Fig. 1.16. Under tensile loading, the matrix multicracking evolution is observed under the optical microscope, and the experimental matrix cracking density versus the applied stress curve is shown in Fig. 1.16a. The first matrix cracking stress of $\sigma_{mc}$ is about 100 MPa, and the saturation matrix cracking stress of $\sigma_{sat}$ is about 240 MPa. The experimental tensile stress–strain curve of unidirectional SiC/CAS composite is shown in Fig. 1.16b. The tensile stress–strain curve of unidirectional SiC/CAS-II composite can be divided into three stages, i.e., the initial linear stage (from initial loading to the first matrix cracking), matrix cracking and fiber/matrix interface debonding non-linear stage (from first matrix cracking to the saturation matrix cracking), and fibers failure stage (from the saturation matrix cracking to the final fracture of the composite). The composite tensile strength is about 350 MPa, and the failure strain is about 0.68%. The material properties of SiC/CAS-II composite are listed in Table 1.1. The predicted tensile stress–strain curve of unidirectional SiC/CAS composite at each damage stage, the composite tensile strength and failure strain agree with experimental data.

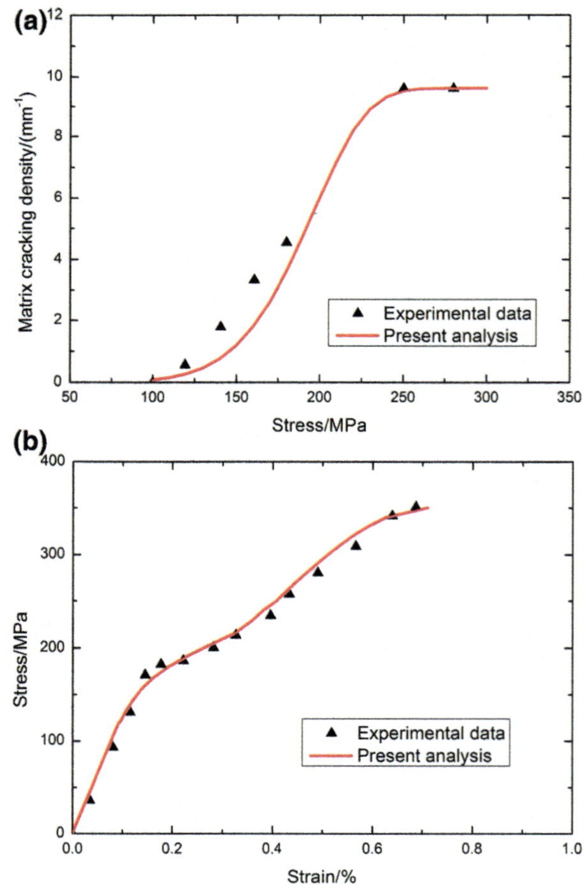


**Fig. 1.16** Experimental and the theoretical **a** matrix crack density versus applied stress curves; and **b** tensile stress–strain curves of unidirectional SiC/CAS-II composite. Reprinted with permission from Ref. [71]. Copyright 2013, Springer Science Business Media New York

#### 1.2.4.3 SiC/Borosilicate Composite

Okabe et al. [57] investigated the tensile behavior of SiC/Borosilicate composite at room temperature. The experimental and theoretical matrix cracking density and tensile stress–strain curves of unidirectional SiC/Borosilicate composite at room temperature are shown in Fig. 1.17. The matrix multicracking evolution of SiC/Borosilicate is observed under the optical microscope, and the experimental matrix cracking density versus the applied stress curve is shown in Fig. 1.17a. The first matrix cracking stress of $\sigma_{mc}$ is about 150 MPa, and the saturation matrix cracking stress of $\sigma_{sat}$ is about 390 MPa. The experimental tensile stress–strain curve of unidirectional SiC/Borosilicate composite is shown in Fig. 1.17b. The tensile stress–strain curve of unidirectional SiC/Borosilicate composite can be divided into three stages, i.e., the initial linear stage (from initial loading to the first matrix cracking), matrix cracking and fiber/matrix interface debonding non-linear stage (from first matrix cracking to the saturation matrix cracking), and fibers failure stage (from the saturation matrix cracking to the final fracture of the composite).

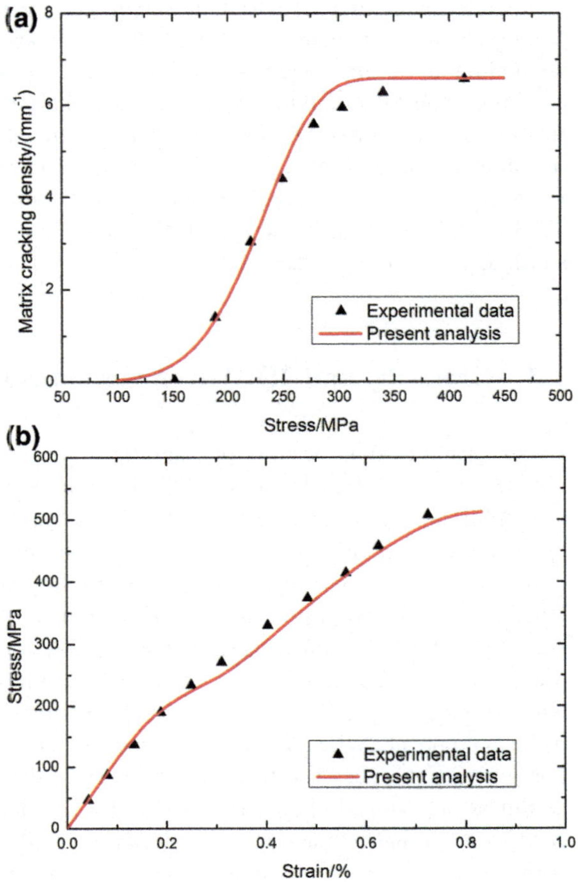


**Fig. 1.17** Experimental and the theoretical **a** matrix crack density versus applied stress curves; and **b** tensile stress–strain curves of unidirectional SiC/Borosilicate composite. Reprinted with permission from Ref. [71]. Copyright 2013, Springer Science Business Media New York

The composite tensile strength is about 512 MPa, and the failure strain is about 0.72%. The material properties of SiC/Borosilicate composite are listed in Table 1.1. The predicted tensile stress–strain curve of unidirectional SiC/Borosilicate composite at each damage stage, the composite tensile strength and failure strain agree with experimental data.

#### 1.2.4.4 SiC/1723 Composite

Zawada et al. [58] investigated the tensile behavior of unidirectional SiC/1723 composite at room temperature. The experimental and theoretical matrix cracking density and tensile stress–strain curves of unidirectional SiC/1723 composite at room temperature are shown in Fig. 1.18. Under tensile loading, matrix multicracking evolution of SiC/1723 is monitored using the acoustic emission and optical microscope. The first matrix cracking stress is about 200 MPa. The experimental tensile stress–strain curve of unidirectional SiC/1723 composite is shown in Fig. 1.18b. The tensile stress–strain curve of unidirectional SiC/1723 composite can be divided into three stages, i.e., the initial linear stage (from initial loading to the first matrix cracking), matrix cracking and fiber/matrix interface debonding non-linear stage (from first matrix cracking to the saturation matrix cracking), and fibers failure stage (from the saturation matrix cracking to the final fracture of the composite). The composite tensile strength is about 680 MPa, and the failure strain is about 0.54%. The material properties of SiC/1723 composite are listed in Table 1.1. The predicted tensile stress–strain curve of unidirectional SiC/1723 composite at each damage stage, the composite tensile strength and failure strain agree with experimental data.

## 1.3 Cross-Ply and 2D Woven Ceramic-Matrix Composites

Under tensile loading of cross-ply and 2D woven CMCs, the first matrix cracking initiates in the 90° plies in cross-ply CMCs or transverse yarns in 2D woven CMCs [59]. With increasing applied stress, transverse multicracking occurs and gradually approaches to the saturation [53]. The term *transverse crack* describes the cracking in the 90° plies in cross-ply CMCs or transverse yarns in 2D woven CMCs. Some transverse cracks propagate into the 0° plies or longitudinal yarns forming major crackings. However, some matrix crackings may form in the matrix-rich region at low stress level. The term *matrix crack* is used to describe the separation of matrix with bridging fibers in 0° plies of a cross-ply composite or longitudinal yarn of a 2D woven composite. When matrix cracking occurs, the cracking may deflect along the fiber/matrix interface. The fiber/matrix interface debonding and sliding affects the nonlinear behavior of fiber-reinforced CMCs. As the saturation of matrix cracking and the fiber/matrix interface completely debonding, the matrix stress remains constant. The extra stress with increasing applied stress is carried by fibers in the 0°

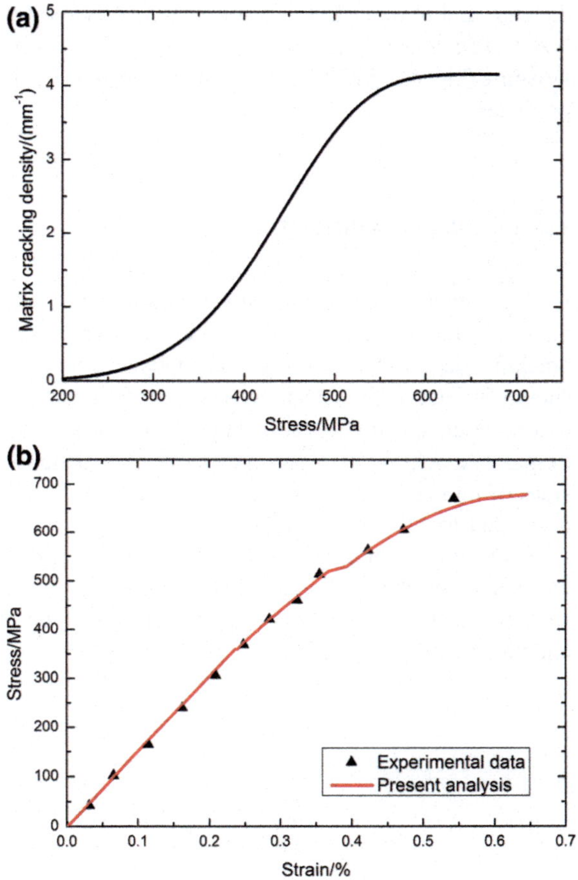


**Fig. 1.18** Experimental and the theoretical **a** matrix crack density versus applied stress curves; and **b** tensile stress–strain curves of unidirectional SiC/1723 composite

plies or longitudinal yarns. The fibers gradually fracture, which affect the tensile modulus and strain of the composite; and when the broken fibers fraction approaches to the critical value, the composite finally fractures. For a cross-ply composite, the tensile stress–strain curve deviates twice for damage accumulation in the 90° and 0° plies.

In this section, the tensile behavior of cross-ply and 2D woven CMCs at room temperature is investigated. The Kuo-Chou shear-lag model is used to analyze the microstress field of the damaged composite considering the damage mechanisms of transverse multicracking, matrix multicracking, fiber/matrix interface debonding, and fibers failure. Combining the shear-lag model with damage models of fracture mechanics transverse cracking model, statistical matrix cracking model, fiber/matrix interface fracture mechanics criterion, and GLS fibers failure criterion, the tensile stress–strain curve of cross-ply and 2D woven CMCs for different damage stages are predicted. The effects of transverse fracture energy, fiber/matrix interface shear stress, fiber/matrix interface debonded energy, and fibers Weibull modulus on damage and tensile stress–strain curves of cross-ply SiC/CAS composite are analyzed. The tensile stress–strain curves of cross-ply SiC/CAS and SiC/1723, 2D woven C/SiC and SiC/SiC composites corresponding to different damage stages are predicted.

### 1.3.1 Stress Analysis

Under tensile loading, the stress–strain curve appears linear-elastically before damage occurred inside of composite. When transverse cracking and matrix cracking occurred in cross-ply composite, the unit cell for cross-ply analysis is shown in Fig. 1.19. The transverse crack spacing in 90° plies is $l_t$, and $l_c$ is the matrix crack spacing in the 0° plies. Due to transverse cracking distributing relative to matrix cracking, the fiber/matrix interface debonded lengths on the left and right sides of unit cell are different, in which $l_{dr}$ and $l_{dl}$ denote the right and left interface debonded lengths, respectively.

Three entities of cross-ply composite are considered, i.e., 90° ply, fiber in 0° ply and matrix in 0° ply. The axial stress of 90° ply, i.e., $\sigma_T^o(x)$, fiber and matrix axial stresses of 0° ply, i.e., $\sigma_f^o(x)$ and $\sigma_m^o(x)$, can be described using the following equations [55].

$$\sigma_T^o(x) = \frac{E_2}{E_c}\sigma + E_2(\alpha_c - \alpha_2)\Delta T \tag{1.45}$$

$$\sigma_f^o(x) = \frac{E_f}{E_c}\sigma + E_f(\alpha_c - \alpha_f)\Delta T \tag{1.46}$$

$$\sigma_m^o(x) = \frac{E_m}{E_c}\sigma + E_m(\alpha_c - \alpha_m)\Delta T \tag{1.47}$$

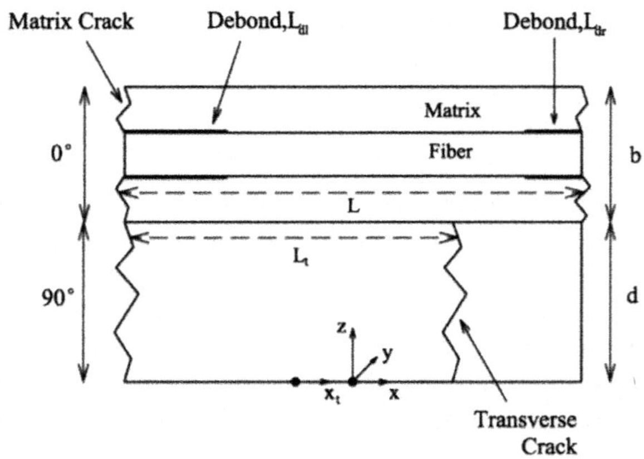


**Fig. 1.19** Unit cell of cross-ply CMCs. Reprinted with permission from Ref. [59]. Copyright 2015, Springer Science Business Media New York

where $E_2$ and $E_c$ denote the 90° ply and composite elastic modulus, respectively; $\alpha_2$ and $\alpha_c$ denote the 90° ply and composite thermal expansion coefficient, respectively.

The composite elastic modulus of $E_c$ can be determined using the following equation.

$$E_c = \frac{bE_1 + dE_2}{b + d} \tag{1.48}$$

where $b$ and $d$ denote the thickness of 0° ply and half of 90° ply; and $E_1$ denotes the 0° ply elastic modulus.

The governing differential equations for the axial stress in the 90° ply, i.e., $\sigma_T(x_t)$, and the axial stress of fiber in the 0° ply, i.e., $\sigma_f(x)$, can be described using the following equations [55].

$$\frac{d^2\sigma_T(x_t)}{dx^2} = \lambda^2\left[\sigma_T(x_t) - \sigma_T^o\right] \tag{1.49}$$

$$\frac{d^2\sigma_f(x)}{dx^2} - \beta^2\sigma_f(x) = -\beta^2\left[\sigma_f^o - \frac{E_f}{E_1}\frac{d}{b}\left(\sigma_T(x) - \sigma_T^o\right)\right] \tag{1.50}$$

where $\lambda$ and $\beta$ are shear-lag parameters; and $x_t$ denotes the local coordinate in the 90° ply as shown in Fig. 1.19.

The solutions of Eqs. (1.49) and (1.50) provide the expression of $\sigma_T(x_t)$ and $\sigma_f(x)$. The matrix axial stress in the 0° ply is determined using the following equation.

$$\sigma_m(x) = \frac{1}{V_m}\left[\frac{b+d}{b}\sigma - \frac{d}{b}\sigma_T\left(x_{t|x}\right)\right] - \frac{V_f}{V_m}\sigma_f(x) \tag{1.51}$$

where $\sigma_T\left(x_{t|x}\right)$ is $\sigma_T(x_t)$ evaluated at $x_t = x$ and $V_f$ and $V_m$ denote the volume fraction of the fiber and the matrix, respectively.

The fiber axial stress distribution of cross-ply CMCs after transverse cracking and matrix cracking is shown in Fig. 1.20. It can be found that the transverse cracking and matrix cracking affect the microstress field of the damaged composite.

## 1.3.2 Damage Models

### 1.3.2.1 Transverse Multicracking

Under tensile loading of cross-ply CMCs, transverse cracking in the 90° layers occurs first, leading to the early non-linear behavior. There are three main approaches to predict the transverse cracking evolution in 90° ply, i.e., the

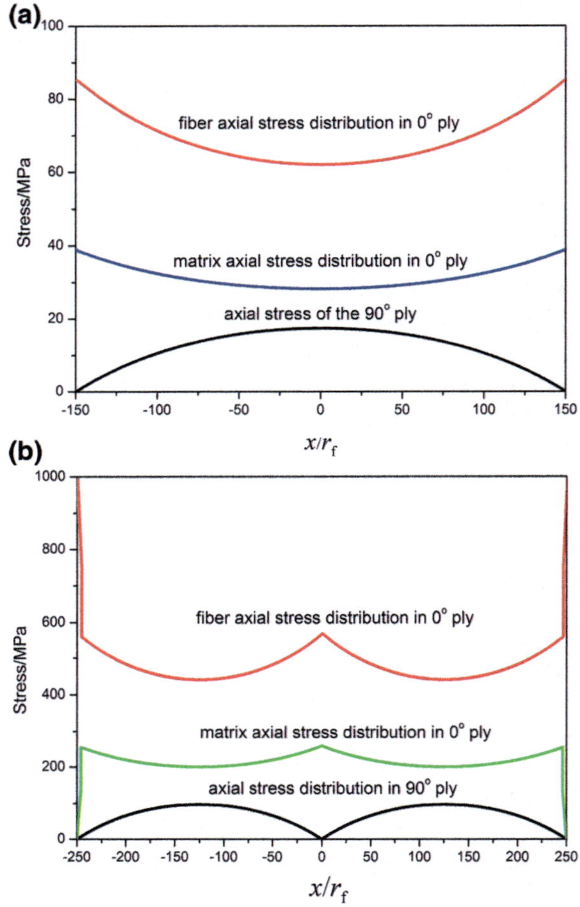


**Fig. 1.20** Fiber axial stress distribution of cross-ply CMCs after **a** transverse cracking; and **b** matrix cracking

maximum stress approach, the fracture mechanics approach, and the statistical failure approach. The maximum stress approach [60] assumes that the matrix of 90° ply exists a single strength, and when the applied stress approaches to the failure strength of the matrix, the transverse cracking occurs. The fracture mechanics approach [61] assumes that when the strain energy release rate in 90° ply approaches to the critical value, the transverse cracking occurs. The statistical failure approach [62] considers the distribution of the internal flaw and failure strength, when the local stress approaches to the failure strength, the transverse cracking occurs.

In the present analysis, the fracture mechanics approach developed by Laws and Dvorak [61] is adopted to determine the transverse multicracking evolution.

$$\zeta_t = \frac{(b+d)E_c}{b\lambda E_1 E_2}\left(\sigma_t^R + \frac{E_2}{E_c}\sigma\right)^2\left[2\tanh\left(\frac{\lambda l_t}{2}\right) - \tanh(\lambda l_t)\right] \tag{1.52}$$

where $\zeta_t$ denotes the transverse fracture energy; and $\sigma_t^R$ denotes the 90° ply thermal residual stress.

The initial transverse cracking stress of $\sigma_{cr}^{90}$ can be determined using the following equation.

$$\sigma_{cr}^{90} = \left(\lambda\frac{b}{b+d}\frac{E_1 E_c}{E_2}\zeta_t\right)^{1/2} - \frac{E_c}{E_2}\sigma_t^R \tag{1.53}$$

#### 1.3.2.2 Matrix Multicracking

The instantaneous matrix crack spacing is determined using the following equation.

$$l_c = l_{sat}\left\{1 - \exp\left[-\left(\frac{\sigma_m}{\sigma_R}\right)^m\right]\right\}^{-1} \tag{1.54}$$

#### 1.3.2.3 Interface Debonding

The fracture mechanics approach is adopted to determine fiber/matrix interface debonding and is given by the following equation [40].

$$\zeta_d = \frac{F}{4\pi r_f}\frac{\partial w_f(x = l_c/2)}{\partial l_{dr}} - \frac{1}{2}\int_{\frac{l_c}{2}-l_{dr}}^{\frac{l_c}{2}} \tau_i \frac{\partial v(x)}{\partial l_{dr}}dx \tag{1.55}$$

where $w_f(x = l_c/2)$ denotes the fiber axial displacement at the matrix crack plane; and $l_{dr}$ denotes the right interface debonded length in the unit cell.

The fiber and matrix axial displacements, i.e., $w_f(x)$ and $w_m(x)$, can be described using the following equations.

$$\begin{aligned} w_f(x) &= \int_{\frac{l_c}{2}-l_{dr}}^{x} \frac{\sigma_f(x)}{E_f}dx \\ &= \frac{1}{V_f E_f}\left[\frac{b+d}{b}\sigma - \frac{d}{b}\sigma_T\left(x_{t|x=l_c/2}\right)\right]\left[x - \left(\frac{l_c}{2} - l_{dr}\right)\right] + \frac{\tau_i}{r_f E_f}\left[\left(\frac{l_c}{2} - x\right)^2 - l_{dr}^2\right] \end{aligned} \tag{1.56}$$

$$
\begin{aligned}
w_{\mathrm{m}}(x) &= \int_{\frac{l_c}{2}-l_{\mathrm{dr}}}^{x} \frac{\sigma_{\mathrm{m}}(x)}{E_{\mathrm{m}}} \mathrm{d}x \\
&= \frac{1}{V_{\mathrm{m}}E_{\mathrm{m}}}\left[\frac{b+d}{b}\sigma - \frac{d}{b}\sigma_{\mathrm{T}}^{\mathrm{o}}\right]\left[x - \left(\frac{l_{\mathrm{c}}}{2} - l_{\mathrm{dr}}\right)\right] \\
&\quad + \frac{d}{b}\frac{\sigma_{\mathrm{T}}^{\mathrm{o}}}{\lambda V_{\mathrm{m}}E_{\mathrm{m}}}\frac{1}{\cosh\left(\frac{\lambda l_{\mathrm{t}}}{2}\right)}\left\{\sinh\left[\lambda\left(x + \frac{l_{\mathrm{c}}}{2} - \frac{l_{\mathrm{t}}}{2}\right)\right] - \sinh\left[\lambda\left(l_{\mathrm{c}} - \frac{l_{\mathrm{t}}}{2} - l_{\mathrm{dr}}\right)\right]\right\} \\
&\quad - \frac{1}{V_{\mathrm{m}}E_{\mathrm{m}}}\left[\frac{b+d}{b}\sigma - \frac{d}{b}\sigma_{\mathrm{T}}\left(x_{t|x=l_c/2}\right)\right]\left[x - \left(\frac{l_{\mathrm{c}}}{2} - l_{\mathrm{dr}}\right)\right] - \frac{V_{\mathrm{f}}}{V_{\mathrm{m}}E_{\mathrm{m}}}\frac{\tau_{\mathrm{i}}}{r_{\mathrm{f}}}\left[\left(\frac{l_{\mathrm{c}}}{2} - x\right)^2 - l_{\mathrm{dr}}^2\right]
\end{aligned}
\tag{1.57}
$$

Using Eqs. (1.56) and (1.57), the relative displacement between the fiber and the matrix of $v(x)$ can be described using the following equation.

$$
\begin{aligned}
v(x) &= \left|w_{\mathrm{f}}(x) - w_{\mathrm{m}}(x)\right| \\
&= \frac{E_1}{V_{\mathrm{f}}V_{\mathrm{m}}E_{\mathrm{f}}E_{\mathrm{m}}}\left[\frac{b+d}{b}\sigma - \frac{d}{b}\sigma_{\mathrm{T}}\left(x_{t|x=\frac{l_c}{2}}\right)\right]\left[x - \left(\frac{l_{\mathrm{c}}}{2} - L_{\mathrm{dr}}\right)\right] \\
&\quad - \frac{d}{b}\frac{\sigma_{\mathrm{T}}^{\mathrm{o}}}{\lambda V_{\mathrm{m}}E_{\mathrm{m}}}\frac{1}{\cosh\left(\lambda\frac{l_{\mathrm{t}}}{2}\right)}\left\{\sinh\left[\lambda\left(x + \frac{l_{\mathrm{c}}}{2} - \frac{l_{\mathrm{t}}}{2}\right)\right] - \sinh\left[\lambda\left(l_{\mathrm{c}} - \frac{l_t}{2} - l_{\mathrm{dr}}\right)\right]\right\} \\
&\quad + \frac{E_1}{V_{\mathrm{m}}E_{\mathrm{f}}E_{\mathrm{m}}}\frac{\tau_{\mathrm{i}}}{r_{\mathrm{f}}}\left[\left(\frac{l_{\mathrm{c}}}{2} - x\right)^2 - l_{\mathrm{dr}}^2\right] - \frac{1}{V_{\mathrm{m}}E_{\mathrm{m}}}\left[\frac{b+d}{b}\sigma - \frac{d}{b}\sigma_{\mathrm{T}}^{\mathrm{o}}\right]\left[x - \left(\frac{l_{\mathrm{c}}}{2} - l_{\mathrm{dr}}\right)\right]
\end{aligned}
\tag{1.58}
$$

Substituting $w_{\mathrm{f}}(x = l_{\mathrm{c}}/2)$ and $v(x)$ into Eq. (1.55), it leads to the following equation for $l_{\mathrm{dr}}$.

$$
\begin{aligned}
& l_{\mathrm{dr}}^2 - \frac{r_{\mathrm{f}}V_{\mathrm{m}}E_{\mathrm{m}}}{2V_{\mathrm{f}}E_1}\frac{\sigma}{\tau_{\mathrm{i}}}l_{\mathrm{dr}} - \frac{r_{\mathrm{f}}}{2V_{\mathrm{f}}\tau_{\mathrm{i}}}\left[\frac{b+d}{b}\sigma - \frac{d}{b}\sigma_{\mathrm{T}}\left(x_{t|x=l_c/2}\right)\right]l_{\mathrm{dr}} \\
&\quad + \frac{r_{\mathrm{f}}E_{\mathrm{f}}}{2E_1\tau_{\mathrm{i}}}\left[\frac{b+d}{b}\sigma - \frac{d}{b}\sigma_{\mathrm{T}}^{\mathrm{o}}\right]l_{\mathrm{dr}} + \frac{r_{\mathrm{f}}E_{\mathrm{f}}}{2E_1\tau_{\mathrm{i}}}\frac{d}{b}\sigma_{\mathrm{T}}^{\mathrm{o}}\frac{\cosh\left[\lambda\left(l_{\mathrm{c}} - \frac{l_{\mathrm{t}}}{2} - l_{\mathrm{dr}}\right)\right]}{\cosh\left(\lambda\frac{l_{\mathrm{t}}}{2}\right)}l_{\mathrm{dr}} \\
&\quad + \frac{r_{\mathrm{f}}^2V_{\mathrm{m}}E_{\mathrm{m}}\sigma}{4V_{\mathrm{f}}^2E_1\tau_{\mathrm{i}}^2}\left[\frac{b+d}{b}\sigma - \frac{d}{b}\sigma_{\mathrm{T}}\left(x_{t|x=l_c/2}\right)\right] - \frac{r_{\mathrm{f}}V_{\mathrm{m}}E_{\mathrm{f}}E_{\mathrm{m}}}{E_1\tau_{\mathrm{i}}^2}\zeta_{\mathrm{d}} = 0
\end{aligned}
\tag{1.59}
$$

The left fiber/matrix interface debonded length of $l_{\mathrm{dl}}$ satisfies the following equation.

$$\begin{aligned} & l_{\mathrm{dl}}^2-\frac{r_{\mathrm{f}} V_{\mathrm{m}} E_{\mathrm{m}}}{2 V_{\mathrm{f}} E_1} \frac{\sigma}{\tau_{\mathrm{i}}} l_{\mathrm{dl}}-\frac{r_{\mathrm{f}}}{2 V_{\mathrm{f}} \tau_{\mathrm{i}}}\left[\frac{b+d}{b} \sigma-\frac{d}{b} \sigma_{\mathrm{T}}\left(x_{t \mid x=-l_c / 2}\right)\right] l_{\mathrm{dl}} \\ & +\frac{r_{\mathrm{f}} E_{\mathrm{f}}}{2 E_1 \tau_{\mathrm{i}}}\left[\frac{b+d}{b} \sigma-\frac{d}{b} \sigma_{\mathrm{T}}^{\mathrm{o}}\right] l_{\mathrm{dl}}+\frac{r_{\mathrm{f}} E_{\mathrm{f}}}{2 E_1 \tau_{\mathrm{i}}} \frac{d}{b} \sigma_{\mathrm{T}}^{\mathrm{o}} \frac{\cosh \left[\lambda\left(\frac{l_{\mathrm{c}}}{2}-l_{\mathrm{dl}}\right)\right]}{\cosh \left(\lambda \frac{L_{\mathrm{t}}}{2}\right)} l_{\mathrm{dl}} \\ & +\frac{r_{\mathrm{f}}^2 V_{\mathrm{m}} E_{\mathrm{m}} \sigma}{4 V_{\mathrm{f}}^2 E_1 \tau_{\mathrm{i}}^2}\left[\frac{b+d}{b} \sigma-\frac{d}{b} \sigma_{\mathrm{T}}\left(x_{t \mid x=-l_c / 2}\right)\right]-\frac{r_{\mathrm{f}} V_{\mathrm{m}} E_{\mathrm{f}} E_{\mathrm{m}}}{E_1 \tau_{\mathrm{i}}^2} \zeta_{\mathrm{d}}=0 \end{aligned} \tag{1.60}$$

### 1.3.2.4 Fiber Failure

The GLS criterion is used to determine the stress sharing between the intact and fracture fibers [45].

$$\frac{b+d}{b} \frac{\sigma}{V_{\mathrm{f}}}=[1-P(T)] T+\frac{2 \tau_i}{r_{\mathrm{f}}}\langle L\rangle P(T) \tag{1.61}$$

and

$$P(T)=1-\exp \left[-\left(\frac{T}{\sigma_{\mathrm{c}}}\right)^{m_{\mathrm{f}}+1}\right] \tag{1.62}$$

The mechanical equilibrium requires the following equation [45].

$$\frac{b+d}{b} \frac{\sigma}{V_{\mathrm{f}}}=[1-P(T)] T+P(T) \frac{T}{2}=\left(1-\frac{1}{2} P(T)\right) T \tag{1.63}$$

To obtain the analytical results, it is to approximate the fiber failure probability as the following equation.

$$P(T)=1-\exp \left[-\left(\frac{T}{\sigma_{\mathrm{c}}}\right)^{m_{\mathrm{f}}+1}\right]=1-\left[1-\left(\frac{T}{\sigma_{\mathrm{c}}}\right)^{m_{\mathrm{f}}+1}\right]=\left(\frac{T}{\sigma_{\mathrm{c}}}\right)^{m_{\mathrm{f}}+1} \tag{1.64}$$

Substituting Eq. (1.64) into Eq. (1.63), it leads to the form of the following equation.

$$\frac{\sigma}{V_{\mathrm{f}}}=\frac{b}{b+d} T\left[1-\frac{1}{2}\left(\frac{T}{\sigma_{\mathrm{c}}}\right)^{m_{\mathrm{f}}+1}\right] \tag{1.65}$$

Calculating the maximum value of $T$, the composite ultimate tensile strength of $\sigma_{\mathrm{UTS}}$ is given by the following equation.

$$\sigma_{\mathrm{UTS}} = V_{\mathrm{fL}}\sigma_{\mathrm{c}}\left(\frac{2}{m_{\mathrm{f}}+2}\right)^{\frac{1}{m_{\mathrm{f}}+1}}\left(\frac{m_{\mathrm{f}}+1}{m_{\mathrm{f}}+2}\right) \tag{1.66}$$

where $V_{\mathrm{fL}}(= V_{\mathrm{f}}b/(b + d))$ denotes the volume fraction of fibers in longitudinal direction.

#### 1.3.2.5 Stress–Strain Relationship

For modeling the tensile stress–strain curves, the linear-elastic region can be described using the following equation.

$$\varepsilon_{\mathrm{c}} = \sigma/E_{\mathrm{c}} \tag{1.67}$$

When damages of transverse cracking, matrix cracking, fiber/matrix interface debonding, and fibers failure occurred inside of CMCs, the composite strain can be determined using the following equation.

$$\varepsilon_{\mathrm{c}} = \frac{2}{E_{\mathrm{f}}l_{\mathrm{c}}}\int\limits_{l_{\mathrm{c}}/2}\sigma_{\mathrm{f}}(x)\mathrm{d}x - (\alpha_{\mathrm{c}} - \alpha_{\mathrm{f}})\Delta T \tag{1.68}$$

The fiber axial stress of $\sigma_{\mathrm{f}}(x)$ depends on the extent of transverse cracking, matrix cracking, fiber/matrix interface debonding, and fibers fracture. The transverse crack spacing is determined by Eq. (1.52), the matrix crack spacing is determined by Eq. (1.54), the fiber/matrix interface debonded length is determined by Eqs. (1.59) and (1.60), and the intact fiber stress at the matrix crack plane is determined by Eq. (1.65). Combining cross-ply shear-lag model and failure criterions, the tensile stress–strain curve of cross-ply CMCs can be predicted for each damage stage.

### *1.3.3 Results and Discussions*

Under tensile loading of cross-ply CMCs, the damage mechanisms of transverse cracking and fibers failure affect the nonlinear behavior of tensile stress–strain curves. The effects of transverse fracture energy and fiber Weibull modulus on the damage and tensile stress–strain curve of cross-ply SiC/CAS composite are analysed. The material properties of cross-ply SiC/CAS composite are given by $E_{\mathrm{f}}$ = 210 GPa, $E_{\mathrm{m}}$ = 90 GPa, $V_{\mathrm{f}}$ = 38%, $r_{\mathrm{f}}$ = 7.5 μm, $\alpha_{\mathrm{f}} = 3.1\times10^{-6}$/°C, $\alpha_{\mathrm{m}} = 4.5\times10^{-6}$/°C, and $\Delta T = -1000$ °C.

#### 1.3.3.1 Effect of Transverse Fracture Energy

The effect of transverse fracture energy on transverse cracking evolution and tensile stress–strain curve is shown in Fig. 1.21. With increasing transverse fracture energy, the critical transverse cracking stress increases, leading to the increase of deviation stress of tensile stress–strain curve. The transverse cracking density decreases with increasing transverse fracture energy, leading to the degradation of composite stiffness.

#### 1.3.3.2 Effect of Fiber Weibull Modulus

The effect of fiber Weibull modulus (i.e., $m_f = 2$ and 4) on the fiber failure and tensile stress–strain curve of cross-ply CMCs is shown in Fig. 1.22. When the fiber Weibull modulus increases, the broken fibers fraction decrease at the same applied stress, and the composite failure strain decreases.

### *1.3.4 Experimental Comparisons*

The damage evolution and tensile stress–strain curves of cross-ply SiC/CAS, SiC/CAS-II, SiC/1723, 2D C/SiC, and SiC/SiC composites at room temperature are predicted using the present analysis.

#### 1.3.4.1 Cross-Ply CMCs

Beyerle et al. [63] investigated the tensile behavior of cross-ply $[0/90]_{4s}$ SiC/CAS composite at room temperature. The experimental and theoretical predicted transverse multicracking evolution, matrix multicracking evolution, and the tensile stress–strain curves of cross-ply SiC/CAS composite are shown in Fig. 1.23. Under tensile loading, the transverse multicracking evolution and matrix multicracking evolution are observed using the optical microscope and acoustic emission. The experimental transverse multicracking evolution and matrix multicracking evolution curves are shown in Fig. 1.23a, b. For transverse cracking, the first transverse cracking stress is about 40 MPa, and the saturation transverse cracking stress is about 80 MPa, and for matrix cracking, the first matrix cracking stress is about 70 MPa and the saturation matrix cracking stress is about 170 MPa. The experimental tensile stress–strain curve of cross-ply SiC/CAS composite is shown in Fig. 1.23c. The tensile stress–strain curve can be divided into four stages, i.e., initial

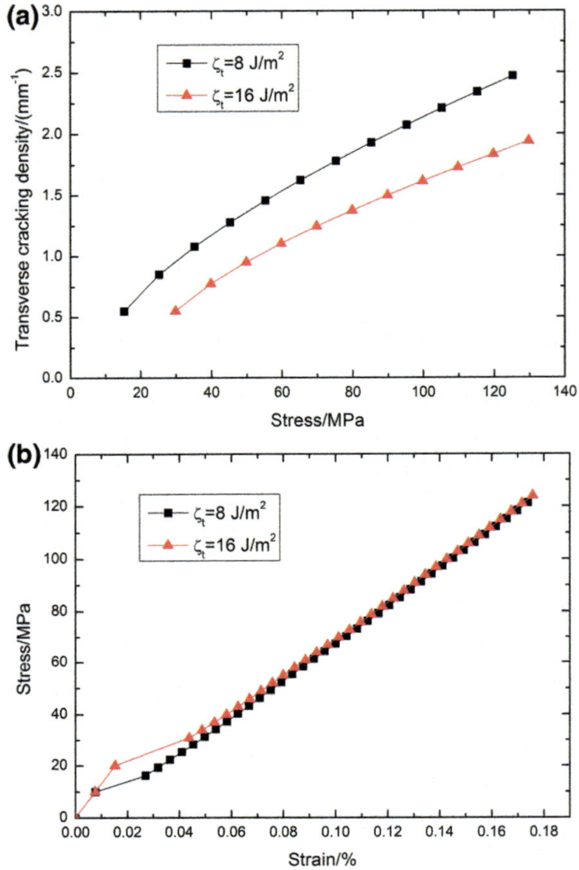


**Fig. 1.21** Effect of transverse fracture energy on **a** transverse cracking; and **b** tensile stress–strain curve

linear-elastic region (i.e., from initial loading transverse cracking), first nonlinear region (i.e., from first transverse cracking to first matrix cracking), second nonlinear region (i.e., from first matrix cracking to matrix cracking saturation), and final fibers failure region (i.e., from the saturation matrix cracking to final composite fracture). The composite tensile strength is about 230 MPa, and the failure strain is about 0.8%. The material properties of cross-ply $[0/90]_{4s}$ SiC/CAS composite are given by Table 1.2. The predicted transverse multicracking, matrix multicracking, and tensile stress–strain curves are shown in Fig. 1.22a–c, respectively, and the predicted stress–strain curve of each damage stage agreed with experimental data.

Pryce and Smith [53] investigated the tensile behavior of cross-ply $[0/90]_{3s}$ SiC/CAS composite at room temperature. The experimental and theoretical predicted transverse multicracking evolution, matrix multicracking evolution, and the tensile stress–strain curves of cross-ply SiC/CAS composite are shown in Fig. 1.24. Under tensile loading, the transverse multicracking evolution and matrix multicracking evolution are observed using the optical microscope and acoustic emission. The experimental transverse multicracking evolution and matrix multicracking

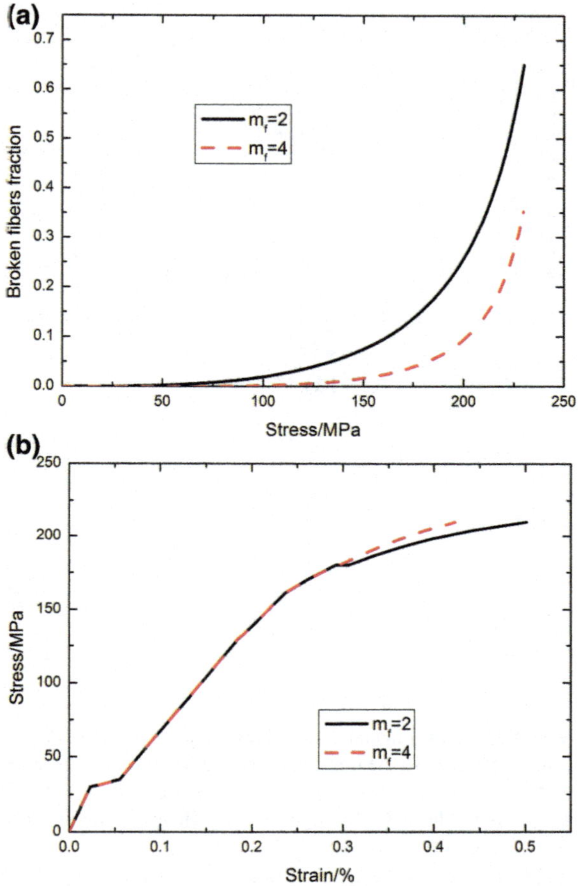


**Fig. 1.22** Effect of fiber Weibull modulus on **a** fibers failure; and **b** tensile stress–strain curve

evolution curves are shown in Fig. 1.24a, b. For transverse cracking, the first transverse cracking stress is about 40 MPa, and the saturation transverse cracking stress is about 80 MPa, and for matrix cracking, the first matrix cracking stress is about 60 MPa and the saturation matrix cracking stress is about 130 MPa. The experimental tensile stress–strain curve of cross-ply SiC/CAS composite is shown in Fig. 1.24c. The tensile stress–strain curve can be divided into four stages, i.e., initial linear-elastic region (i.e., from initial loading to transverse cracking), first nonlinear region (i.e., from first transverse cracking to first matrix cracking), second nonlinear region (i.e., from first matrix cracking to matrix cracking saturation), and final fibers failure region (i.e., from the saturation matrix cracking to final composite fracture). The composite tensile strength is about 143 MPa, and the failure strain is about 0.5%. The material properties of cross-ply $[0/90]_{3s}$ SiC/CAS composite are given by Table 1.2. The predicted transverse multicracking, matrix multicracking, and tensile stress–strain curves are shown in Fig. 1.24a–c,

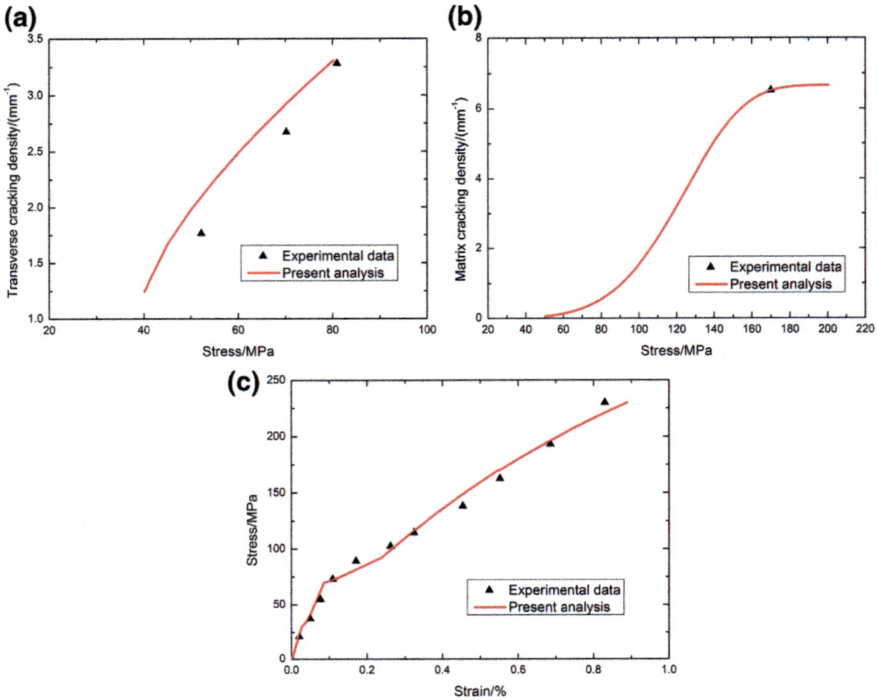


**Fig. 1.23** Experimental data and the theoretical analysis of cross-ply $[0/90]_{4s}$ SiC/CAS composite corresponding to **a** the transverse crack density versus the applied stress; **b** the matrix crack density versus the applied stress; and **c** the tensile stress–strain curve. Reprinted with permission from Ref. [59]. Copyright 2015, Springer Science Business Media New York

**Table 1.2** Material properties of cross-ply SiC/CAS and SiC/1723 composites

| Items | $[0/90]_{4s}$-SiC/CAS [63] | $[0/90]_{3s}$-SiC/CAS [53] | $[0_2/90_4]_s$-SiC/CAS [53] | $[0_3/90/0_3]$-SiC/CAS [64] | $[0_3/90_3/0_3]$-SiC/CAS [64] | $[0_3/90_2/0_3]$-SiC/CAS [65] | $[0/90]_{3s}$-SiC/1723 [58] |
|---|---|---|---|---|---|---|---|
| $E_f$ (GPa) | 200 | 190 | 190 | 210 | 210 | 193 | 190 |
| $E_m$ (GPa) | 85 | 90 | 90 | 95.5 | 95.5 | 98 | 88 |
| $V_f$ | 0.37 | 0.34 | 0.34 | 0.35 | 0.35 | 0.38 | 0.35 |
| $r_f$ (μm) | 7.5 | 8 | 8 | 7.5 | 7.5 | 7.5 | 7.5 |
| $\nu_f$ | 0.2 | 0.2 | 0.2 | 0.2 | 0.2 | 0.2 | 0.2 |
| $\nu_m$ | 0.2 | 0.2 | 0.2 | 0.2 | 0.2 | 0.2 | 0.2 |
| $\alpha_f$ ($10^{-6}$ ×°C$^{-1}$) | 4 | 3.3 | 3.3 | 4 | 3.1 | 4 | 3.1 |
| $\alpha_m$ ($10^{-6}$ × °C$^{-1}$) | 5 | 4.6 | 4.6 | 5 | 4.5 | 5 | 4.36 |
| $\Delta T$ (°C) | −1000 | −1000 | −1000 | −1000 | −1000 | −1000 | −1000 |
| $\zeta_m$ (J·m$^{-2}$) | 25 | 6 | 6 | 25 | 25 | 25 | 25 |
| $\tau_i$ (MPa) | 13 | 10 | 10 | 12 | 12 | 15 | 35 |
| $m$ | 5 | 5 | 6 | 6 | 6 | 6 | 6 |
| $m_f$ | 3.6 | 3.6 | 3.6 | 2.0 | 2.0 | 2.0 | 3.6 |
| $\sigma_{cr}^{90}$ (MPa) | 40 | 20 | 20 | 40 | 35 | 25 | 45 |
| $\sigma_{cr}^{0}$ (MPa) | 70 | 40 | 40 | 80 | 60 | 60 | 80 |
| $\sigma_R$ (MPa) | 130 | 95 | 78 | 200 | 145 | 135 | 150 |
| $\sigma_{UTS}$ (MPa) | 230 | 146 | 107 | 336 | 260 | 290 | 284 |

respectively, and the predicted stress–strain curve of each damage stage agreed with experimental data.

Pryce and Smith [53] investigated the tensile behavior of cross-ply $[0_2/90_4]_s$ SiC/CAS composite at room temperature. The experimental and theoretical predicted transverse multicracking evolution, matrix multicracking evolution, and the tensile stress–strain curves of cross-ply SiC/CAS composite are shown in Fig. 1.25. Under tensile loading, the transverse multicracking evolution and matrix multicracking evolution are observed using the optical microscope and acoustic emission. The experimental transverse multicracking evolution and matrix multicracking evolution curves are shown in Fig. 1.25a, b. For transverse cracking, the first transverse cracking stress is about 30 MPa, and the saturation transverse cracking stress is about 65 MPa, and for matrix cracking, the first matrix cracking stress is about 50 MPa and the saturation matrix cracking stress is about 100 MPa. The experimental tensile stress–strain curve of cross-ply SiC/CAS composite is shown in Fig. 1.25c. The tensile stress–strain curve can be divided into four stages, i.e., initial linear-elastic region (i.e., from initial loading to transverse cracking), first nonlinear region (i.e., from first transverse cracking to first matrix cracking), second nonlinear

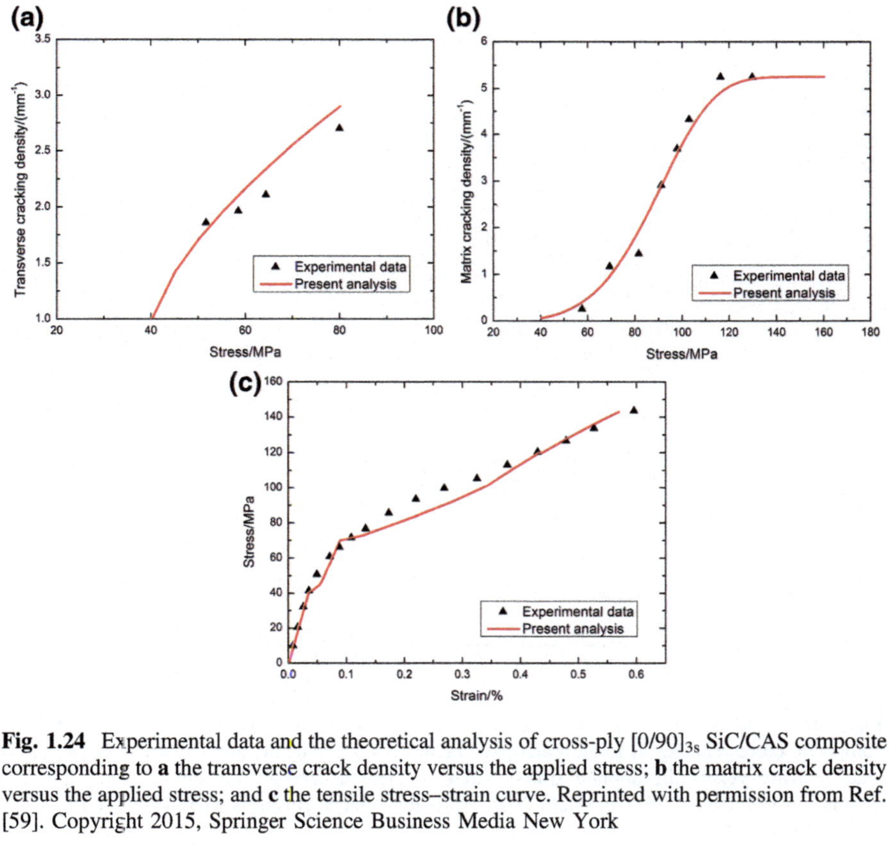


**Fig. 1.24** Experimental data and the theoretical analysis of cross-ply $[0/90]_{3s}$ SiC/CAS composite corresponding to **a** the transverse crack density versus the applied stress; **b** the matrix crack density versus the applied stress; and **c** the tensile stress–strain curve. Reprinted with permission from Ref. [59]. Copyright 2015, Springer Science Business Media New York

region (i.e., from first matrix cracking to matrix cracking saturation), and final fibers failure region (i.e., from the saturation matrix cracking to final composite fracture). The composite tensile strength is about 107 MPa, and the failure strain is about 0.6%. The material properties of cross-ply $[0_2/90_4]_s$ SiC/CAS composite are given by Table 1.2. The predicted transverse multicracking, matrix multicracking, and tensile stress–strain curves are shown in Fig. 1.25a–c, respectively, and the predicted stress–strain curve of each damage stage agreed with experimental data.

Karandikar and Chou [64] investigated the tensile behavior of cross-ply $[0_3/90/0_3]$ SiC/CAS composite at room temperature. The experimental and theoretical predicted transverse multicracking evolution, matrix multicracking evolution and the tensile stress–strain curves of cross-ply SiC/CAS composite are shown in Fig. 1.26. Under tensile loading, the transverse multicracking evolution and matrix multicracking evolution are observed using the optical microscope and acoustic emission. The experimental transverse multicracking evolution and matrix multicracking evolution curves are shown in Fig. 1.26a, b. For transverse cracking, the first transverse cracking stress is about 20 MPa, and the saturation transverse

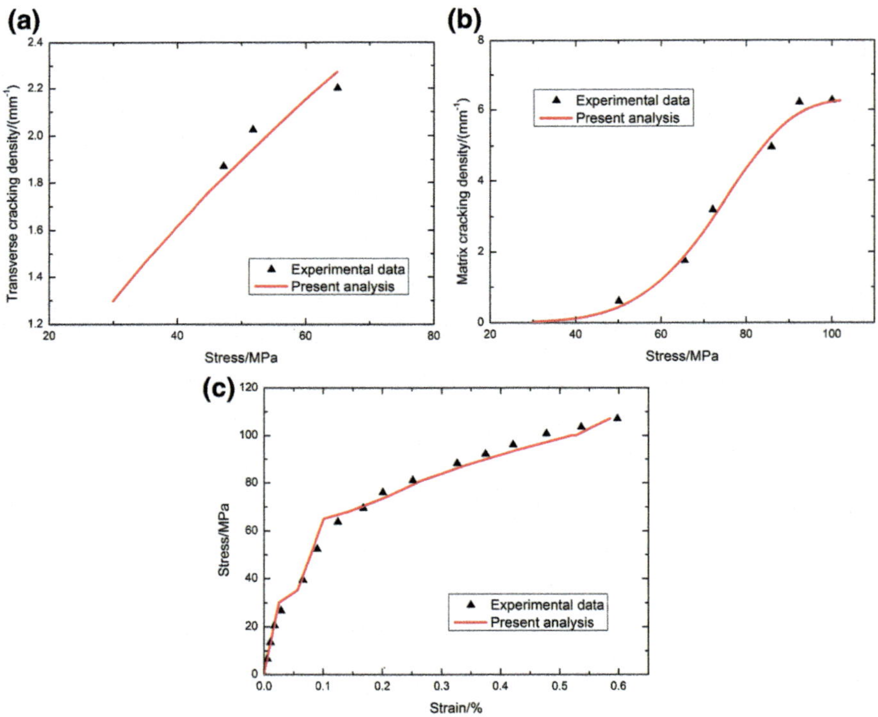


**Fig. 1.25** Experimental data and the theoretical analysis of cross-ply $[0_2/90_4]_s$ SiC/CAS composite corresponding to **a** the transverse crack density versus the applied stress; **b** the matrix crack density versus the applied stress; and **c** the tensile stress–strain curve. Reprinted with permission from Ref. [59]. Copyright 2015, Springer Science Business Media New York

cracking stress is about 80 MPa; and for matrix cracking, the first matrix cracking stress is about 100 MPa and the saturation matrix cracking stress is about 260 MPa. The experimental tensile stress–strain curve of cross-ply SiC/CAS composite is shown in Fig. 1.26c. The tensile stress–strain curve can be divided into four stages, i.e., initial linear-elastic region (i.e., from initial loading to transverse cracking), first nonlinear region (i.e., from first transverse cracking to first matrix cracking), second nonlinear region (i.e., from first matrix cracking to matrix cracking saturation), and final fibers failure region (i.e., from the saturation matrix cracking to final composite fracture). The composite tensile strength is about 334 MPa, and the failure strain is about 0.7%. The material properties of cross-ply $[0_3/90/0_3]$ SiC/CAS composite are given by Table 1.2. The predicted transverse multicracking, matrix multicracking, and tensile stress–strain curves are shown in Fig. 1.26a–c, respectively, and the predicted stress–strain curve of each damage stage agreed with experimental data.

Karandikar and Chou [64] investigated the tensile behavior of cross-ply $[0_3/90_3/0_3]$ SiC/CAS composite at room temperature. The experimental and theoretical predicted transverse multicracking evolution, matrix multicracking evolution and

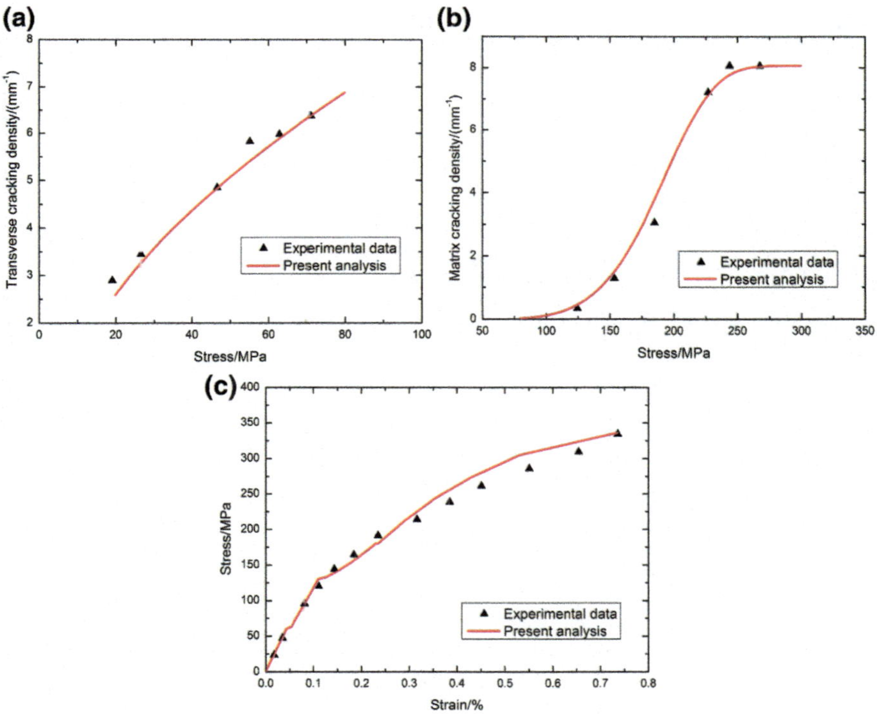


**Fig. 1.26** Experimental data and the theoretical analysis of cross-ply $[0_3/90/0_3]$ SiC/CAS composite corresponding to **a** the transverse crack density versus the applied stress; **b** the matrix crack density versus the applied stress; and **c** the tensile stress–strain curve. Reprinted with permission from Ref. [59]. Copyright 2015, Springer Science Business Media New York

the tensile stress–strain curves of cross-ply SiC/CAS composite are shown in Fig. 1.27. Under tensile loading, the transverse multicracking evolution and matrix multicracking evolution are observed using the optical microscope and acoustic emission. The experimental transverse multicracking evolution and matrix multicracking evolution curves are shown in Fig. 1.27a, b. For transverse cracking, the first transverse cracking stress is about 10 MPa, and the saturation transverse cracking stress is about 60 MPa; and for matrix cracking, the first matrix cracking stress is about 50 MPa and the saturation matrix cracking stress is about 200 MPa. The experimental tensile stress–strain curve of cross-ply SiC/CAS composite is shown in Fig. 1.27c. The tensile stress–strain curve can be divided into four stages, i.e., initial linear-elastic region (i.e., from initial loading to transverse cracking), first nonlinear region (i.e., from first transverse cracking to first matrix cracking), second nonlinear region (i.e., from first matrix cracking to matrix cracking saturation), and final fibers failure region (i.e., from the saturation matrix cracking to final composite fracture). The composite tensile strength is about 261 MPa, and the failure strain is about 0.75%. The material properties of cross-ply $[0_3/90_3/0_3]$ SiC/CAS composite are given by Table 1.2. The predicted transverse multicracking, matrix

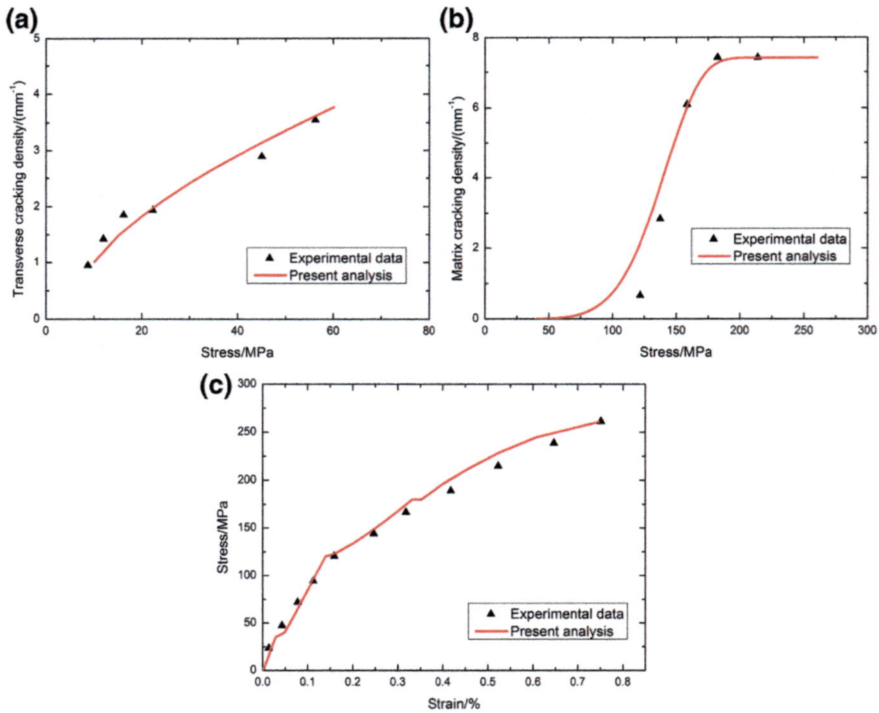


**Fig. 1.27** Experimental data and the theoretical analysis of cross-ply $[0_3/90_3/0_3]$ SiC/CAS composite corresponding to **a** the transverse crack density versus the applied stress; **b** the matrix crack density versus the applied stress; and **c** the tensile stress–strain curve. Reprinted with permission from Ref. [59]. Copyright 2015, Springer Science Business Media New York

multicracking, and tensile stress–strain curves are shown in Fig. 1.27a–c, respectively, and the predicted stress–strain curve of each damage stage agreed with experimental data.

Wang and Parvizi-Majidi [65] investigated the tensile behavior of cross-ply $[0_3/90_2/0_3]$ SiC/CAS composite at room temperature. The experimental and theoretical predicted transverse multicracking evolution, matrix multicracking evolution and the tensile stress–strain curves of cross-ply SiC/CAS composite are shown in Fig. 1.28. Under tensile loading, the transverse multicracking evolution and matrix multicracking evolution are observed using the optical microscope and acoustic emission. The experimental transverse multicracking evolution and matrix multicracking evolution curves are shown in Fig. 1.28a, b. For transverse cracking, the first transverse cracking stress is about 15 MPa, and the saturation transverse cracking stress is about 40 MPa; and for matrix cracking, the first matrix cracking stress is about 60 MPa and the saturation matrix cracking stress is about 160 MPa. The experimental tensile stress–strain curve of cross-ply SiC/CAS composite is shown in Fig. 1.28c. The tensile stress–strain curve can be divided into four stages, i.e., initial linear-elastic region (i.e., from initial loading to transverse cracking), first nonlinear region (i.e., from first transverse cracking to first matrix cracking), second nonlinear region (i.e., from first matrix cracking to matrix cracking saturation), and final fibers failure region (i.e., from the saturation matrix cracking to final composite fracture). The composite tensile strength is about 290 MPa, and the failure strain is about 0.7%. The material properties of cross-ply $[0_3/90_2/0_3]$ SiC/CAS composite are given by Table 1.2. The predicted transverse multicracking, matrix multicracking, and tensile stress–strain curves are shown in Fig. 1.28a–c, respectively, and the predicted stress–strain curve of each damage stage agreed with experimental data.

Zawada et al. [58] investigated the tensile behavior of cross-ply $[0/90]_{3s}$ SiC/1723 composite at room temperature. The experimental and theoretical predicted transverse multicracking evolution, matrix multicracking evolution and the tensile stress–strain curves of cross-ply SiC/CAS composite are shown in Fig. 1.29. For transverse cracking, the first transverse cracking stress is about 40 MPa, and the saturation transverse cracking stress is about 130 MPa; and for matrix cracking, the first matrix cracking stress is about 80 MPa and the saturation matrix cracking stress is about 200 MPa. The experimental tensile stress–strain curve of cross-ply SiC/CAS composite is shown in Fig. 1.29c. The tensile stress–strain curve can be divided into four stages, i.e., initial linear-elastic region (i.e., from initial loading to transverse cracking), first non-linear region (i.e., from first transverse cracking to first matrix cracking), second non-linear region (i.e., from first matrix cracking to matrix cracking saturation), and final fibers failure region (i.e., from the saturation matrix cracking to final composite fracture). The composite tensile strength is about 284 MPa, and the failure strain is about 0.6%. The material properties of cross-ply $[0/90]_{3s}$ SiC/1723 composite are given by Table 1.2. The predicted transverse multicracking, matrix multicracking, and tensile stress–strain curves are shown in Fig. 1.29a–c, respectively, and the predicted stress–strain curve of each damage stage agreed with experimental data.

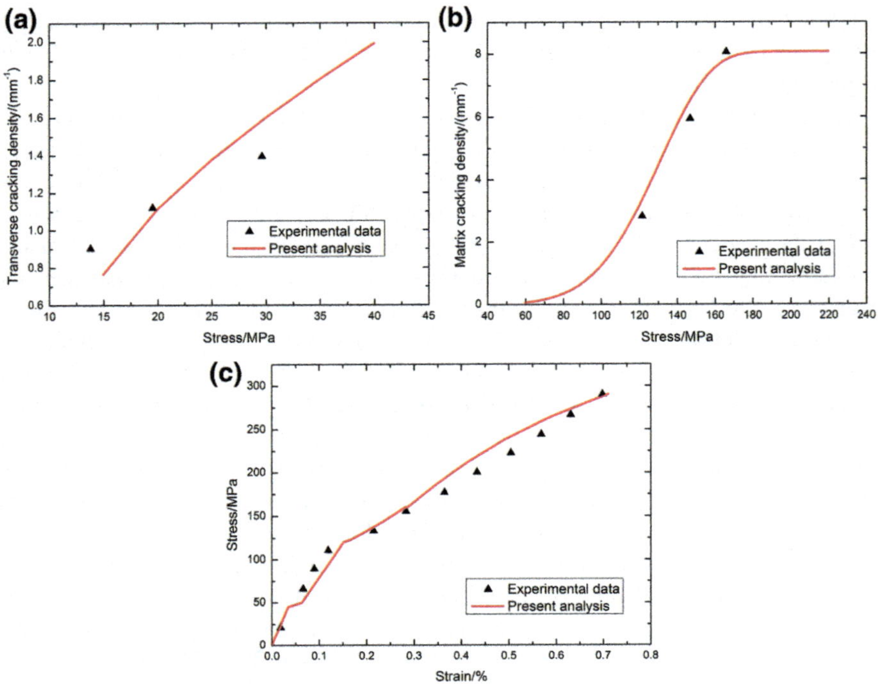


**Fig. 1.28** Experimental data and the theoretical analysis of cross-ply $[0_3/90_2/0_3]$ SiC/CAS composite corresponding to **a** the transverse crack density versus the applied stress; **b** the matrix crack density versus the applied stress; and **c** the tensile stress–strain curve. Reprinted with permission from Ref. [59]. Copyright 2015, Springer Science Business Media New York

#### 1.3.4.2 2D Woven CMCs

Wang et al. [66] investigated the tensile behavior of 2D C/SiC composite at room temperature. The monotonic tensile stress–strain curves of 2D C/SiC composite are predicted and compared with experimental data, as shown in Fig. 1.30. The material properties are given as $E_f = 230$ GPa, $E_m = 350$ GPa, $r_f = 3.5$ μm, $V_f = 42\%$, $m = 4$, $\tau_i = 8$ MPa, $m_f = 6$, and $\sigma_{uts} = 245$ MPa. The tensile stress–strain curves of 2D C/SiC composite can be divided into three regions, i.e., (1) region I, the initial linear-elastic region; (2) region II, the matrix multi-cracking and fiber/matrix interface debonding region; and (3) region III, the fiber failure region. The predicted each damage stage of tensile stress–strain curves agreed with experimental data.

Morscher [67] investigated the tensile behavior of 2D MI-Sylramic-SiC/SiC composite at room temperature. The monotonic tensile stress–strain curves of 2D MI-Sylramic-SiC/SiC composite are predicted and compared with experimental data, as shown in Fig. 1.31. The material properties of 2D MI-Sylramic-SiC/SiC composite are given in Table 1.3. The tensile stress–strain curves of 2D

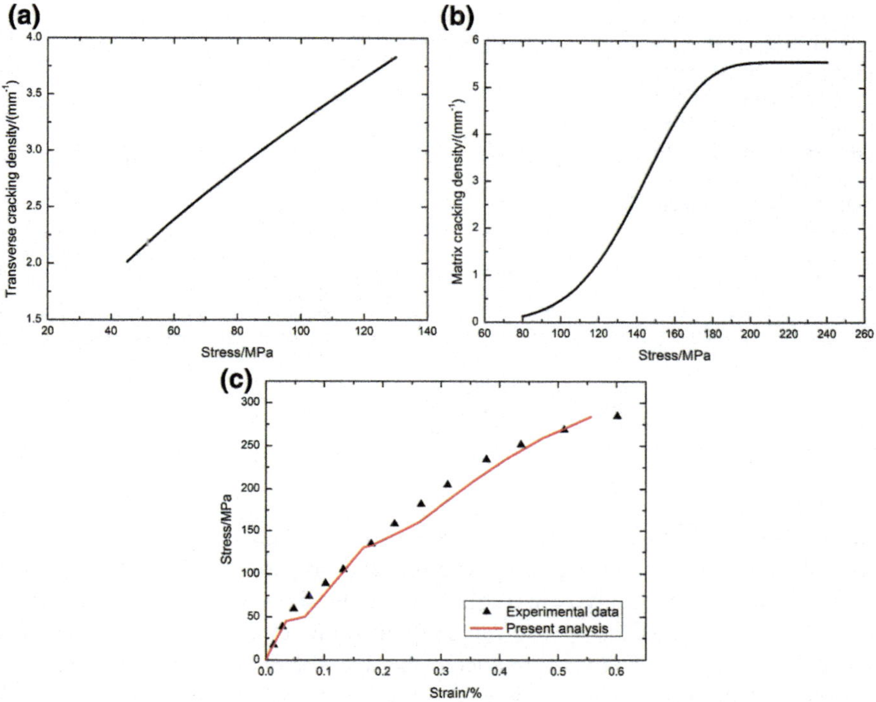


**Fig. 1.29** Experimental data and the theoretical analysis of cross-ply $[0/90]_{3s}$ SiC/1723 composite corresponding to **a** the transverse crack density versus the applied stress; **b** the matrix crack density versus the applied stress; and **c** the tensile stress–strain curve. Reprinted with permission from Ref. [59]. Copyright 2015, Springer Science Business Media New York

MI-Sylramic-SiC/SiC composite can be divided into three regions, i.e., the initial linear-elastic region; the matrix multicracking and fiber/matrix interface debonding region; and the fiber failure region. The predicted each damage stage of tensile stress–strain curves agreed with experimental data.

Morscher et al. [68] investigated the tensile behavior of 2D CVI-Hi-Nicalon-SiC/SiC composite at room temperature. The monotonic tensile stress–strain curves of 2D CVI-Hi-Nicalon-SiC/SiC composite are predicted and compared with experimental data, as shown in Fig. 1.32. The material properties of 2D CVI-Hi-Nicalon-SiC/SiC composite are given in Table 1.4. The tensile stress–strain curves of 2D CVI-Hi-Nicalon-SiC/SiC composite can be divided into three regions, i.e., the initial linear-elastic region; the matrix multicracking and fiber/matrix interface debonding region; and the fiber failure region. The predicted each damage stage of tensile stress–strain curves agreed with experimental data.

Morscher et al. [68] investigated the tensile behavior of 2D CVI-Sylramic-SiC/SiC composite at room temperature. The monotonic tensile stress–strain curves of

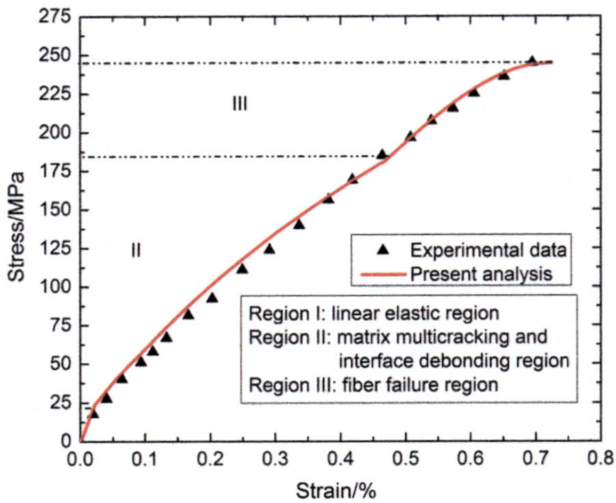


**Fig. 1.30** Experimental data and the theoretical analysis of monotonic tensile stress–strain curve of 2D C/SiC composite. Reprinted with permission from Ref. [69]. Copyright 2018, Springer Science Business Media New York

2D CVI-Sylramic-SiC/SiC composite are predicted and compared with experimental data, as shown in Fig. 1.33. The material properties of 2D CVI-Sylramic-SiC/SiC composite are given in Table 1.5. The tensile stress–strain curves of 2D CVI-Sylramic-SiC/SiC composite can be divided into three regions, i.e., the initial linear-elastic region; the matrix multicracking and fiber/matrix interface debonding region; and the fiber failure region. The predicted each damage stage of tensile stress–strain curves agreed with experimental data.

## 1.4 2.5D Woven Ceramic-Matrix Composites

Compared with tensile behavior of unidirectional, cross-ply and 2D woven CMCs, the damage evolution processes of 2.5D woven CMCs are even more complicated [69]. Dalmaz et al. [70] investigated damage evolution in 2.5D C/SiC composite under tensile loading conducted in situ in a scanning electron microscope. The cracking growth form preexisting cracks and undamaged zones in the transverse and longitudinal directions and sliding at the fiber/matrix interface or at the interface with adjacent fibers were observed under tensile loading. Wang et al. [66] investigated tensile behavior of 2.5D C/SiC composite using the acoustic emission. The composite shows a slightly linear region followed by a highly nonlinear region and finally a quasi-linear region up to fracture.

In this section, the tensile damage evolution and stress–strain behavior of 2.5D CMCs are investigated. Based on the microstructure analysis, the 2.5D woven CMCs can be divided into four elements, including longitudinal warp yarn, transverse weft yarn, matrix outside of yarns and open porosity. The damage mechanisms of matrix multicracking, fiber/matrix interface debonding, and fibers failure are determined using the matrix statistical cracking model, fiber/matrix

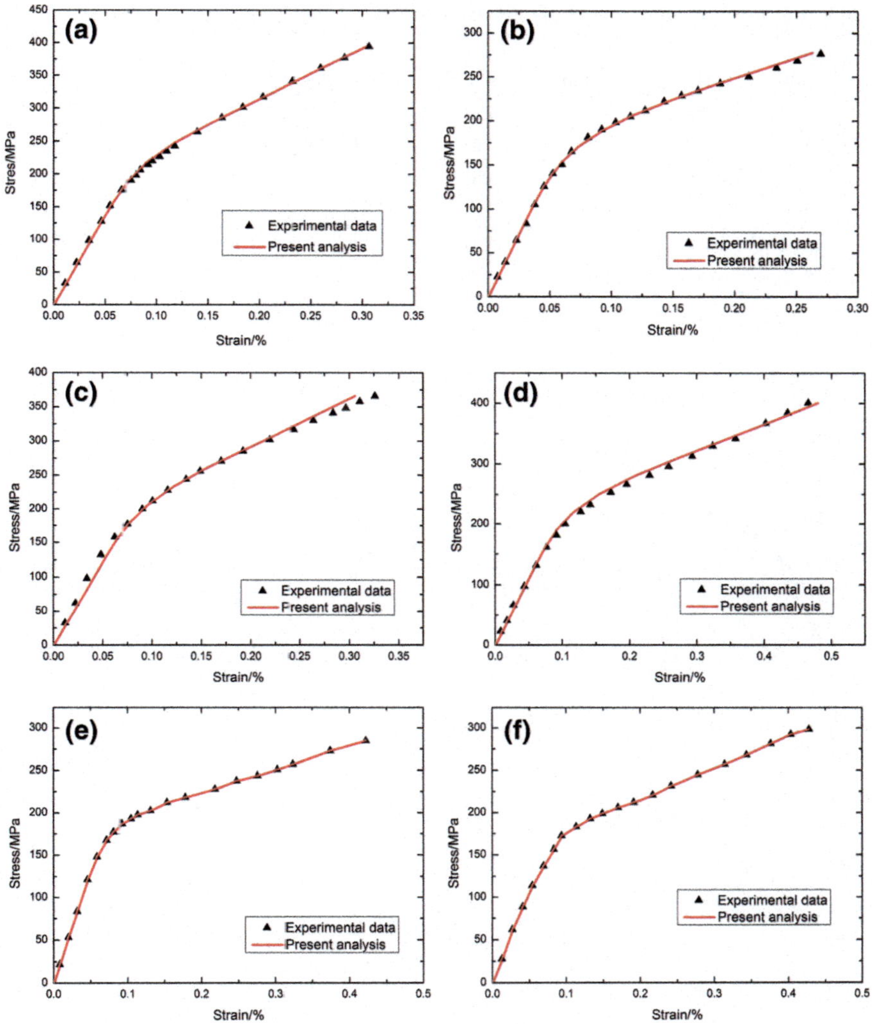


**Fig. 1.31** Experimental and theoretical tensile stress–strain curve of 2D MI-Sylramic-SiC/SiC composite

**Table 1.3** Material properties of 2D MI-Sylramic-SiC/SiC composite

| | $r_f$ (μm) | $V_f$ (%) | $E_f$ (GPa) | $E_m$ (GPa) | $E_c$ (GPa) | $\sigma_R$ (MPa) | $m$ | $\tau_i$ (MPa) |
|---|---|---|---|---|---|---|---|---|
| a | 5 | 40 | 380 | 450 | 277 | 150 | 4 | 25 |
| b | 5 | 28 | 380 | 450 | 289 | 150 | 4 | 30 |
| c | 5 | 36 | 380 | 450 | 246 | 150 | 4 | 20 |
| d | 5 | 34 | 380 | 450 | 224 | 150 | 4 | 27 |
| e | 5 | 24 | 380 | 450 | 279 | 150 | 4 | 40 |
| f | 5 | 30 | 380 | 450 | 219 | 150 | 4 | 11 |

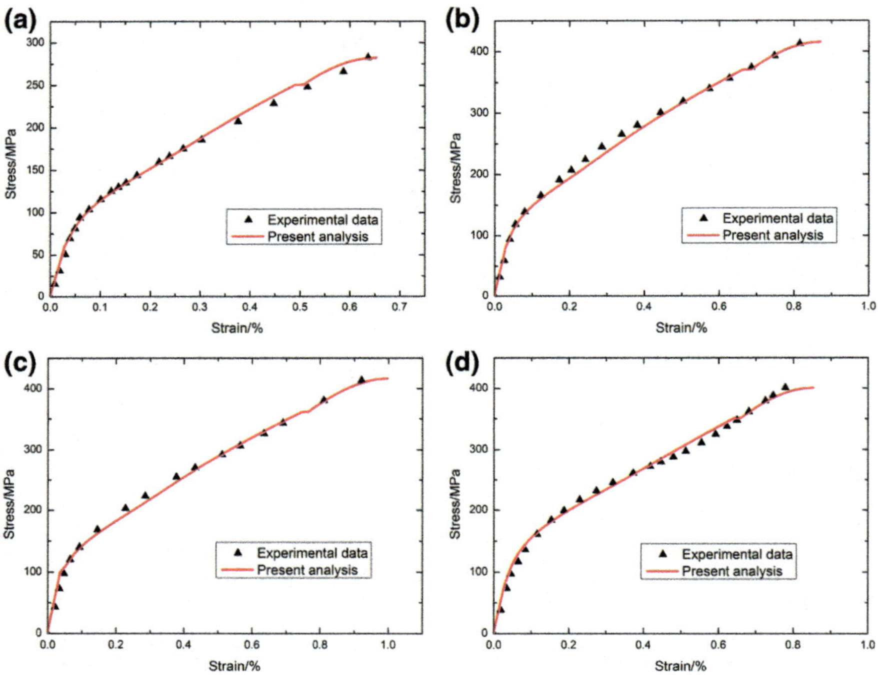


**Fig. 1.32** Experimental and theoretical tensile stress–strain curve of 2D CVI-Hi-Nicalon-SiC/SiC composite

**Table 1.4** Material properties of 2D CVI-Hi-Nicalon-SiC/SiC composite

| | $r_f$ (μm) | $V_f$ (%) | $E_f$ (GPa) | $E_m$ (GPa) | $E_c$ (GPa) | $\sigma_R$ (MPa) | $m$ | $\tau_i$ (MPa) | $m_f$ | $\sigma_{uts}$ (MPa) |
|---|---|---|---|---|---|---|---|---|---|---|
| a | 6.85 | 28 | 280 | 425 | 199 | 130 | 4 | 17 | 6 | 282 |
| b | 6.85 | 33 | 280 | 425 | 258 | 180 | 4 | 67 | 6 | 415 |
| c | 6.85 | 31 | 280 | 425 | 243 | 165 | 4 | 47 | 6 | 416 |
| d | 6.85 | 34 | 280 | 425 | 231 | 200 | 4 | 30 | 6 | 400 |

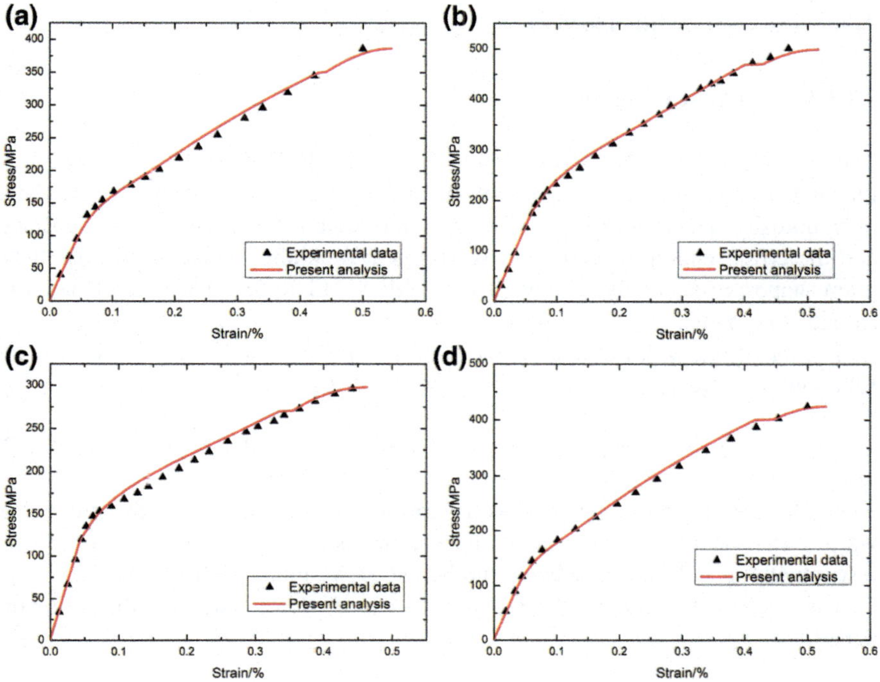


**Fig. 1.33** Experimental and theoretical tensile stress–strain curve of 2D CVI-Sylramic-SiC/SiC C composite

**Table 1.5** Material properties of 2D CVI-Hi-Sylramic-SiC/SiC composite

| | $r_f$ (μm) | $V_f$ (%) | $E_f$ (GPa) | $E_m$ (GPa) | $E_c$ (GPa) | $\sigma_R$ (MPa) | $m$ | $\tau_i$ (MPa) | $m_f$ | $\sigma_{uts}$ (MPa) |
|---|---|---|---|---|---|---|---|---|---|---|
| a | 3.5 | 40 | 230 | 350 | 81.2 | 50 | 4 | 6 | 6 | 252 |
| b | 3.5 | 42 | 230 | 350 | 108 | 50 | 4 | 8 | 6 | 245 |

interface fracture mechanics criterion, and fibers GLS failure criterion. The effects of fiber volume fraction and fiber Weibull modulus on the damage evolution and tensile stress–strain curve are analyzed. The tensile stress–strain curves of 2.5D C/SiC composites are predicted by combining damage models and shear-lag model.

## 1.4.1 Theoretical Models

### 1.4.1.1 Geometric Model

The unit cell of 2.5D CMCs is shown in Fig. 1.34. The 2.5D CMCs can be divided into four elements, i.e., 0° warp yarn, 90° weft yarn, matrix outside of yarns, and open porosity. The cross section of warp yarn is assumed to exhibit rectangle shape, and the cross section of weft warp is assumed to exhibit lenticular and parallelogram shapes, respectively. The warp yarn is divided into four 1/4 arc parts and two straight line parts.

The length $L_x$, width $L_y$ and height $L_z$ of the unit cell can be described using the following equation.

$$L_x = 10(N_w - 1)/e_{90},\ L_y = 10N_j/e_0,\ L_z = 2\left(h_j + h_w\right) \tag{1.69}$$

where $N_w$ denotes the number of weft yarn in unit cell; $N_j$ denotes the number of warp yarn in unit cell; $e_{90}$ and $e_0$ denote the density of weft and warp yarns, respectively; and $h_j$ and $h_w$ denote the height of warp and weft yarns.

The cross-sectional area of warp yarn $A_j$ can be described using the following equation.

$$A_j = h_j w_j \tag{1.70}$$

where $w_j$ denotes the width of warp yarn cross section.

The cross-sectional area of the lenticular weft yarn can be described using the following equation.

$$A_{wl} = \frac{1}{8}h_w^2\left(k_w^2 + 1\right)^2 \arcsin\left(\frac{2k_w}{k_w^2 + 1}\right) - \frac{1}{4}h_w^2 k_w\left(k_w^2 - 1\right) \tag{1.71}$$

where $k_w$ is the ratio of major axis to minor axis of lenticulated cross section.

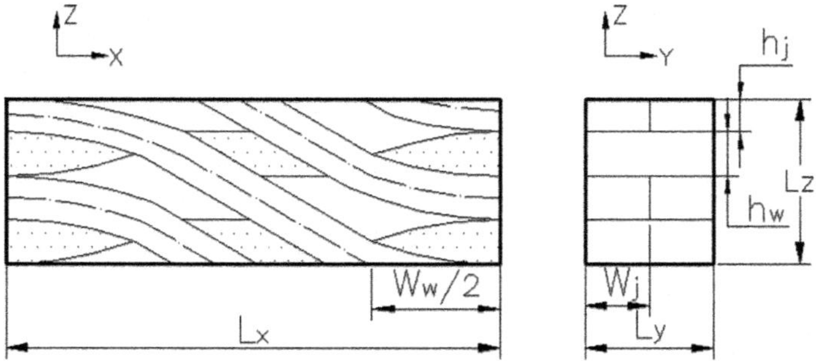


**Fig. 1.34** Schematic of unit cell for 2.5D CMCs. Reprinted with permission from Ref. [69]. Copyright 2018, Springer Science Business Media New York

$$k_w = w_w / h_w \tag{1.72}$$

The cross-sectional area of parallelogram weft yarn can be described using the following equation.

$$A_{wp} = h_{wp} w_{wp} \tag{1.73}$$

where $h_{wp}$ and $w_{wp}$ denote the height and width of parallelogram weft yarn.

The warp yarn is divided into four 1/4 arc parts and two straight line parts. The length of the warp yarn $l_j$ can be described using the following equation.

$$l_j = 4l_c + 2l_l \tag{1.74}$$

where

$$l_c = (R + h_j/2)\theta, l_l = \sqrt{[h_j(1 + \cos\theta) + h_w]^2 + [L_x - (2R + h_j)\sin\theta]^2} \tag{1.75}$$

where $R$ and $\theta$ denote the curvature radius and half chord angle of lenticulated cross section.

$$R = \frac{w_w^2 + h_w^2}{4h_w} = \frac{h_w}{4}(1 + k_w^2), \ \theta = \arcsin\left(\frac{2w_w h_w}{w_w^2 + h_w^2}\right) = \arcsin\left(\frac{2k_w}{1 + k_w^2}\right) \tag{1.76}$$

#### 1.4.1.2 Volume Content Model

The volume content of warp and weft yarns, i.e., $V_{inner0}$ and $V_{inner90}$, in the unit cell can be described using the following equation.

$$V_{inner0} = \frac{n_j l_j A_j}{L_x L_y L_z}, V_{inner90} = \frac{n_{wl} L_y A_w + n_{wp} L_y A_{wp}}{L_x L_y L_z} \tag{1.77}$$

where $n_j$, $n_{wl}$, and $n_{wp}$ denote the number of warp yarn, lenticulated and parallelogram weft yarn, respectively.

The fiber volume content inside of composite $V_f$ can be described using the following equation.

$$V_f = \frac{n_j l_j A_j \varphi_j + n_{wl} L_y A_w \varphi_w + n_{wp} L_y A_{wp} \varphi_{wp}}{L_x L_y L_z} \tag{1.78}$$

where $\varphi_j$, $\varphi_{wl}$, and $\varphi_{wp}$ denote the fiber bundle packing density for rectangle, lenticular, and parallelogram shapes, respectively.

The interphase content inside of composite $V_i$ can be described using the following equation.

$$V_i = \frac{\pi(r_f + t_i)^2 - \pi r_f^2}{\pi r_f^2} V_f = \frac{2r_f t_i + t_i^2}{r_f^2} V_f \tag{1.79}$$

where $t_i$ denotes the interphase thickness.

The matrix volume content of $V_m$ can be described using the following equation.

$$V_m = \frac{d_{spec} - d_f V_f - d_i V_i}{d_m} \tag{1.80}$$

where $d_{spec}$, $d_f$, $d_i$, and $d_m$ denote the density of specimen, fiber, interphase, and matrix, respectively.

The porosity volume content inside of composite $V_p$ can be described using the following equation.

$$V_p = 1 - V_f - V_i - V_m \tag{1.81}$$

The fiber volume content along the warp and weft yarns direction, i.e., $V_{fl}$ and $V_{ft}$, can be described using the following equation.

$$V_{fl} = \frac{e_0}{e_0 + e_{90}} V_f,\ V_{ft} = \frac{e_{90}}{e_0 + e_{90}} V_f \tag{1.82}$$

where $e_{90}$ and $e_0$ denote the density of weft and warp yarns, respectively.

The fiber volume content inside of warp yarn $V_{fli}$ can be described using the following equation.

$$V_{fli} = \frac{V_{fl}}{V_{inner0}} = \frac{e_0}{e_0 + e_{90}} \frac{n_j l_j A_j \varphi_j + n_w L_y A_w \varphi_w + n_{wp} L_y A_{wp} \varphi_{wp}}{n_j l_j A_j} \tag{1.83}$$

The fiber volume content inside of weft yarn $V_{fti}$ can be described using the following equation.

$$V_{fti} = \frac{V_{ft}}{V_{inner90}} = \frac{e_{90}}{e_0 + e_{90}} \frac{n_j l_j A_j \varphi_j + n_w L_y A_w \varphi_w + n_{wp} L_y A_{wp} \varphi_{wp}}{n_w L_y A_w + n_{wp} L_y A_{wp}} \tag{1.84}$$

The interphase volume contents inside of warp and weft yarns, i.e., $V_{il}$ and $V_{it}$ can be described using the following equation.

$$V_{il} = \frac{2r_f t_i + t_i^2}{r_f^2} V_{fli},\ V_{it} = \frac{2r_f t_i + t_i^2}{r_f^2} V_{fti} \tag{1.85}$$

The matrix volume contents inside of warp and weft yarns, i.e., $V_{\mathrm{ml}}$ and $V_{\mathrm{mt}}$, can be described using the following equations.

$$\begin{aligned} V_{\mathrm{ml}} &= 1 - V_{\mathrm{fli}} - V_{\mathrm{il}} - V_{\mathrm{pl}} \\ V_{\mathrm{mt}} &= 1 - V_{\mathrm{fti}} - V_{\mathrm{it}} - V_{\mathrm{pt}} \end{aligned} \tag{1.86}$$

where $V_{\mathrm{pl}}, V_{\mathrm{pt}}$ denote the porosity volume fraction inside of the warp and weft yarns, respectively.

The matrix volume content outside of yarns $V_{\text{m-outer}}$ can be described using the following equation.

$$V_{\mathrm{m-outer}} = V_{\mathrm{m}} - V_{ml} V_{\mathrm{inner0}} - V_{mt} V_{\mathrm{inner90}} \tag{1.87}$$

#### 1.4.1.3 Stiffness Model

The elastic modulus of 2.5D CMCs can be determined using the stiffness average method. The stiffness matrix of the unit cell can be described using the following equation.

$$[C] = V_{\mathrm{m-outer}}[C]_{\mathrm{m}} + V_{\mathrm{inner90}}[C]_{\mathrm{w}} + V_{\mathrm{inner0}}[C]_{\mathrm{j}} \tag{1.88}$$

where $[C]_{\mathrm{m}}$, $[C]_{\mathrm{w}}$, and $[C]_{\mathrm{j}}$ denote the matrix, weft and warp yarns stiffness matrix, respectively.

Substituting the matrix, weft and warp yarns stiffness matrix and corresponding volume fraction into Eq. (1.88), the stiffness matrix of the unit cell can be obtained. The compliance matrix is given by $[S] = [C]^{-1}$, and the elastic constants of the composite can be described using the following equation.

$$\begin{cases} E_x = 1/S_{11},\ E_y = 1/S_{22},\ E_z = 1/S_{33} \\ G_{yz} = 1/S_{44},\ G_{xz} = 1/S_{55},\ G_{xy} = 1/S_{66} \\ \mu_{xy} = -S_{12}/S_{11},\ \mu_{xz} = -S_{13}/S_{11},\ \mu_{yz} = -S_{23}/S_{22} \end{cases} \tag{1.89}$$

where $E$ denotes the elastic modulus; $G$ denotes the shear modulus; and $\mu$ denotes the Poisson ratio.

#### 1.4.1.4 Matrix Multicracking

The matrix cracking density in 2.5D CMCs can be described using the following equation [37].

$$\rho_{\mathrm{m}} = \rho_{\mathrm{m-sat}}\left\{1 - \exp\left[-\left(\frac{\sigma_{\mathrm{minimatrix}}}{\sigma_{\mathrm{cr}}}\right)\right]^{m}\right\} \tag{1.90}$$

where $\rho_{\text{m-sat}}$ denotes the saturation matrix cracking density; and $\sigma_{\text{minimatrix}}$ denotes the stress carried by weft yarn and matrix outside of yarns.

$$\sigma_{\mathrm{minimatrix}} = \frac{(\sigma_{\mathrm{c}} + \sigma_{\mathrm{th}})}{E_{\mathrm{c}}}\left(\frac{E_{\mathrm{c}} - V_{\mathrm{inner0}}E_{\mathrm{inner0}}}{1 - V_{\mathrm{inner0}}}\right) \tag{1.91}$$

where $E_{\text{inner0}}$ and $E_{\mathrm{c}}$ denote the elastic modulus of warp yarn and composite.

### 1.4.1.5 Interface Debonding

The fracture mechanics approach is used to determine the fiber/matrix interface debonding [40].

$$\zeta_{\mathrm{d}} = \frac{F}{4\pi r_{\mathrm{f}}}\frac{\partial w_{\mathrm{f}}(0)}{\partial L_{\mathrm{d}}} - \frac{1}{2}\int_{0}^{l_{\mathrm{d}}} \tau_{\mathrm{i}}\frac{\partial v(x)}{\partial l_{\mathrm{d}}}\mathrm{d}x \tag{1.92}$$

The axial displacements of the fiber and the matrix, i.e., $w_{\mathrm{f}}(x)$ and $w_{\mathrm{m}}(x)$, can be described using the following equation.

$$\begin{aligned} w_{\mathrm{f}}(x) &= \int_{x}^{l_{\mathrm{c}}/2} \frac{\sigma_{\mathrm{f}}}{E_{\mathrm{f}}}\mathrm{d}x \\ &= \frac{\sigma}{V_{\mathrm{fl}}E_{\mathrm{f}}}(l_{\mathrm{d}} - x) - \frac{\tau_{\mathrm{i}}}{r_{\mathrm{f}}E_{\mathrm{f}}}\left(l_{\mathrm{d}}^{2} - x^{2}\right) - \frac{2\tau_{\mathrm{i}}}{\rho E_{\mathrm{f}}}l_{\mathrm{d}} + \frac{r_{\mathrm{f}}V_{\mathrm{m-outer}}E_{\mathrm{m-mini}}}{\rho V_{\mathrm{fl}}E_{\mathrm{f}}E_{\mathrm{c}}}\sigma + \frac{\sigma}{E_{\mathrm{c}}}(l_{\mathrm{c}}/2 - l_{\mathrm{d}}) \end{aligned} \tag{1.93}$$

$$\begin{aligned} w_{\mathrm{m}}(x) &= \int_{x}^{l_{\mathrm{c}}/2} \frac{\sigma_{\mathrm{m}}}{E_{\mathrm{m}}}\mathrm{d}x \\ &= \frac{V_{\mathrm{fl}}\tau_{i}}{V_{\mathrm{m-outer}}E_{\mathrm{m-mini}}r_{\mathrm{f}}}\left(l_{\mathrm{d}}^{2} - x^{2}\right) + \frac{2V_{\mathrm{fl}}\tau_{\mathrm{i}}}{\rho V_{\mathrm{m-outer}}E_{\mathrm{m-mini}}}l_{\mathrm{d}} - \frac{r_{\mathrm{f}}}{\rho E_{\mathrm{c}}}\sigma + \frac{\sigma}{E_{\mathrm{c}}}(l_{\mathrm{c}}/2 - l_{\mathrm{d}}) \end{aligned} \tag{1.94}$$

where $E_{\text{m-mini}}$ denotes the elastic modulus of weft yarn and matrix outside of yarns.

Using Eqs. (1.93) and (1.94), the relative displacement between fiber and matrix of $v(x)$ can be described using the following equation.

$$
\begin{aligned}
v(x) &= |w_{\mathrm{f}}(x) - w_{\mathrm{m}}(x)| \\
&= \frac{\sigma}{V_{\mathrm{fl}}E_{\mathrm{f}}}(l_{\mathrm{d}} - x) - \frac{\tau_{\mathrm{i}}E_{\mathrm{c}}}{V_{\mathrm{m-outer}}E_{\mathrm{m-mini}}E_{\mathrm{f}}r_{\mathrm{f}}}\left(l_{\mathrm{d}}^2 - x^2\right) - \frac{2\tau_{\mathrm{i}}E_{\mathrm{c}}l_{\mathrm{d}}}{\rho V_{\mathrm{m-outer}}E_{\mathrm{m-mini}}E_{\mathrm{f}}} + \frac{r_{\mathrm{f}}}{\rho V_{\mathrm{fl}}E_{\mathrm{f}}}\sigma
\end{aligned} \tag{1.95}
$$

Substituting $w_{\mathrm{f}}(x = 0)$ and $v(x)$ into Eq. (1.92), it leads to the form of the following equation.

$$
\begin{aligned}
&\frac{E_{\mathrm{c}}\tau_{\mathrm{i}}^2}{V_{\mathrm{m-outer}}E_{\mathrm{m-mini}}E_{\mathrm{f}}r_{\mathrm{f}}}l_{\mathrm{d}}^2 + \left(\frac{E_{\mathrm{c}}\tau_{\mathrm{i}}^2}{\rho V_{\mathrm{m-outer}}E_{\mathrm{m-mini}}E_{\mathrm{f}}} - \frac{\tau_{\mathrm{i}}\sigma}{V_{\mathrm{fl}}E_{\mathrm{f}}}\right)l_{\mathrm{d}} \\
&+ \frac{r_{\mathrm{f}}V_{\mathrm{m-outer}}E_{\mathrm{m-mini}}\sigma^2}{4(V_{\mathrm{fl}})^2E_{\mathrm{f}}E_{\mathrm{c}}} - \frac{r_{\mathrm{f}}\tau_{\mathrm{i}}}{2\rho V_{\mathrm{fl}}E_{\mathrm{f}}}\sigma - \zeta_{\mathrm{d}} = 0
\end{aligned} \tag{1.96}
$$

Solve Eq. (1.96), the fiber/matrix interface debonded length $l_{\mathrm{d}}$ in the longitudinal yarns can be described using the following equation.

$$
l_{\mathrm{d}} = \frac{r_{\mathrm{f}}}{2}\left(\frac{V_{\mathrm{m-outer}}E_{\mathrm{m-mini}}\sigma}{V_{\mathrm{fl}}E_{\mathrm{c}}\tau_{\mathrm{i}}} - \frac{1}{\rho}\right) - \sqrt{\left(\frac{r_{\mathrm{f}}}{2\rho}\right)^2 + \frac{r_{\mathrm{f}}V_{\mathrm{m-outer}}E_{\mathrm{m-mini}}E_{\mathrm{f}}}{E_{\mathrm{c}}\tau_{\mathrm{i}}^2}\zeta_{\mathrm{d}}} \tag{1.97}
$$

#### 1.4.1.6 Fibers Failure

The GLS criterion is used to determine the load carried by intact and fracture fibers along the longitudinal yarns [45].

$$
\frac{\sigma}{V_{\mathrm{fl}}} = [1 - P(T)]T + \frac{2\tau_i}{r_{\mathrm{f}}}\langle L\rangle P(T) \tag{1.98}
$$

Then $\sigma$ is given by the following equation:

$$
\frac{\sigma}{V_{\mathrm{fl}}} = T\left[1 - \frac{1}{2}\left(\frac{T}{\sigma_{\mathrm{c}}}\right)^{m_{\mathrm{f}}+1}\right] \tag{1.99}
$$

The composite ultimate tensile strength of $\sigma_{\mathrm{UTS}}$ can be described using the following equation.

$$
\sigma_{\mathrm{UTS}} = V_{\mathrm{fl}}\sigma_{\mathrm{c}}\left(\frac{2}{m_{\mathrm{f}}+2}\right)^{\frac{1}{m_{\mathrm{f}}+1}}\left(\frac{m_{\mathrm{f}}+1}{m_{\mathrm{f}}+2}\right) \tag{1.100}
$$

#### 1.4.1.7 Stress–Strain Relationship

For modeling the tensile stress–strain curves of 2.5D woven CMCs, the linear-elastic region can be described using the following equation.

$$\varepsilon_c = \sigma / E_c \tag{1.101}$$

Considering damage mechanisms of matrix multicracking and fiber/matrix interface debonding, the stress–strain relationship can be described using the following equation.

$$\varepsilon_c = \frac{\sigma}{E_c} + \frac{(1 - V_{fl})E_{m-mini}l_d(\sigma)\rho_m}{V_{fl}E_f E_{inner0}}(\sigma + \sigma_{th}), \quad \rho_m^{-1} > 2l_d(\sigma) \tag{1.102}$$

After saturation of matrix cracking, the applied stress is mainly carried by fibers inside of warp yarns. As fibers gradually failure, the load carried by intact fibers increases. The extra strain caused by fibers failure can be described using the following equation.

$$\Delta\varepsilon_c = \frac{1}{E_f}\left(T - \frac{\sigma}{V_{fl}}\right), \quad l_d(\sigma) > \rho_m^{-1} \tag{1.103}$$

### 1.4.2 Results and Discussions

The effects of warp yarn density, weft yarn density, yarn cross-sectional shape on the volume fraction of 2.5D C/SiC composite are discussed. The material properties are given by $E_f$ = 230 GPa, $E_m$ = 450 GPa, $E_i$ = 6.9 GPa, $d_{spec}$ = 2.02 g/cm$^3$, $d_f$ = 1.74 g/cm$^3$, $d_i$ = 2.2 g/cm$^3$, $d_m$ = 3.2 g/cm$^3$, $r_f$ = 3.5 μm, $t_i$ = 0.2 μm.

#### 1.4.2.1 Effect of Warp Yarn Density on Volume Fraction

The effect of warp yarn density (i.e., $e_0$ = 0.5–10.5/cm) on the volume fraction of warp yarn, weft yarn, matrix outside of yarn, porosity, fiber, and matrix is shown in Fig. 1.35. With increasing the density of warp yarn, the volume fraction of warp yarn increases, the volume fraction of weft yarn remains unchanged, the matrix outside of yarns decreases, the porosity volume fraction decreases, the fiber volume fraction increases, and the matrix volume fraction decrease.

### 1.4.2.2 Effect of Weft Yarn Density on Volume Fraction

The effect of weft yarn density (i.e., $e_{90}$ = 0.5–5.5/cm) on the volume fraction of warp yarn, weft yarn, matrix outside of yarn, porosity, fiber, and matrix is shown in Fig. 1.36. With increasing the density of weft yarn, the volume fraction of warp yarn increases, the volume fraction of weft yarn increases, the matrix outside of yarns decreases, the porosity volume fraction decreases, the fiber volume fraction increases, and the matrix volume fraction decrease.

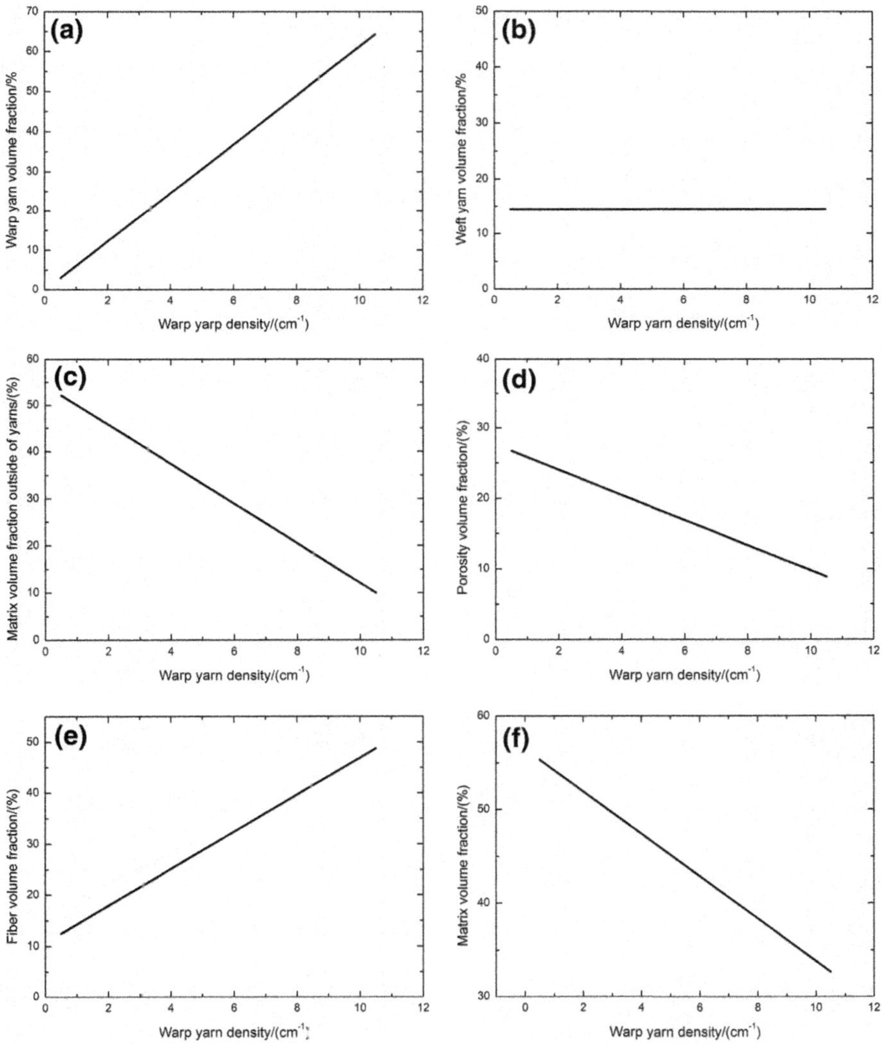


**Fig. 1.35** Effect of warp yarn density on the volume fraction of **a** warp yarn; **b** weft yarn; **c** matrix outside of yarn, **d** porosity; **e** fiber; and **f** matrix

#### 1.4.2.3 Effect of Yarn Cross-Sectional Shape on Volume Fraction

The effect of yarn cross-sectional shape (i.e., $k_w$ = 1–10) on the volume fraction of warp yarn, weft yarn, matrix outside of yarn, porosity, fiber, and matrix is shown in Fig. 1.37. With increasing of $k_w$, the volume fraction of warp yarn increases, the volume fraction of weft yarn increases, the matrix outside of yarns decreases, the porosity volume fraction decreases, the fiber volume fraction increases, and the matrix volume fraction decrease.

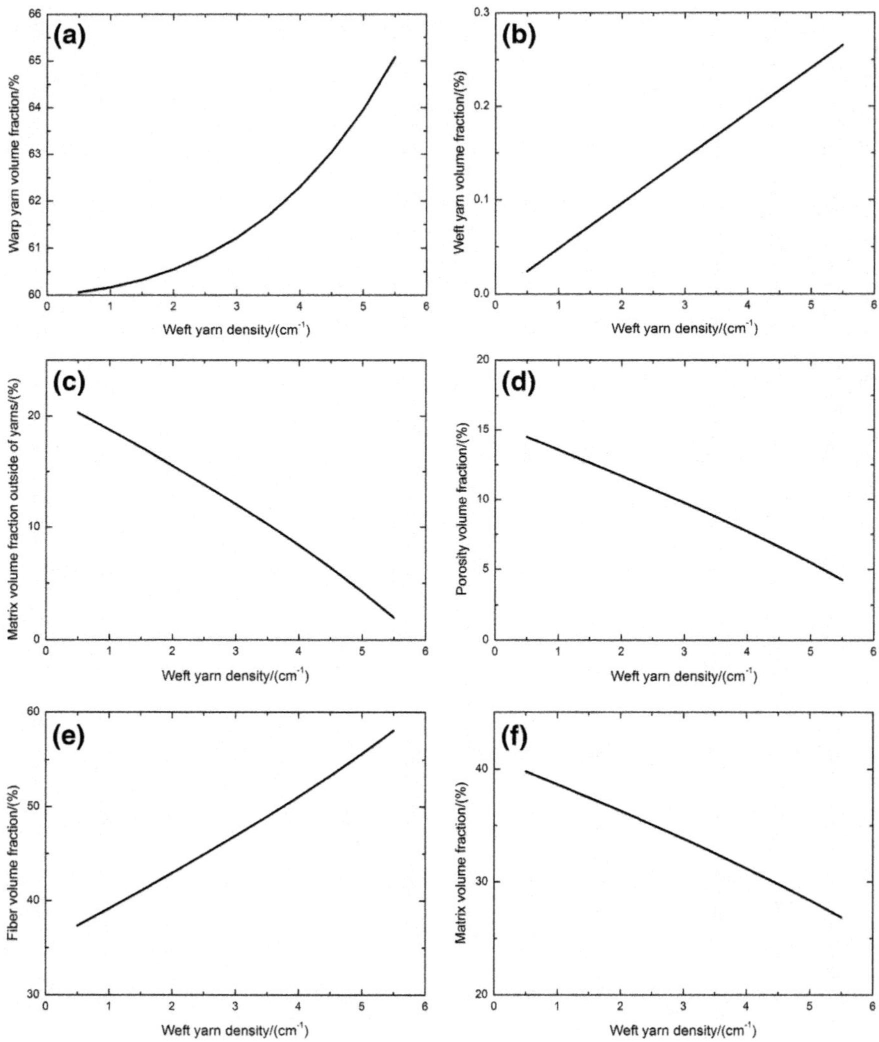


**Fig. 1.36** Effect of weft yarn density on the volume fraction of **a** warp yarn; **b** weft yarn; **c** matrix outside of yarn; **d** porosity; **e** fiber; and **f** matrix

#### 1.4.2.4 Effect of Warp Yarn Density on Elastic Modulus

The effect of warp yarn density ($e_0$ = 0.5–10.5/cm) on the elastic modulus of 2.5D CMCs is shown in Fig. 1.38. When the warp yarn density increases, the composite elastic modulus along the warp yarn direction of $E_x$ increases, and the composite elastic modulus along the weft direction and the thickness direction of $E_y$ ane $E_z$ decreases; the weft yarn elastic modulus along the warp direction, weft directional,

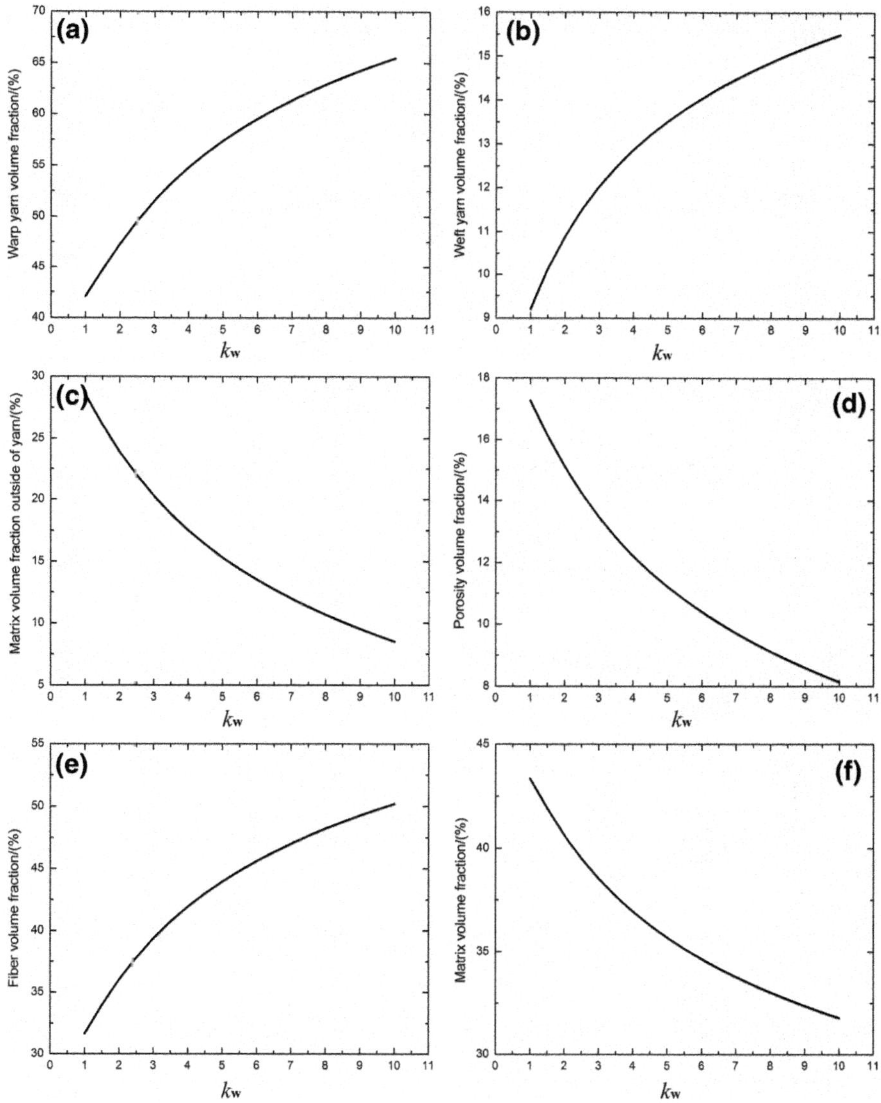


**Fig. 1.37** Effect of yarn cross-sectional shape on the volume fraction of **a** warp yarn; **b** weft yarn; **c** matrix outside of yarn; **d** porosity; **e** fiber; and **f** matrix

and thickness direction of $E_{xw}$, $E_{yw}$, and $E_{zw}$ remain unchanged; the warp yarn elastic modulus along the warp direction, weft direction and thickness direction of $E_{xj}$, $E_{yj}$, and $E_{zj}$ all increase; the matrix elastic modulus along the warp direction, weft direction and thickness direction of $E_{xm}$, $E_{ym}$, and $E_{zm}$ all decrease; the composite shear modulus of $G_{xy}$, $G_{xz}$, $G_{yz}$ all decrease; the composite Poisson ratio of $v_{xy}$ decreases, $v_{xz}$ increases, and $v_{yz}$ increases first and then decreases.

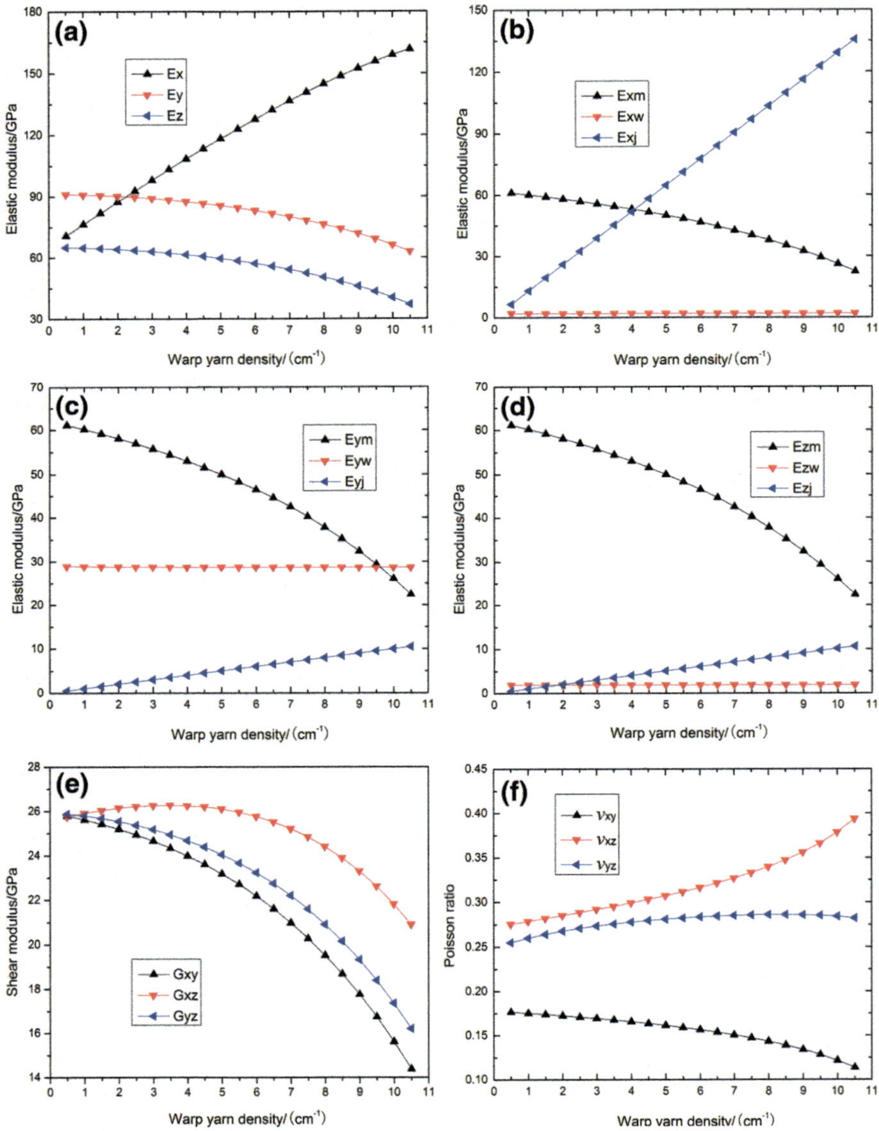


**Fig. 1.38** Effect of warp yarn density on elastic modulus of 2.5D CMCs

#### 1.4.2.5 Effect of Weft Yarn Density on Elastic Modulus

The effect of weft yarn density ($e_0$ = 0.5–5.5/cm) on the elastic modulus of 2.5D CMCs is shown in Fig. 1.39. When the weft yarn density increases, the composite elastic modulus along the warp yarn direction of $E_x$ decreases, and the composite elastic modulus along the weft direction of $E_y$ increases, and the composite elastic modulus along the thickness direction of $E_z$ decreases; the weft yarn elastic modulus along the warp direction, weft directional and thickness direction of $E_{xw}$, $E_{yw}$, and $E_{zw}$ all increases; the matrix elastic modulus along the warp direction, weft direction and thickness direction of $E_{xm}$, $E_{ym}$, and $E_{zm}$ all decrease; the warp yarn elastic modulus along the warp direction of $E_{xj}$ decreases, and the warp yarn elastic modulus along the weft direction and the thickness direction of $E_{yj}$ and $E_{zj}$ increase; the composite shear modulus of $G_{xy}$ and $G_{yz}$ decrease, and $G_{xz}$ increases; and the composite Poisson ratio of $v_{xy}$ and $v_{yz}$ decreases, and $v_{xz}$ increases.

#### 1.4.2.6 Effect of Yarn Cross-Section Shape on Elastic Modulus

The effect of yarn cross-sectional shape (i.e., $k_w$ = 1–10) on the elastic modulus of 2.5D CMCs is shown in Fig. 1.40. When $k_w$ increases, the composite elastic modulus along the warp direction of $E_x$ increases, the composite elastic modulus along the weft direction of $E_y$ increases when $k_w < 2$, and decreases when $k_w > 2$, and the composite elastic modulus along the thickness direction of $E_z$ decreases; the weft yarn elastic modulus along the warp direction, weft direction and thickness direction of $E_{xw}$, $E_{yw}$, and $E_{zw}$ increase; the matrix elastic modulus along the warp direction, weft direction and thickness direction of $E_{xm}$, $E_{ym}$, and $E_{zm}$ all decrease; and the warp yarn elastic modulus along the warp direction and weft direction of $E_{xj}$, $E_{yj}$ increase, and along the thickness direction of $E_{zj}$ decreases when $k_w < 2$ and increases when $k_w > 2$; the composite shear modulus of $G_{xy}$ and $G_{yz}$ decrease, and $G_{xz}$ increases when $k_w < 2$ and decrease when $k_w > 2$; and the composite Poisson ratio of $v_{xy}$ decreases, $v_{xz}$ increases rapidly when $k_w < 2$ and increases slowly when $k_w > 2$, and $v_{yz}$ increases when $k_w < 2$ and decreases when $k_w > 2$.

#### 1.4.2.7 Effect of Warp Yarn Density on Damage Evolution

The effect of warp yarn density ($e_0$ = 8 and 10/cm) on matrix multicracking and tensile stress–strain curve is shown in Fig. 1.41. When the warp yarn density increases, the composite elastic modulus along the warp direction increases, the matrix cracking characteristic stress increases, and the matrix cracking saturation stress increases, and the tensile stress–strain curve deviates at higher stress level, and the strain decreases at the same applied stress.

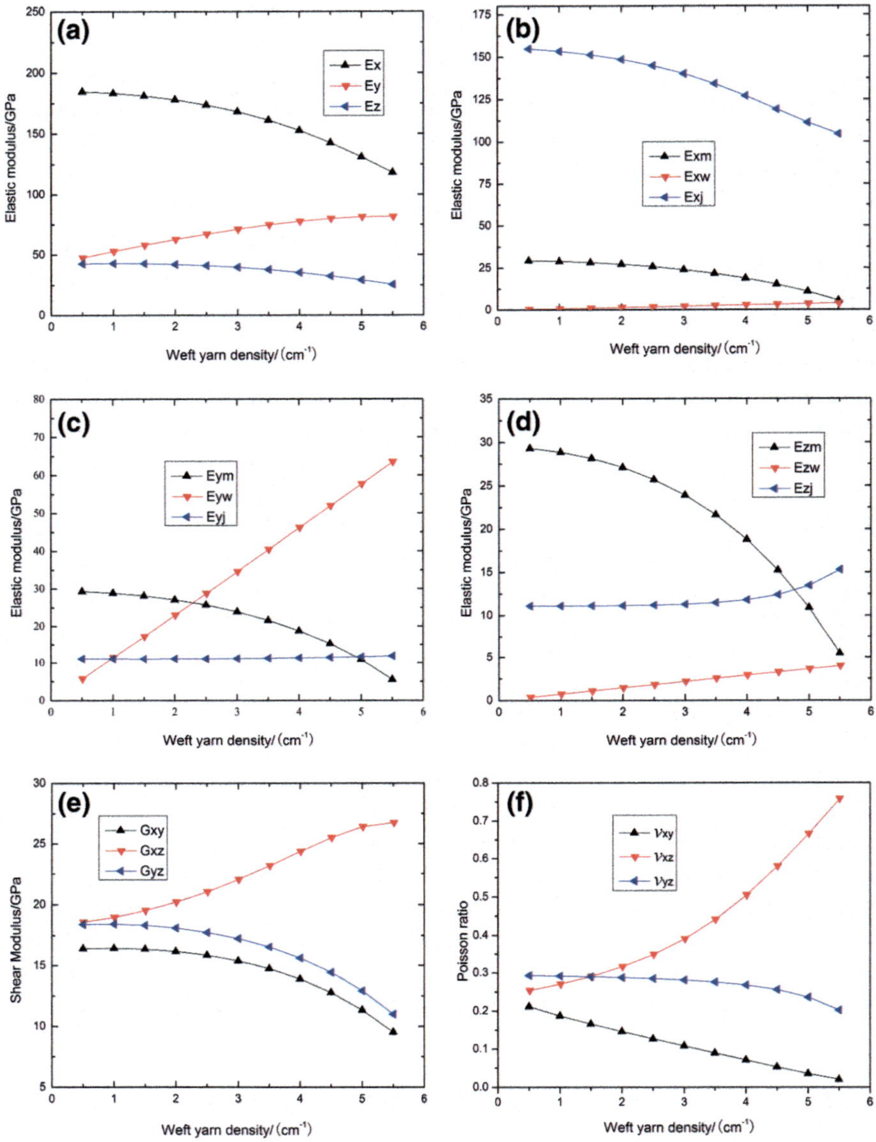


**Fig. 1.39** Effect of weft yarn density on elastic modulus of 2.5D CMCs

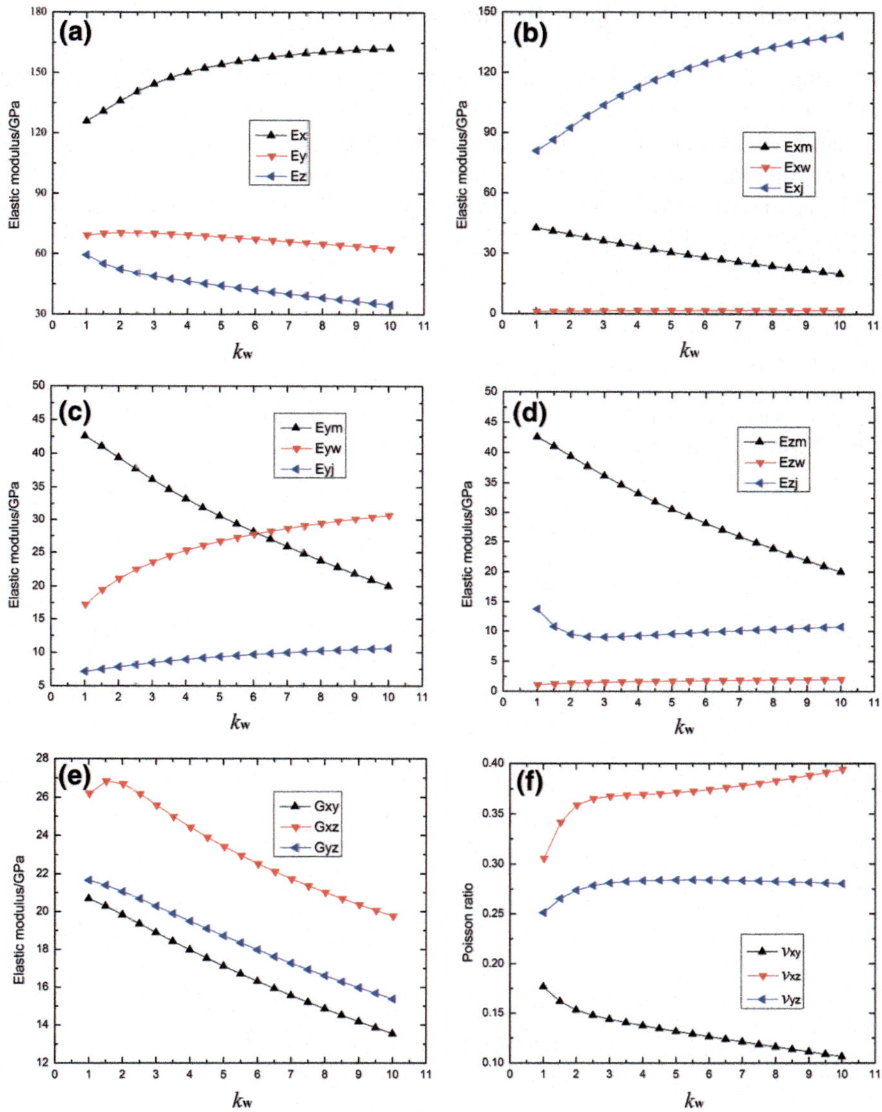


**Fig. 1.40** Effect of yarn cross-sectional shape on elastic modulus of 2.5D CMCs

#### 1.4.2.8 Effect of Weft Yarn Density on Damage Evolution

The effect of weft yarn density ($e_{90}$ = 3 and 5/cm) on the multicracking evolution and tensile stress–strain curve is shown in Fig. 1.42. When the weft yarn density increases, the composite elastic modulus along the warp direction decreases, the matrix cracking characteristic stress decreases, and the matrix cracking approaches to the saturation at low applied stress, and the tensile stress–strain deviates at low applied stress, and increases the strain at the damage stage of tensile stress–strain curve.

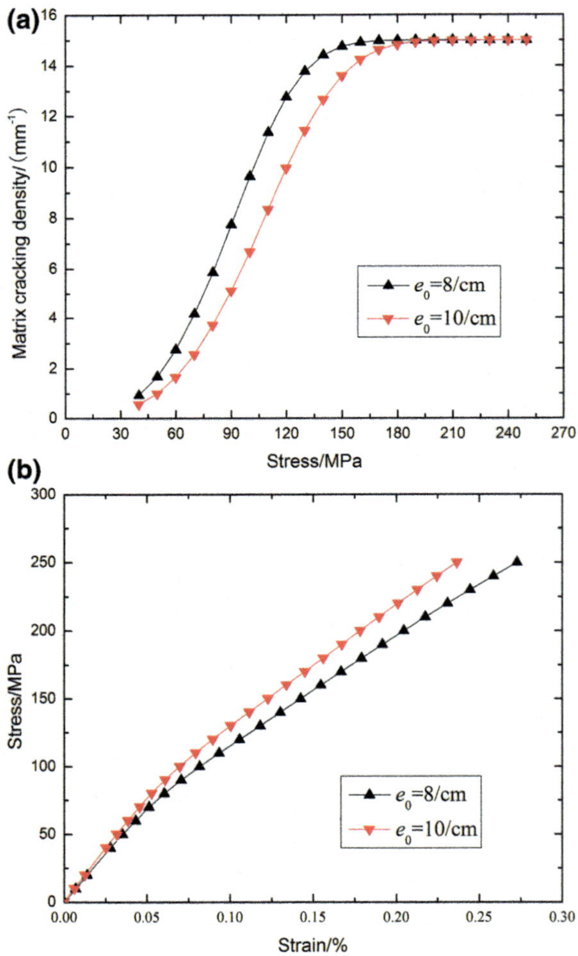


**Fig. 1.41** Effect of warp yarn density ($e_0$ = 8 and 10/cm) on **a** matrix multicracking; and **b** tensile stress–strain curve

### 1.4.3 Experimental Comparisons

The monotonic tensile stress–strain curve of 2.5D C/SiC composite is predicted and compared with experimental data, as shown in Fig. 1.43. The material properties are given as $E_f$ = 230 GPa, $E_m$ = 350 GPa, $r_f$ = 3.5 μm, $e_0$ = 10/cm, $N_f$ = 1000, $t$ = 3 mm, $V_f$ = 40%, $m$ = 4, $\tau_i$ = 7 MPa, $m_f$ = 6, and $\sigma_{uts}$ = 312 MPa [66]. The predicted tensile stress–strain curves of 2.5D C/SiC composite can be divided into three regions, i.e., the linear elastic region, matrix multicracking and fiber/matrix interface debonding nonlinear region, and the fiber failure region, which agreed with experimental data.

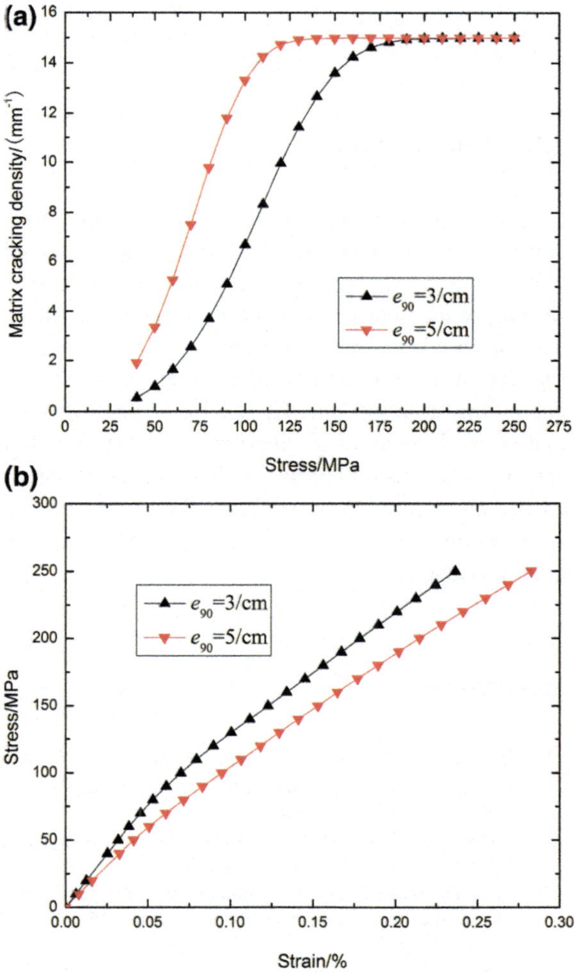


**Fig. 1.42** Effect of weft yarn density ($e_{90}$ = 3 and 5/cm) on **a** matrix multicracking; and **b** tensile stress–strain curve

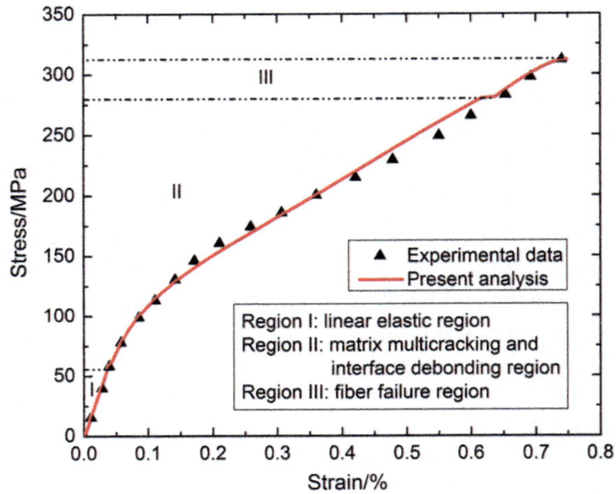


**Fig. 1.43** Experimental and theoretical predicted tensile stress–strain curve of 2.5D C/SiC composite. Reprinted with permission from Ref. [69]. Copyright 2018, Springer Science Business Media New York

## 1.5 Conclusions

In this chapter, the micromechanical approach to predict the tensile stress–strain curves of fiber-reinforced CMCs has been developed. When matrix cracking, fiber/matrix interface debonding and fibers failure occur, the shear-lag model is adopted to analyze the microstress field of the damaged fiber-reinforced CMCs, i.e., the fiber and matrix axial stress distributions. Combining the shear-lag model with damage models of matrix statistical cracking, fracture mechanics fiber/matrix interface debonding criterion and Global Load Sharing (GLS) fibers failure criterion, the matrix cracking spacing, fiber/matrix interface debonding length and fibers broken fraction have been determined. The tensile stress–strain curves of fiber-reinforced CMCs corresponding to different damage stages have been modeled. The tensile stress–strain curves of unidirectional, cross-ply, 2D and 2.5D woven CMCs have been predicted.

## References

1. Naslain R (2004) Design, preparation and properties of non-oxide CMCs for application in engines and nuclear reactors: an overview. Compos Sci Technol 64(2):155–170. https://doi.org/10.1016/S0266-3538(03)00230-6
2. Padture NP (2016) Advanced structural ceramics in aerospace propulsion. Nat Mater 15:804–809. https://doi.org/10.1038/nmat4687
3. Ji FS, Dharani LR (1998) Non-axisymmetric matrix cracking and interface debonding with friction in ceramic composites. Appl Compos Mater 5(6):379–397. https://doi.org/10.1023/A:1008820315282
4. Curtin WA (2000). Stress–strain behavior of brittle matrix composites. In: Comprehensive composite materials, vol 4. Elsevier Science Ltd., pp 47–76. https://doi.org/10.1016/B0-08-042993-9/00088-7

5. Li LB, Song YD (2010) An approach to estimate interface shear stress of ceramic matrix composites from hysteresis loops. Appl Compos Mater 17(3):309–328. https://doi.org/10.1007/s10443-009-9122-6
6. Li LB, Song YD, Sun ZG (2008) Influence of fiber Poisson contraction on matrix cracking development of ceramic matrix composites. J Aerosp Power 23(12):2196–2201
7. Li LB, Song YD, Sun ZG (2008) Uniaxial tensile behavior of unidirectional fiber reinforced ceramic matrix composites. Acta Mater Compos Sinica 25(4):154–160
8. Cox HL (1952) The elasticity and strength of paper and other fibrous materials. British J Appl Phys 3(3):72–79. https://doi.org/10.1088/0508-3443/3/3/302
9. Hedgepeth JM, Dyke PD (1967) Local stress concentration in imperfect filamentary composite materials. J Compos Mater 1(3):294–304. https://doi.org/10.1177/002199836700100305
10. Dyke PD, Hedgepeth JM (1969) Stress concentration from single-filament failure in composite materials. Text Res J 39:618–626. https://doi.org/10.1177/004051756903900702
11. Zweben C (1974) An approximate method of analysis for notched unidirectional composites. Eng Fract Mech 6(1):1–10. https://doi.org/10.1016/0013-7944(74)90042-3
12. Budiansky B, Hutchinson JW, Evans AG (1986) Matrix fracture in fiber-reinforced ceramics. J Mech Phys Solids 34(2):167–189. https://doi.org/10.1016/0022-5096(86)90035-9
13. Marshall DB, Cox BN, Evans AG (1985) The mechanics of matrix cracking in brittle-matrix fiber composites. Acta Metall 33(11):2013–2021. https://doi.org/10.1016/0001-6160(85)90124-5
14. Aveston J, Cooper GA, Kelly A (1971) Single and multiple fracture. In: Properties of fiber composites: conference on proceedings. National Physical Laboratory, IPC, England, pp 15–26
15. Aveston J, Kelly A (1973) Theory of multiple fracture of fibrous composites. J Mater Sci 8(3):352–362. https://doi.org/10.1007/BF00550155
16. Chiang YC (2001) On fiber debonding and matrix cracking in fiber-reinforced ceramics. Compos Sci Technol 61(2):1743–1756. https://doi.org/10.1016/S0266-3538(01)00078-1
17. Chiang YC (2007) On a matrix cracking model using Coulomb's friction law. Eng Fract Mech 74(10):1602–1616. https://doi.org/10.1016/j.engfracmech.2006.09.006
18. Li LB (2017) Synergistic effects of temperature and oxidation on matrix cracking in fiber-reinforced ceramic-matrix composites. Appl Compos Mater 24:691–715. https://doi.org/10.1007/s10443-016-9535-y
19. Li LB (2017) Modeling first matrix cracking stress of fiber-reinforced ceramic-matrix composites considering fiber fracture. Theor Appl Fract Mech 92:24–32. https://doi.org/10.1016/j.tafmec.2017.05.004
20. Li LB (2017) Modeling matrix cracking of fiber-reinforced ceramic-matrix composites under oxidation environment at elevated temperature. Theor Appl Fract Mech 87:110–119. https://doi.org/10 1016/j.tafmec.2016.11.003
21. Li LB (2017) Synergistic effects of fiber debonding and fracture on matrix cracking in fiber-reinforced ceramic-matrix composites. Mater Sci Eng A 682:482–490. https://doi.org/10.1016/j.msea.2016.11.077
22. Cox BN, Marshall DB, Thouless MD (1989) Influence of statistical fiber strength distribution on matrix cracking in fiber composites. Acta Metall 37(7):1933–1943. https://doi.org/10.1016/0001-6160(89)90078-3
23. Marshall DB, Evans AG (1988) Influence of residual stress on the toughness of reinforced brittle materials. Mater Forum 11(1):304–312
24. Cox BN, Marshall DB (1991) Stable and unstable solutions for bridged cracks in various specimens. Acta Metall Mater 39(4):579–589. https://doi.org/10.1016/0956-7151(91)90126-L
25. Cox BN (1991) Extrinsic factors in the mechanics of bridged cracks. Acta Metall Mater 39(6):1189–1201. https://doi.org/10.1016/0956-7151(91)90207-H
26. Chiang YC, Wang ASD, Chou TW (1993) On matrix cracking in fiber reinforced ceramics. J Mech Phys Solids 41(7):1137–1154. https://doi.org/10.1016/0022-5096(93)90087-V
27. Chiang YC (2000) Tensile failure in fiber reinforced ceramic matrix composites. J Mater Sci 35(21):5449–5455. https://doi.org/10.1023/A:1004868928586

28. Thouless MD, Evans AG (1988) Effects of pull-out on the mechanical properties of ceramic-matrix-composites. Acta Metall 36(3):517–522. https://doi.org/10.1016/0001-6160(88)90083-1
29. Cao HC, Bischoff E, Sbaizero O, Ruhle M, Evans AG, Marshall DB, Brennan J (1990) Effect of interfaces on the mechanical performance of fiber-reinforced brittle materials. J Am Ceram Soc 73(6):1691–1699. https://doi.org/10.1111/j.1151-2916.1990.tb09814.x
30. Danchaivijit SD, Shetty DK (1993) Matrix cracking in ceramic matrix composites. J Am Ceram Soc 76(10):2497–2504. https://doi.org/10.1111/j.1151-2916.1993.tb03972.x
31. Kim RY, Pagano NJ (1991) Crack initiation in unidirectional brittle-matrix composites. J Am Ceram Soc 74(5):1082–1090. https://doi.org/10.1111/j.1151-2916.1991.tb04346.x
32. Barsoum MW, Kangutkar P, Wang ASD (1992) Matrix crack initiation in ceramic matrix composites part I: experiments and test results. Compos Sci Technol 44(3):257–269. https://doi.org/10.1016/0266-3538(92)90016-V
33. Lee JW, Daniel IM (1992) Deformation and failure of longitudinally loaded brittle-matrix composites. In: Grimes GC (ed) Proceedings of the tenth symposium on composite materials: testing and design, ASTM, pp 204–221. https://doi.org/10.1520/stp20156s
34. Zok FW, Spearing SM (1992) Matrix crack spacing in brittle matrix composites. Acta Metall Mater 40(8):2033–2043. https://doi.org/10.1016/0956-7151(92)90189-L
35. Weitsman Y, Zhu H (1993) Multi-fracture of ceramic composites. J Mech Phys Solids 41(2):351–388. https://doi.org/10.1016/0022-5096(93)90012-5
36. Solti JP, Mall S, Robertson DD (1997) Modeling of matrix failure in ceramic matrix composites. J Compos Technol Research 19(1):29–40. https://doi.org/10.1520/CTR10011J
37. Curtin WA (1993) Multiple matrix cracking in brittle matrix composites. Acta Metall Mater 41(5):1369–1377. https://doi.org/10.1016/0956-7151(93)90246-O
38. Curtin WA (1999) Stochastic damage evolution and failure in fiber-reinforced composites. Adv Appl Mech 36:164–248. https://doi.org/10.1016/S0065-2156(08)70186-8
39. Hsueh CH (1996) Crack-wake interface debonding criterion for fiber-reinforced ceramic composites. Acta Mater 44(6):2211–2216. https://doi.org/10.1016/1359-6454(95)00369-X
40. Gao YC, Mai YW, Cotterell B (1988) Fracture of fiber-reinforced materials. Zeitschrift für Angewandte Mathematik und Physik (ZAMP) 39(4):550–572. https://doi.org/10.1007/BF00948962
41. Sun YJ, Singh RN (1998) The generation of multiple matrix cracking and fiber-matrix interfacial debonding in a glass composite. Acta Mater 46(5):1657–1667. https://doi.org/10.1016/S1359-6454(97)00347-9
42. Cao HC, Thouless MD (1990) Tensile tests of ceramic-matrix composites: theory and experiment. J Am Ceram Soc 73(7):2091–2094. https://doi.org/10.1111/j.1151-2916.1990.tb05273.x
43. Sutcu M (1989) Weibull statistics applied to fiber failure in ceramic composites and work of fracture. Acta Metall 37(2):651–661. https://doi.org/10.1016/0001-6160(89)90249-6
44. Schwietert HR, Steif PS (1990) A theory for the ultimate strength of a brittle-matrix composites. J Mech Phys Solids 38(3):325–343. https://doi.org/10.1016/0022-5096(90)90002-L
45. Curtin WA (1991) Theory of mechanical properties of ceramic matrix composites. J Am Ceram Soc 74(11):2837–2845. https://doi.org/10.1111/j.1151-2916.1991.tb06852.x
46. Solti JP, Mall S, Robertson DD (1995) Modeling damage in unidirectional ceramic matrix composites. Compos Sci Technol 54(1):55–66. https://doi.org/10.1016/0266-3538(95)00041-0
47. Liao K, Reifsnider KL (2000) A tensile strength model for unidirectional fiber-reinforced brittle matrix composite. Int J Fract 106(2):95–115. https://doi.org/10.1023/A:1007645817753
48. Zhou SJ, Curtin WA (1995) Failure of fiber composites: a lattice Green function model. Acta Metall Mater 43(8):3093–3104. https://doi.org/10.1016/0956-7151(95)00003-E
49. Ibnabdeljalil M, Curtin WA (1997) Strength and reliability of fiber reinforced composites: localized load-sharing and associated size effects. Int J Solids Struct 34(21):2649–2668. https://doi.org/10.1016/S0020-7683(96)00179-5

50. Beyerlein IJ, Phoenix SL (1996) Stress concentrations around multiple fiber breaks in an elastic matrix with local yielding or de-bonding using quadratic influence superposition. J Mech Phys Solids 44(12):1997–2039. https://doi.org/10.1016/S0022-5096(96)00068-3
51. Dutton RE, Pagano NJ, Kim RY (2000) Modeling the ultimate tensile strength of unidirectional glass-matrix composites. J Am Ceram Soc 83(1):166–174. https://doi.org/10.1111/j.1151-2916.2000.tb01166.x
52. Xia ZH, Curtin WA (2000) Toughness-to-brittle transitions in ceramic-matrix composites with increasing interfacial shear stress. Acta Mater 48(20):4879–4892. https://doi.org/10.1016/S1359-6454(00)00291-3
53. Pryce AW, Smith PA (1992) Behavior of unidirectional and crossply ceramic matrix composites under quasi-static tensile loading. J Mater Sci 27(10):2695–2704. https://doi.org/10.1007/BF00540692
54. Beyerle DS, Spearing SM, Zok FW, Evans AG (1992) Damage and failure in unidirectional ceramic-matrix composites. J Am Ceram Soc 75(10):2719–2725. https://doi.org/10.1111/j.1151-2916.1992.tb05495.x
55. Kuo WS, Chou TW (1995) Multiple cracking of unidirectional and cross-ply ceramic matrix composites. J Am Ceram Soc 78(3):745–755. https://doi.org/10.1111/j.1151-2916.1995.tb08242.x
56. Sørensen BF, Holmes JW (1996) Effects of loading rate on the monotonic tensile behavior of a continuous-fiber-reinforced glass-ceramic matrix composite. J Am Ceram Soc 79(2):313–320. https://doi.org/10.1111/j.1151-2916.1996.tb08122.x
57. Okabe T, Komotori J, Shimizu M, Takeda N (1999) Mechanical behavior of sic fiber reinforced brittle-matrix composites. J Mater Sci 34(14):3405–3412. https://doi.org/10.1023/A:1004637200310
58. Zawada LP, Butkus LM, Hartman GA (1991) Tensile and fatigue behavior of silicon carbide fiber-reinforced aluminosilicate glass. J Am Ceram Soc 74(11):2851–2858. https://doi.org/10.1111/j.1151-2916.1991.tb06854.x
59. Li LB, Song YD, Sun YC (2015) Modeling tensile behavior of cross-ply C/SiC ceramic matrix composites. Mech Compos Mater 51(3):359–376. https://doi.org/10.1007/s11029-015-9507-6
60. Garrett KW, Bailey JE (1977) Multiple transverse fracture in 90° cross-ply laminates of a glass fiber-reinforced polyester. J Mater Sci 12(1):157–168. https://doi.org/10.1007/BF00738481
61. Laws N, Dvorak GJ (1988) Progressive transverse cracking in composite laminates. J Compos Mater 22(10):900–916. https://doi.org/10.1177/002199838802201001
62. Fukunaga H, Chou TW, Peters PWM, Schulte K (1984) Probabilistic failure strength analysis of graphite/epoxy cross-ply laminates. J Compos Mater 18(4):339–356. https://doi.org/10.1177/002199838401800403
63. Beyerle DB, Spearing SM, Evans AG (1992) Damage mechanisms and the mechanical properties of a laminated 0/90 ceramic matrix composite. J Am Ceram Soc 75(12):3321–3330. https://doi.org/10.1111/j.1151-2916.1992.tb04428.x
64. Karandikar P, Chou TW (1993) Characterization and modeling of microcracking and elastic moduli changes in Nicalon/CAS composites. Compos Sci Technol 46(3):253–263. https://doi.org/10.1016/0266-3538(93)90159-E
65. Wang SW, Parvizi-Majidi A (1992) Experimental characterization of the tensile behavior of nicalon fiber-reinforced calcium aluminosilicate composites. J Mater Sci 27(20):5483–5496. https://doi.org/10.1007/BF00541610
66. Wang YQ, Zhang LT, Cheng LF. (2008). Tensile behavior of a 2D and 2.5D C/SiC composites fabricated by chemical vapor infiltration. J Chin Ceram Soc 36(8):1062–1068
67. Morscher GN (2004) Stress-dependent matrix cracking in 2d woven SiC-fiber reinforced melt-infiltrated SiC matrix composites. Compos Sci Technol 64(9):1311–1319. https://doi.org/10.1016/j.compscitech.2003.10.022

68. Morscher GN, Singh M, Kiser JD, Freedman M, Bhatt R (2007) Modeling stress-dependent matrix cracking and stress–strain behavior in 2D woven SiC fiber reinforced CVI SiC matrix. Compos Sci Technol 67:1009–1017. https://doi.org/10.1016/j.compscitech.2006.06.007
69. Li LB (2018) Modeling the monotonic and cyclic tensile stress–strain behaviour of 2D and 2.5D woven C/SiC ceramic-matrix composites. Mech Compos Mater 54(2):165–178. https://doi.org/10.1007/s11029-018-9729-5
70. Dalmaz A, Reynaud P, Rouby D, Fantozzi G (1996) Damage propagation in carbon/silicon carbide composites during tensile tests under the SEM. J Mater Sci 31:4213–4219. https://doi.org/10.1007/BF00356441
71. Li LB, Song YD, Sun YC (2014) Modeling tensile behavior of unidirectional C/SiC ceramic matrix composites. Mech Compos Mater 49(6):659–672. https://doi.org/10.1007/s11029-013-9382-y

# Chapter 2
# Fatigue Hysteresis Behavior of Ceramic-Matrix Composites

## 2.1 Introduction

Under cyclic fatigue loading, the basic damage mechanisms include matrix cracking and fiber/matrix interface debonding and closure/open of matrix crack upon unloading/reloading [1]. The fatigue stress–strain hysteresis loops appear due to the frictional sliding between the fiber and the matrix in the matrix crack spacing [2]. The cyclic fatigue stress–strain hysteresis loops can be used as a useful tool to reveal the damage evolution inside of fiber-reinforced CMCs [3–14]. Marshall and Evans [15] observed through experiments for the fatigue stress–strain hysteresis loop of unidirectional fiber-reinforced CMC at room temperature. Holmes and Cho [16] investigated the characteristic of fatigue hysteresis loops area and fatigue hysteresis modulus, and the relationship between the fatigue hysteresis and damage mechanisms of matrix cracking and fiber/matrix interface debonding/sliding, and observed the surface temperature rising under cyclic fatigue loading of unidirectional SiC/CAS-II composite at room temperature. Fantozzi and Reynaud [17] investigated the fatigue hysteresis loops development characteristic of two different CMCs, SiC/[Si–B–C], and C/[Si–B–C] at elevated temperature. The fibers type will affect the fatigue hysteresis evolution due to the thermal expansion coefficient mismatch between the fibers and the matrix. Kotil et al. [18], Cho et al. [19], Pryce and Smith [20], and Ahn and Curtin [21] investigated the cyclic loading/unloading fatigue stress–strain hysteresis loops of unidirectional CMCs for the fiber/matrix interface without chemical bonding. Solti et al. [22, 23] developed the fatigue stress–strain hysteresis loops models for fiber-reinforced CMCs with the chemical bonding fiber/matrix interface shear stress and frictional interface shear stress at the debonded interface. Vagaggini et al. [24] divided the fiber/matrix interface debonded energy to small and large condition when the fiber/matrix interface debonding for the fatigue stress–strain hysteresis based on the Hutchinson–Jenson fiber pullout model [25]. Keith and Kedward [26] investigated the fatigue hysteresis loops in 2D woven CMC for the partially debonding and completely debonding interface based on Pryce–Smith



L. Li, *Damage, Fracture, and Fatigue of Ceramic-Matrix Composites*,
https://doi.org/10.1007/978-981-13-1783-5_2

model [20]. Li et al. investigated the effect of interface debonding [27, 28], fibers Poisson contraction [29], matrix multicracking [30–33], fibers fracture [34, 35], interface wear [36–38], interface oxidation [39], loading type [40], and fiber preforms [41] on the fatigue stress–strain hysteresis loops of different CMCs and compared the difference of hysteresis behavior between C/SiC and SiC/SiC composites [42].

In this chapter, the fatigue hysteresis behavior of unidirectional, 2D cross-ply and woven, and 2.5D woven fiber-reinforced ceramic-matrix composites (CMCs) are analyzed. Based on the fiber/matrix interface debonding and sliding behavior, the fiber/matrix interface debonding and sliding lengths are determined using the fracture mechanics approach. The fiber/matrix interface debonding ratio and interface sliding ratio are determined for different interface slip cases. The effects of fiber volume fraction, peak stress, matrix crack spacing, interface shear stress, interface debonded energy, fibers failure, fiber Poisson contraction, fiber strength, fiber Weibull modulus, matrix cracking mode, applied cycle number and fiber/matrix interface wear on the fatigue stress–strain hysteresis loops and the fiber/matrix interface debonding and sliding are discussed. The experimental cyclic fatigue stress–strain hysteresis loops of unidirectional SiC/CAS, SiC/1723 and C/SiC, 2D cross-ply SiC/CAS and woven SiC/SiC, and 2.5D woven C/SiC composites under cyclic loading/unloading tensile and tension–tension fatigue loading are predicted.

## 2.2 Unidirectional Ceramic-Matrix Composites

In this section, the cyclic loading/unloading and tension–tension fatigue hysteresis behavior of unidirectional fiber-reinforced CMCs at room and elevated temperatures are investigated. When matrix cracking and fiber/matrix interface debonding occurred upon first loading to the peak stress, the microstress field in the damaged fiber-reinforced CMCs is described using the BHE shear-lag model [43]. Upon continually unloading/reloading, the debonding and sliding between the fibers and the matrix are the main reason for the stress–strain hysteresis behavior of fiber-reinforced CMCs. The fracture mechanics approach is used to determine the fiber/matrix interface debonding length for first loading to peak stress, unloading interface counter slip length and reloading interface new slip length for cyclic loading. Based on the fiber/matrix interface debonding and sliding condition in matrix crack spacing, the fatigue stress–strain hysteresis loops are divided into four different cases. The effects of fiber/matrix interface shear stress, fiber/matrix interface debonded energy, fibers failure, fatigue peak stress, applied cycle number, fibers strength and fiber Weibull modulus on the debonding and sliding in the fiber/matrix interface and the fatigue stress–strain hysteresis loops of fiber-reinforced CMCs are discussed. The cyclic loading/unloading and tension–tension fatigue hysteresis loops of SiC/CAS, SiC/1723, and C/SiC composites are predicted and compared with experimental data.

## 2.2.1 Stress Analysis

### 2.2.1.1 Initial Loading

Upon loading to the fatigue peak stress of $\sigma_{\max}$, matrix cracking and fiber/matrix interface debonding occur, but no fibers fail. The Budiansky–Hutchinson–Evans shear-lag model [43] is used to describe the microstress distributions of the fiber, matrix, and the fiber/matrix interface shear stress inside of CMCs. The unit cell of BHE shear-lag model is shown in Fig. 1.1. The fiber, matrix, and fiber/matrix interface axial distributions can be described using the following equations.

$$\sigma_{\mathrm{f}}(x)=\begin{cases}\frac{\sigma}{V_{\mathrm{f}}}-\frac{2\tau_{\mathrm{i}}}{r_{\mathrm{f}}}x, & x\in(0,l_{\mathrm{d}})\\ \sigma_{\mathrm{fo}}+\left(\frac{V_{\mathrm{m}}}{V_{\mathrm{f}}}\sigma_{\mathrm{mo}}-2\frac{l_{\mathrm{d}}}{r_{\mathrm{f}}}\tau_{\mathrm{i}}\right)\exp\left(-\rho\frac{x-l_{\mathrm{d}}}{r_{\mathrm{f}}}\right), & x\in(l_{\mathrm{d}},l_{\mathrm{c}}/2)\end{cases}\tag{2.1}$$

$$\sigma_{\mathrm{m}}(x)=\begin{cases}2\tau_{\mathrm{i}}\frac{V_{\mathrm{f}}}{V_{\mathrm{m}}}\frac{x}{r_{\mathrm{f}}}, & x\in(0,l_{\mathrm{d}})\\ \sigma_{\mathrm{mo}}-\left(\sigma_{\mathrm{mo}}-2\tau_{\mathrm{i}}\frac{V_{\mathrm{f}}}{V_{\mathrm{m}}}\frac{l_{\mathrm{d}}}{r_{\mathrm{f}}}\right)\exp\left[-\frac{\rho(x-l_{\mathrm{d}})}{r_{\mathrm{f}}}\right], & x\in(l_{\mathrm{d}},l_{\mathrm{c}}/2)\end{cases}\tag{2.2}$$

$$\tau_{\mathrm{i}}(x)=\begin{cases}\tau_{\mathrm{i}}, & x\in(0,l_{\mathrm{d}})\\ \frac{\rho}{2}\left(\frac{V_{\mathrm{m}}}{V_{\mathrm{f}}}\sigma_{\mathrm{mo}}-2\tau_{\mathrm{i}}\frac{l_{\mathrm{d}}}{r_{\mathrm{f}}}\right)\exp\left[-\frac{\rho(x-l_{\mathrm{d}})}{r_{\mathrm{f}}}\right], & x\in(l_{\mathrm{d}},l_{\mathrm{c}}/2)\end{cases}\tag{2.3}$$

where $V_{\mathrm{f}}$ and $V_{\mathrm{m}}$ denote the fiber and matrix volume fraction, respectively; $r_{\mathrm{f}}$ denotes the fiber radius; $l_{\mathrm{d}}$ and $l_{\mathrm{c}}$ denote the fiber/matrix interface debonded length and the matrix crack spacing, respectively; $\tau_i$ denotes the interface frictional shear stress in the debonded region; $\rho$ denotes the shear-lag model parameter; and $\sigma_{\mathrm{fo}}$ and $\sigma_{\mathrm{mo}}$ denote the fiber and matrix axial stress in the bonded region, respectively.

Considering fibers failure upon cyclic loading, the fiber axial stress distribution can be described using the following equation.

$$\sigma_{\mathrm{f}}(x)=\begin{cases}T-\frac{2\tau_{\mathrm{i}}}{r_{\mathrm{f}}}x, & x\in(0,l_{\mathrm{d}})\\ \sigma_{\mathrm{fo}}+\left(T-\sigma_{\mathrm{fo}}-2\frac{l_{\mathrm{d}}}{r_{\mathrm{f}}}\tau_{\mathrm{i}}\right)\exp\left(-\rho\frac{x-l_{\mathrm{d}}}{r_{\mathrm{f}}}\right), & x\in(l_{\mathrm{d}},l_{\mathrm{c}}/2)\end{cases}\tag{2.4}$$

where $T$ denotes intact fiber axial stress on the matrix crack plane.

$$\frac{\sigma}{V_{\mathrm{f}}}=T\left(\frac{\sigma_{\mathrm{c}}}{T}\right)^{m_{\mathrm{f}}+1}\left\{1-\exp\left[-\left(\frac{T}{\sigma_{\mathrm{c}}}\right)^{m_{\mathrm{f}}+1}\right]\right\}\tag{2.5}$$

where $m_{\mathrm{f}}$ denotes the fiber Weibull modulus, and $\sigma_{\mathrm{c}}$ denotes the fiber characteristic strength.

#### 2.2.1.2 Unloading

Upon unloading, the counter slip occurs in the debonded region, and the fiber/matrix interface debonded region can be divided into the counter slip region and slip region. The fiber and matrix axial stress distributions can be described using the following equations.

$$\begin{cases} \sigma_f(x) = \frac{\sigma}{V_f} + \frac{2\tau_i}{r_f}x, & x \in (0, y) \\ \sigma_f(x) = \frac{\sigma}{V_f} + \frac{2\tau_i}{r_f}(2y - x), & x \in (y, l_d) \\ \sigma_f(x) = \sigma_{fo} + \left[\frac{V_m}{V_f}\sigma_{mo} - \frac{2\tau_i}{r_f}(l_d - 2y)\right]\exp\left(-\rho\frac{x - l_d}{r_f}\right), & x \in (l_d, l_c/2) \end{cases} \tag{2.6}$$

$$\begin{cases} \sigma_m(x) = -\frac{V_f}{V_m}\frac{2\tau_i}{r_f}x, & x \in (0, y) \\ \sigma_m(x) = -\frac{V_f}{V_m}\frac{2\tau_i}{r_f}(2y - x), & x \in (y, l_d) \\ \sigma_m(x) = \sigma_{mo} - \left[\sigma_{mo} - \frac{2V_f\tau_i}{r_f V_m}(l_d - 2y)\right]\exp\left(-\rho\frac{x - l_d}{r_f}\right), & x \in (l_d, l_c/2) \end{cases} \tag{2.7}$$

where $y$ denotes the interface counter slip length.

Considering fibers failure, the fiber axial stress distribution upon unloading can be described using the following equation.

$$\begin{cases} \sigma_f(x) = T^U + \frac{2\tau_i}{r_f}x, & x \in (0, y_f) \\ \sigma_f(x) = T^U + \frac{2\tau_i}{r_f}(2y_f - x), & x \in (y_f, l_{df}) \\ \sigma_f(x) = \sigma_{fo} + \left[T^U - \sigma_{fo} - 2\frac{\tau_i}{r_f}(l_{df} - 2y_f)\right]\exp\left(-\rho\frac{x - l_{df}}{r_f}\right), & x \in (l_{df}, l_c/2) \end{cases} \tag{2.8}$$

where $l_{df}$ and $y_f$ denote the fiber/matrix interface debonded length and unloading interface counter slip length considering fibers failure, and $T^U$ denotes the stress carried by the intact fibers upon unloading.

$$\frac{\sigma}{V_f} = 2T\left(\frac{\sigma_c}{T}\right)^{m_f+1}\left\{\exp\left[-\left(\frac{T - T^U}{2T}\right)\left(\frac{T}{\sigma_c}\right)^{m_f+1}\right] - 1 + \frac{1}{2}P(T)\right\} \tag{2.9}$$

#### 2.2.1.3 Reloading

Upon reloading, new slip occurs in the interface counter slip region. The fiber and matrix axial stress distributions upon reloading can be described using the following equations.

$$\begin{cases} \sigma_f(x) = \frac{\sigma}{V_f} - \frac{2\tau_i}{r_f}x, & x \in (0, z) \\ \sigma_f(x) = \frac{\sigma}{V_f} + \frac{2\tau_i}{r_f}(x - 2z), & x \in (z, y) \\ \sigma_f(x) = \frac{\sigma}{V_f} - \frac{2\tau_i}{r_f}(x - 2y + 2z), & x \in (y, l_d) \\ \sigma_f(x) = \sigma_{fo} + \left[\frac{V_m}{V_f}\sigma_{mo} - \frac{2\tau_i}{r_f}(l_d - 2y + 2z)\right]\exp\left(-\rho\frac{x - l_d}{r_f}\right), & x \in (l_d, l_c/2) \end{cases} \tag{2.10}$$

$$\begin{cases} \sigma_m(x) = \frac{V_f}{V_m}\frac{2\tau_i}{r_f}x, & x \in (0, z) \\ \sigma_m(x) = -\frac{V_f}{V_m}\frac{2\tau_i}{r_f}(x - 2z), & x \in (z, y) \\ \sigma_m(x) = \frac{V_f}{V_m}\frac{2\tau_i}{r_f}(x - 2y + 2z), & x \in (y, l_d) \\ \sigma_m(x) = \sigma_{mo} - \left[\sigma_{mo} - \frac{2V_f\tau_i}{r_f V_m}(l_d - 2y + 2z)\right]\exp\left(-\rho\frac{x - l_d}{r_f}\right), & x \in \left(l_d, \frac{l_c}{2}\right) \end{cases} \tag{2.11}$$

where $z$ denotes the interface new slip length.

Considering fibers failure, the fiber axial stress distribution upon reloading can be described using the following equation.

$$\begin{cases} \sigma_f(x) = T^R - \frac{2\tau_i}{r_f}x, & x \in (0, z_f) \\ \sigma_f(x) = T^R + \frac{2\tau_i}{r_f}(x - 2z_f), & x \in (z_f, y_f) \\ \sigma_f(x) = T^R - \frac{2\tau_i}{r_f}(x - 2y_f + 2z_f), & x \in (y_f, l_{df}) \\ \sigma_f(x) = \sigma_{fo} + \left[T^R - \sigma_{fo} - 2\frac{\tau_i}{r_f}(l_{df} - 2y_f + 2z_f)\right]\exp\left(-\rho\frac{x - l_{df}}{r_f}\right), & x \in (l_{df}, l_c/2) \end{cases} \tag{2.12}$$

where $z_f$ denotes the interface new slip length considering fibers failure, and $T^R$ denotes stress carried by the intact fibers upon reloading.

$$\begin{aligned} \frac{\sigma}{V_f} = 2T\left(\frac{\sigma_c}{T}\right)^{m_f+1}\Bigg\{ & \exp\left[-\left(\frac{T_m}{2T}\right)\left(\frac{T}{\sigma_c}\right)^{m_f+1}\right] \\ & - \exp\left[-\left(\frac{T^R - T + T_m}{2T}\right)\left(\frac{T}{\sigma_c}\right)^{m_f+1}\right] + \frac{1}{2}P(T)\Bigg\} \end{aligned} \tag{2.13}$$

where.

$$0 = 2T\left(\frac{\sigma_c}{T}\right)^{m_f+1}\left\{\exp\left[-\left(\frac{T_m}{2T}\right)\left(\frac{T}{\sigma_c}\right)^{m_f+1}\right] - 1 + \frac{1}{2}P(T)\right\} \tag{2.14}$$

### 2.2.2 Interface Debonding and Sliding

The fracture mechanics fiber/matrix interface debonding criterion is used to determine the debonding and sliding lengths, which can be described using the following equation.

$$\zeta_{\mathrm{d}} = \frac{F}{4\pi r_{\mathrm{f}}}\frac{\partial w_{\mathrm{f}}(0)}{\partial l_{\mathrm{d}}} - \frac{1}{2}\int_{0}^{l_{\mathrm{d}}} \tau_{\mathrm{i}}\frac{\partial v(x)}{\partial l_{\mathrm{d}}}\mathrm{d}x \tag{2.15}$$

where $F$ denotes fiber loads at matrix crack plane; $\zeta_{\mathrm{d}}$ denotes the fiber/matrix interface debonded energy; $w_{\mathrm{f}}(0)$ denotes fiber axial displacement at matrix crack plane; and $v(x)$ denotes relative displacement between the fiber and the matrix.

The fiber/matrix interface debonded length of $l_{\mathrm{d}}$ can be described using the following equation.

$$l_{\mathrm{d}} = \frac{r_{\mathrm{f}}}{2}\left(\frac{V_{\mathrm{m}}E_{\mathrm{m}}\sigma}{V_{\mathrm{f}}E_{\mathrm{c}}\tau_{\mathrm{i}}} - \frac{1}{\rho}\right) - \sqrt{\left(\frac{r_{\mathrm{f}}}{2\rho}\right)^2 + \frac{r_{\mathrm{f}}V_{\mathrm{m}}E_{\mathrm{m}}E_{\mathrm{f}}}{E_{\mathrm{c}}\tau_{\mathrm{i}}^2}\zeta_{\mathrm{d}}} \tag{2.16}$$

The initial fiber/matrix interface debonding stress can be described using the following equation.

$$\sigma_{\mathrm{d}} = \frac{V_{\mathrm{f}}E_{\mathrm{c}}\tau_{\mathrm{i}}}{\rho V_{\mathrm{m}}E_{\mathrm{m}}}\left[1 + \sqrt{1 + 4\frac{V_{\mathrm{m}}E_{\mathrm{m}}E_{\mathrm{f}}}{r_{\mathrm{f}}E_{\mathrm{c}}}\frac{\rho^2}{\tau_{\mathrm{i}}^2}\zeta_{\mathrm{d}}}\right] \tag{2.17}$$

The completely interface debonding stress can be described using the following equation.

$$\sigma_{\mathrm{b}} = \frac{V_{\mathrm{f}}E_{\mathrm{c}}\tau_{\mathrm{i}}}{\rho V_{\mathrm{m}}E_{\mathrm{m}}}\left[1 + \rho\frac{l_{\mathrm{c}}}{r_{\mathrm{f}}} + \sqrt{1 + 4\frac{V_{\mathrm{m}}E_{\mathrm{m}}E_{\mathrm{f}}}{r_{\mathrm{f}}E_{\mathrm{c}}}\frac{\rho^2}{\tau_{\mathrm{i}}^2}\zeta_{\mathrm{d}}}\right] \tag{2.18}$$

Considering fibers failure, the fiber/matrix interface debonded length can be described using the following equation.

$$l_{\mathrm{df}} = \frac{r_{\mathrm{f}}}{2}\left(\frac{V_{\mathrm{m}}E_{\mathrm{m}}}{E_{\mathrm{c}}\tau_{i}}T - \frac{1}{\rho}\right) - \sqrt{\left(\frac{r_{\mathrm{f}}}{2\rho}\right)^2 - \frac{r_{\mathrm{f}}^2V_{\mathrm{f}}V_{\mathrm{m}}E_{\mathrm{f}}E_{\mathrm{m}}T}{4E_{\mathrm{c}}^2\tau_{\mathrm{i}}^2}\left(T - \frac{\sigma}{V_{\mathrm{f}}}\right) + \frac{r_{\mathrm{f}}V_{\mathrm{m}}E_{\mathrm{m}}E_{\mathrm{f}}}{E_{\mathrm{c}}\tau_{\mathrm{i}}^2}\zeta_{\mathrm{d}}} \tag{2.19}$$

The interface counter slip length of $y$ upon unloading can be described using the following equation.

$$y = \frac{1}{2}\left\{ l_{\rm d}(\sigma_{\max}) - \left[\frac{r_{\rm f}}{2}\left(\frac{V_{\rm m}E_{\rm m}\sigma}{V_{\rm f}E_{\rm c}\tau_{\rm i}} - \frac{1}{\rho}\right) - \sqrt{\left(\frac{r_{\rm f}}{2\rho}\right)^2 + \frac{r_{\rm f}V_{\rm m}E_{\rm m}E_{\rm f}}{E_{\rm c}\tau_{\rm i}^2}\zeta_{\rm d}}\right]\right\} \tag{2.20}$$

Considering fibers failure, the interface counter slip length of $y_{\rm f}$ can be described using the following equation.

$$\begin{aligned} y_{\rm f} = \frac{1}{2}\Bigg\{ l_{\rm df} - \Bigg[\frac{r_{\rm f}}{2}\left(\frac{V_{\rm m}E_{\rm m}}{E_{\rm c}\tau_{\rm i}}T^{\rm U} - \frac{1}{\rho}\right) \\ -\sqrt{\left(\frac{r_{\rm f}}{2\rho}\right)^2 - \frac{r_{\rm f}^2V_{\rm f}V_{\rm m}E_{\rm f}E_{\rm m}T^{\rm U}}{4E_{\rm c}^2\tau_{\rm i}^2}\left(T^{\rm U} - \frac{\sigma}{V_{\rm f}}\right) + \frac{r_{\rm f}V_{\rm m}E_{\rm m}E_{\rm f}}{E_{\rm c}\tau_{\rm i}^2}\zeta_{\rm d}}\Bigg]\Bigg\} \end{aligned} \tag{2.21}$$

The interface new slip length of $z$ upon reloading can be described using the following equation.

$$z = y(\sigma_{\min}) - \frac{1}{2}\left\{ l_{\rm d} - \left[\frac{r_{\rm f}}{2}\left(\frac{V_{\rm m}E_{\rm m}\sigma}{V_{\rm f}E_{\rm c}\tau_{\rm i}} - \frac{1}{\rho}\right) - \sqrt{\left(\frac{r_{\rm f}}{2\rho}\right)^2 + \frac{r_{\rm f}V_{\rm m}E_{\rm m}E_{\rm f}}{E_{\rm c}\tau_{\rm i}^2}\zeta_{\rm d}}\right]\right\} \tag{2.22}$$

Considering fibers failure, the interface new slip length of $z_{\rm f}$ upon reloading can be described using the following equation.

$$\begin{aligned} z_{\rm f} = y_{\rm f} - \frac{1}{2}\Bigg\{ l_{\rm df} - \Bigg[\frac{r_{\rm f}}{2}\left(\frac{V_{\rm m}E_{\rm m}}{E_{\rm c}\tau_{\rm i}}T^{R} - \frac{1}{\rho}\right) \\ -\sqrt{\left(\frac{r_{\rm f}}{2\rho}\right)^2 - \frac{r_{\rm f}^2V_{\rm f}V_{\rm m}E_{\rm f}E_{\rm m}T^{R}}{4E_{\rm c}^2\tau_{\rm i}^2}\left(T^{R} - \frac{\sigma}{V_{\rm f}}\right) + \frac{r_{\rm f}V_{\rm m}E_{\rm m}E_{\rm f}}{E_{\rm c}\tau_{\rm i}^2}\zeta_{\rm d}}\Bigg]\Bigg\} \end{aligned} \tag{2.23}$$

### 2.2.3 Stress–Strain Hysteresis Loops

When the fiber/matrix interface partially debonds, the stress–strain hysteresis loops without and with considering fibers failure upon unloading can be described using the following equations.

$$\varepsilon_{\rm c\_pu} = \frac{\sigma}{V_{\rm f}E_{\rm f}} + 4\frac{\tau_{\rm i}}{E_{\rm f}}\frac{y^2}{r_{\rm f}L} - 2\frac{\tau_{\rm i}}{E_{\rm f}}\frac{(2y - l_{\rm d})(2y + l_{\rm d} - l_{\rm c})}{r_{\rm f}l_{\rm c}} - (\alpha_{\rm c} - \alpha_{\rm f})\Delta{\rm T} \tag{2.24}$$

$$\left(\varepsilon_{\rm c\_pu}\right)_{\rm f} = \frac{T^{\rm U}}{E_{\rm f}} + 4\frac{\tau_{\rm i}}{E_{\rm f}}\frac{y_{\rm f}^2}{r_{\rm f}l_{\rm c}} - \frac{\tau_{\rm i}}{E_{\rm f}}\frac{1}{r_{\rm f}l_{\rm c}}(2y_{\rm f} - l_{\rm df})(2y_{\rm f} + l_{\rm df} - l_{\rm c}) - (\alpha_{\rm c} - \alpha_{\rm f})\Delta{\rm T} \tag{2.25}$$

When the fiber/matrix interface partially debonds, the stress–strain hysteresis loops without and with considering fibers failure upon reloading can be described using the following equations.

$$\begin{aligned}\varepsilon_{c\_pr} = & \frac{\sigma}{V_f E_f} - 4\frac{\tau_i}{E_f}\frac{z^2}{r_f l_c} + 4\frac{\tau_i}{E_f}\frac{(y-2z)^2}{r_f l_c} \\ & + 2\frac{\tau_i}{E_f}\frac{(l_d - 2y + 2z)(l_d + 2y - 2z - l_c)}{r_f l_c} - (\alpha_c - \alpha_f)\Delta T\end{aligned} \tag{2.26}$$

$$\begin{aligned}\left(\varepsilon_{c\_pr}\right)_f = & \frac{T^R}{E_f} - 4\frac{\tau_i}{E_f}\frac{z_f^2}{r_f l_c} + 4\frac{\tau_i}{E_f}\frac{(y_f-2z_f)^2}{r_f l_c} \\ & + 2\frac{\tau_i}{E_f}\frac{(l_{df} - 2y_f + 2z_f)(l_{df} + 2y_f - 2z_f - l_c)}{r_f l_c} - (\alpha_c - \alpha_f)\Delta T\end{aligned} \tag{2.27}$$

When the fiber/matrix interface completely debonds, the stress–strain hysteresis loops without and with considering fiber failure upon unloading can be described using the following equations.

$$\varepsilon_{c\_fu} = \frac{\sigma}{V_f E_f} + 4\frac{\tau_i}{E_f}\frac{y^2}{r_f l_c} - 2\frac{\tau_i}{E_f}\frac{(2y - l_c/2)^2}{r_f l_c} - (\alpha_c - \alpha_f)\Delta T \tag{2.28}$$

$$(\varepsilon_{c\_fu})_f = \frac{T^U}{E_f} + 4\frac{\tau_i}{E_f}\frac{y_f^2}{r_f l_c} - 2\frac{\tau_i}{E_f}\frac{(2y_f - l_c/2)^2}{r_f l_c} - (\alpha_c - \alpha_f)\Delta T \tag{2.29}$$

When the fiber/matrix interface completely debonds, the stress–strain hysteresis loops without and with considering fibers failure upon reloading can be described using the following equations.

$$\varepsilon_{c\_fr} = \frac{\sigma}{V_f E_f} - 4\frac{\tau_i}{E_f}\frac{z^2}{r_f l_c} + 4\frac{\tau_i}{E_f}\frac{(y-2z)^2}{r_f l_c} - 2\frac{\tau_i}{E_f}\frac{(l_c/2 - 2y + 2z)^2}{r_f l_c} - (\alpha_c - \alpha_f)\Delta T \tag{2.30}$$

$$\begin{aligned}(\varepsilon_{c\_fr})_f = & \frac{T^R}{E_f} - 4\frac{\tau_i}{E_f}\frac{z_f^2}{r_f l_c} + 4\frac{\tau_i}{E_f}\frac{(y_f-2z_f)^2}{r_f l_c} - 2\frac{\tau_i}{E_f}\frac{(l_c/2 - 2y_f + 2z_f)^2}{r_f l_c} \\ & - (\alpha_c - \alpha_f)\Delta T\end{aligned} \tag{2.31}$$

The area of fatigue hysteresis loops can be described using the following equation.

$$S = \int_{\sigma_{min}}^{\sigma_{max}} [\varepsilon_{unload}(\sigma) - \varepsilon_{reload}(\sigma)]d\sigma \tag{2.32}$$

where $\varepsilon_{unload}$ denotes the unloading strain and $\varepsilon_{reload}$ denotes the reloading strain.

### 2.2.4 Results and Discussions

Based on the debonding and sliding condition in the matrix crack spacing, the fatigue stress–strain hysteresis loops under cyclic loading can be divided into four different cases, as following:

(1) Case I, the fiber/matrix interface partially debonding ($l_d < l_c/2$), and the fiber sliding completely relative to the matrix in the interface debonded region ($y/l_d = z/l_d = 1$), as shown in Fig. 2.1.
(2) Case II, the fiber/matrix interface partially debonding ($l_d < l_c/2$), and the fiber sliding partially relative to the matrix in the interface debonded region ($y/l_d = z/l_d < 1$), as shown in Fig. 2.2.
(3) Case III, the fiber/matrix interface completely debonding ($l_d = l_c/2$), and the fiber sliding partially relative to the matrix in the interface debonded region ($y/l_d = z/l_d < 1$), as shown in Fig. 2.3.
(4) Case IV, the fiber/matrix interface completely debonding ($l_d = l_c/2$), and the fiber sliding completely relative to the matrix in the interface debonded region ($y/l_d = z/l_d = 1$), as shown in Fig. 2.4.

The evolution of the fatigue stress–strain hysteresis loops, area of the hysteresis loops, fiber/matrix interface debonding ratio and sliding ratio versus the fiber/matrix interface shear stress curves of unidirectional SiC/CAS composite are shown in Fig. 2.5. The stress–strain hysteresis loops corresponding to the interface slip Case I, II, III, and IV are shown in Fig. 2.5a. The area, residual strain, and peak strain of the stress–strain hysteresis loops depend on the fiber/matrix interface debonding and slip

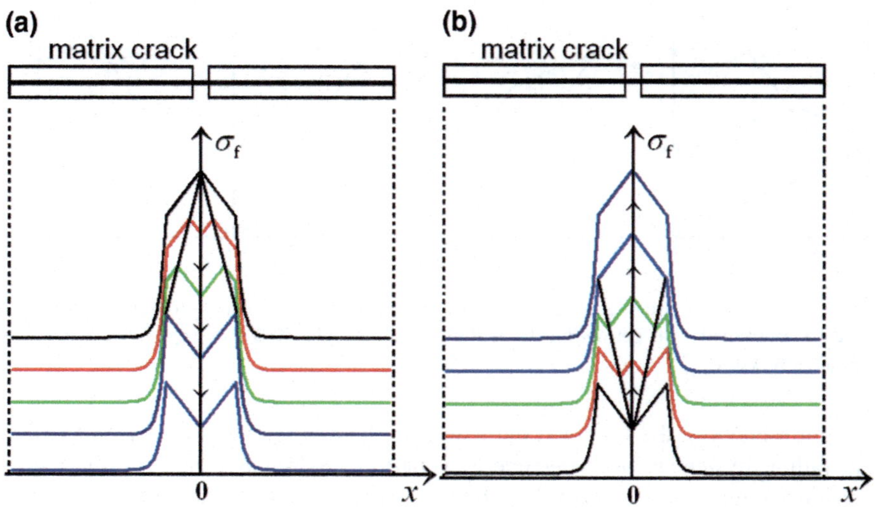


**Fig. 2.1** Fiber axial stress distribution during **a** unloading; **b** reloading for the interface slip Case I. Reprinted with permission from Ref. [6]. Copyright 2009, Springer Science Business Media New York

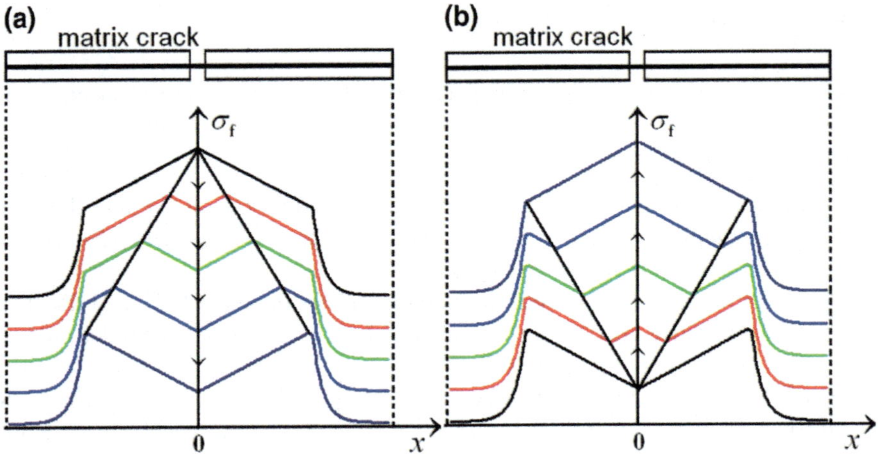


**Fig. 2.2** Fiber axial stress distribution during **a** unloading; **b** reloading for the interface slip Case II. Reprinted with permission from Ref. [6]. Copyright 2009, Springer Science Business Media New York

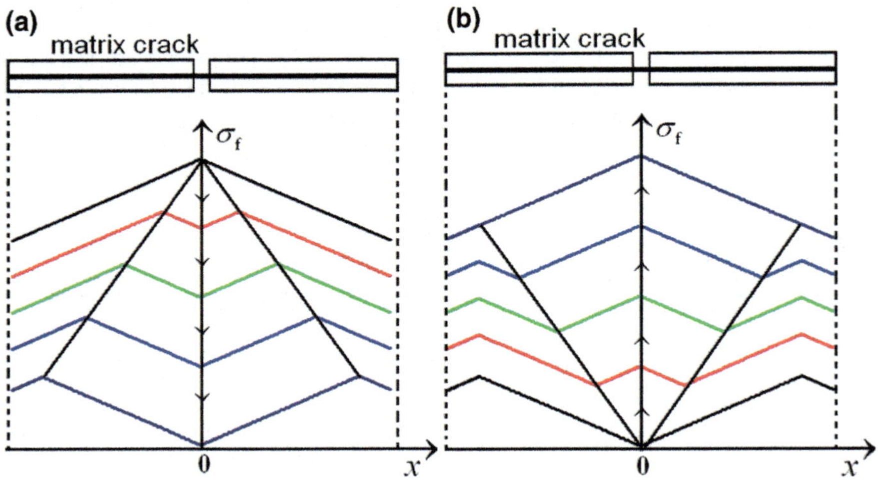


**Fig. 2.3** Fiber axial stress distribution during **a** unloading; **b** reloading for the interface slip Case III. Reprinted with permission from Ref. [6]. Copyright 2009, Springer Science Business Media New York

condition between the fiber and the matrix in the matrix cracking space of CMCs. The fatigue hysteresis dissipated energy versus the fiber/matrix interface shear stress curve is shown in Fig. 2.5b. When the fiber/matrix interface shear stress increases, the fatigue hysteresis dissipated energy increases first and then decreases. When the fiber/matrix interface shear stress increases from $\tau_i$ = 1 MPa to $\tau_i$ = 8.3 MPa, the fatigue hysteresis dissipated energy increases to the peak value, then decreases with

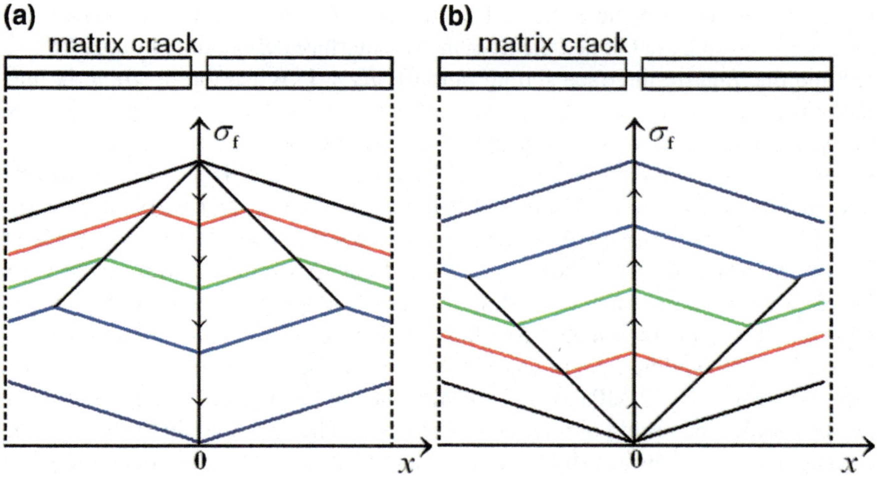


**Fig. 2.4** Fiber axial stress during **a** unloading; **b** reloading for the interface slip Case IV. Reprinted with permission from Ref. [6]. Copyright 2009, Springer Science Business Media New York

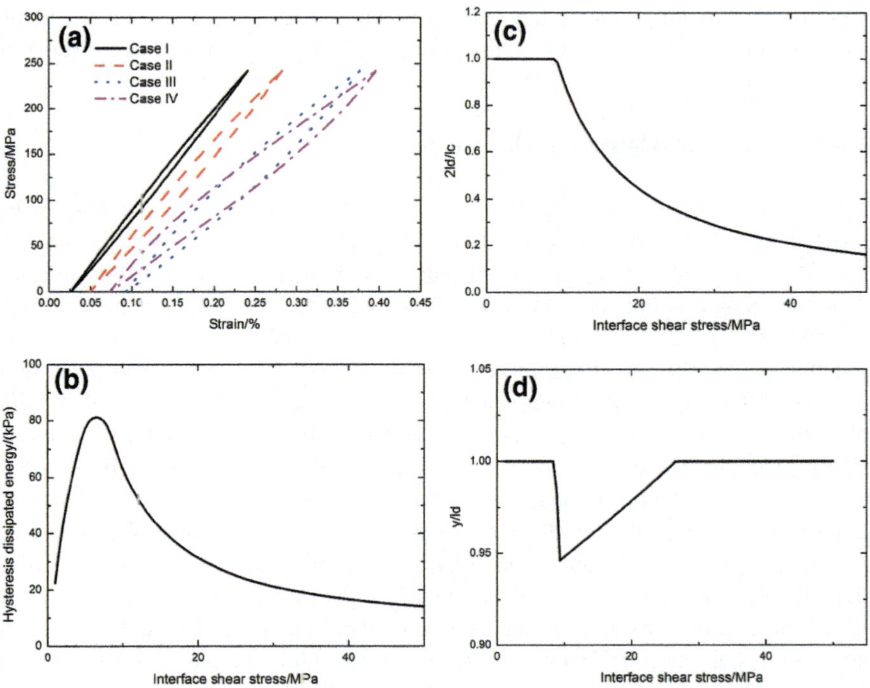


**Fig. 2.5** **a** Fatigue stress–strain hysteresis loops of different fiber/matrix interface slip cases; **b** the fatigue hysteresis dissipated energy versus the fiber/matrix interface shear stress curve; **c** the fiber/matrix interface debonding ratio versus the interface shear stress curve; and **d** the fiber/matrix interface slip ratio versus the interface shear stress curve of unidirectional SiC/CAS composite. Reprinted with permission from Ref. [6]. Copyright 2009, Springer Science Business Media New York

increasing of interface shear stress. The stress–strain hysteresis loops correspond to the interface slip Case IV; i.e., the fiber/matrix interface debonding ratio of $2l_d/l_c = 1$ and the fiber/matrix interface sliding ratio of $y/l_d = 1$. When the fiber/matrix interface shear stress increases from $\tau_i = 8.3$ MPa to $\tau_i = 8.8$ MPa, the fatigue hysteresis dissipated energy decreases with increasing of interface shear stress. The stress–strain hysteresis loops correspond to the interface slip Case III; i.e., the fiber/matrix interface debonding ratio of $2l_d/l_c = 1$ and the fiber/matrix interface sliding ratio of $y/l_d < 1$. When the fiber/matrix interface shear stress increases from $\tau_i = 8.8$ MPa to $\tau_i = 27$ MPa, the fatigue hysteresis dissipated energy decreases with increasing of interface shear stress. The stress–strain hysteresis loops correspond to the interface slip Case II; i.e., the fiber/matrix interface debonding ratio of $2l_d/l_c < 1$ and the fiber/matrix interface sliding ratio of $y/l_d < 1$. When the fiber/matrix interface shear stress increases from $\tau_i = 27$ MPa to $\tau_i = 50$ MPa, the fatigue hysteresis dissipated energy decreases with increasing of interface shear stress. The stress–strain hysteresis loops correspond to the interface slip Case I; i.e., the fiber/matrix interface debonding ratio of $2l_d/l_c < 1$ and the fiber/matrix interface sliding ratio of $y/l_d = 1$.

Based on the developed fatigue hysteresis loops models, the effects of fiber/matrix interface shear stress, fiber/matrix interface debonded energy, fibers failure, fatigue peak stress, applied cycle number, fibers strength and fibers Weibull modulus on the fiber/matrix interface debonding and slip, and then the fatigue stress–strain hysteresis loops of unidirectional SiC/CAS composite are discussed.

#### 2.2.4.1 Effect of Matrix Crack Spacing

The effect of matrix crack spacing (i.e., $l_c = 30r_f$ and $40r_f$) on the fiber/matrix interface debonding and slip and fatigue stress–strain hysteresis loops is shown in Fig. 2.6. When the matrix crack spacing is $l_c = 30r_f$, the fiber/matrix interface debonded length at the peak stress occupies 26% of matrix cracking space; i.e., the fiber/matrix interface debonding ratio of $2l_d/l_c = 0.26$; when the matrix crack spacing is $l_c = 40r_f$, the fiber/matrix interface debonded length at the peak stress occupies 19.6% of matrix crack spacing; i.e., the fiber/matrix interface debonding ratio of $2l_d/l_c = 0.196$. Upon unloading and reloading, the fiber/matrix interface sliding occupies the entire debonded region; i.e., the fiber/matrix interface sliding ratio of $y/l_d = 1$. Upon unloading to the transition stress of $\sigma_{tr\_pu} = 32$ MPa, the unloading interface counter slip length approaches to the interface debonding tip, i.e., $y(\sigma_{tr\_pu})/l_d = 1$; and upon reloading to the transition stress of $\sigma_{tr\_pr} = 168$ MPa, the new slip length approaches to the interface debonding tip, i.e., $z(\sigma_{tr\_pr})/l_d = 1$. When the matrix crack spacing increases, the fiber/matrix interface debonding ratio decreases, and the fiber/matrix interface sliding ratio remains unchanged. The fatigue stress–strain hysteresis area decreases, and the hysteresis modulus increases, and the unloading residual strain also decreases.

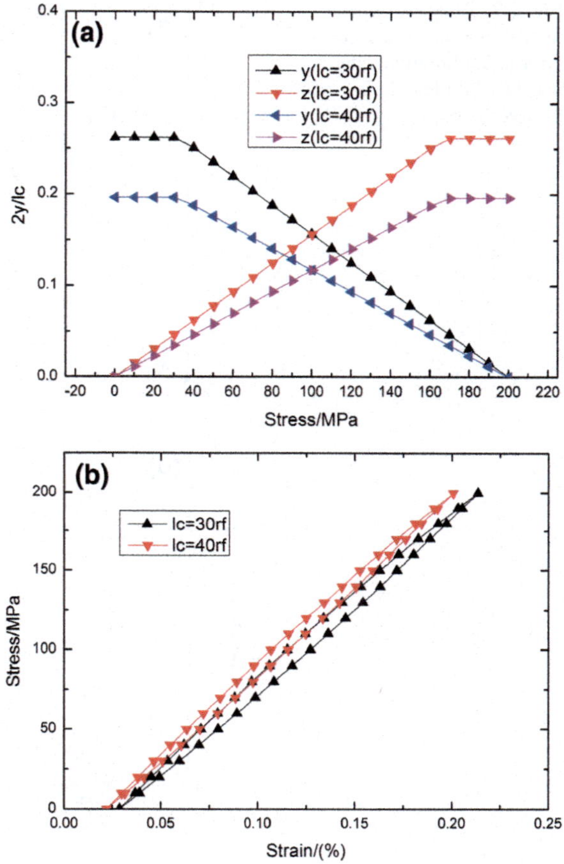


**Fig. 2.6** Effects of matrix crack spacing on **a** unloading/reloading interface slip length; **b** fatigue hysteresis loops

#### 2.2.4.2 Effect of Interface Shear Stress

The effects of fiber/matrix interface shear stress (i.e., $\tau_i$ = 10 MPa and 20 MPa) on the fiber/matrix interface debonding and slip and fatigue stress–strain hysteresis loops are shown in Fig. 2.7. The fiber/matrix interface partially debonds, and interface debonded length occupies 40% of entire matrix crack spacing when the interface shear stress is $\tau_i$ = 10 MPa; i.e., the fiber/matrix interface debonding ratio is $2l_d/l_c$ = 0.4. Upon unloading to the transition stress of $\sigma_{tr\_pu}$ = 28.7 MPa, the interface counter slip length reaches the interface debonded length, i.e., $y(\sigma_{tr\_pu})/l_d$ = 1. Upon reloading to the transition stress of $\sigma_{tr\_pr}$ = 171.2 MPa, the interface new slip length reaches the interface debonded length, i.e., $z(\sigma_{tr\_pr})/l_d$ = 1. When the fiber/matrix interface shear stress is $\tau_i$ = 20 MPa, the interface partially debonds and the interface debonded length occupies 19% of entire matrix crack spacing; i.e., the fiber/matrix interface debonding ratio is $2l_d/l_c$ = 0.19. Upon unloading to the transition stress of $\sigma_{tr\_pu}$ = 36.6 MPa, the interface counter slip length approaches to the interface debonded length, $y(\sigma_{tr\_pu})/l_d$ = 1. Upon reloading to the transition

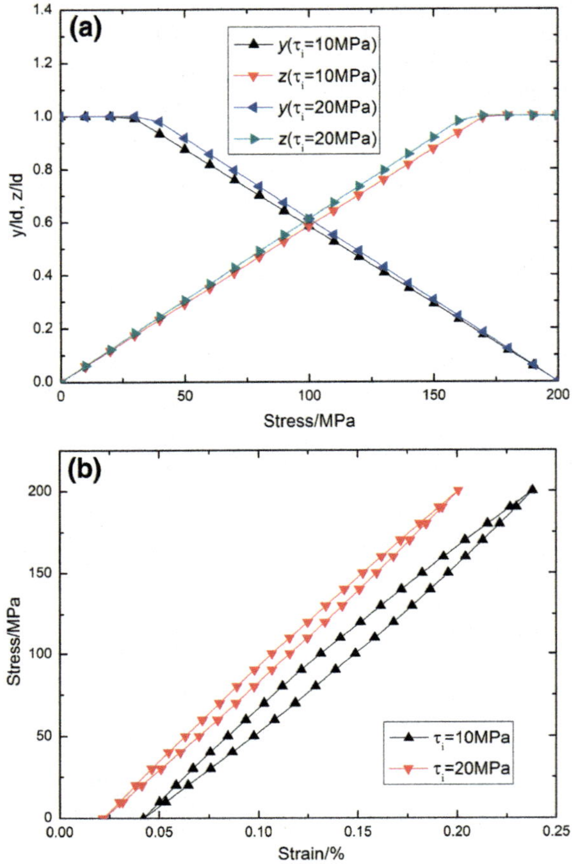


**Fig. 2.7** Effects of fiber/matrix interface shear stress on **a** unloading/reloading interface slip length; **b** fatigue hysteresis loops

stress of $\sigma_{tr\_pr}$ = 163.4 MPa, the interface new slip length approaches to the interface debonded length, i.e., $z(\sigma_{tr\_pr})/l_d = 1$. The unloading transition stress $\sigma_{tr\_pu}$ increases and reloading transition stress $\sigma_{tr\_pr}$ decreases with increasing interface shear stress. The proportion of unloading/reloading interface slip increases with increasing of interface shear stress. The residual strain and the stress–strain fatigue hysteresis loop area decrease, and the fatigue hysteresis modulus increases with increasing of interface shear stress.

#### 2.2.4.3 Effect of Interface Debonded Energy

The effects of fiber/matrix interface debonded energy (i.e., $\zeta_d$ = 0.1 J/m$^2$ and 0.5 J/m$^2$) on the interface debonding and slip and fatigue stress–strain hysteresis loops are shown in Fig. 2.8. The fiber/matrix interface partially debonds, and the interface debonded length occupies 45% of entire matrix crack spacing when the interface debonded energy is $\zeta_d$ = 0.1 J/m$^2$; i.e., the fiber/matrix interface debonding ratio is

$2l_d/l_c = 0.45$. The fiber partially slides relative to the matrix in interface debonded region, and interface counter slip length upon completely unloading occupies 69.1% of the interface debonded length, i.e., $y(\sigma_{min})/l_d = 69.1\%$. When the fiber/matrix interface debonding energy is $\zeta_d = 0.5$ J/m$^2$, the interface partially debonds and the interface debonded length occupies 26% of entire matrix crack spacing; i.e., the fiber/matrix interface debonding ratio is $2l_d/l_c = 0.26$. Upon unloading to the transition stress of $\sigma_{tr\_pu} = 32.7$ MPa, the interface counter slip length approaches to the interface debonded length, i.e., $y(\sigma_{tr\_pu})/l_d = 1$. Upon reloading to the transition stress of $\sigma_{tr\_pr} = 167.3$ MPa, the interface new slip length approaches to the interface debonded length, i.e., $z(\sigma_{tr\_pr})/l_d = 1$. The unloading transition stress $\sigma_{tr\_pu}$ increases, and reloading transition stress $\sigma_{tr\_pr}$ decreases with increasing of the fiber/matrix interface debonded energy. The proportion of unloading/reloading interface slip increases with increasing interface debonded energy. The residual strain and the fatigue hysteresis loop area decrease, and the fatigue hysteresis modulus increases with increasing of the fiber/matrix interface debonded energy.

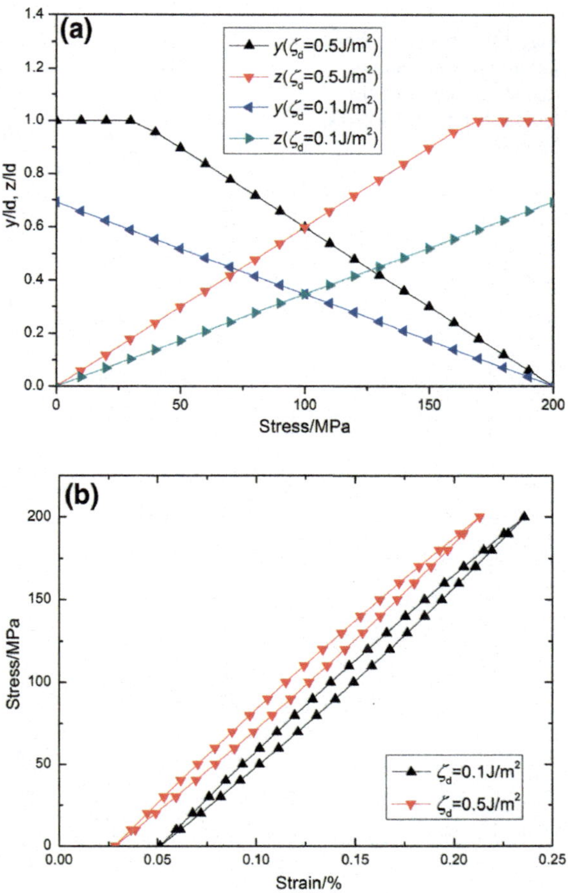


**Fig. 2.8** Effects of fiber/matrix interface debonded energy on **a** unloading/reloading interface slip length; **b** fatigue hysteresis loops

#### 2.2.4.4 Effect of Fibers Failure

The effects of fibers failure on the fiber/matrix interface debonding and slip and fatigue stress–strain hysteresis loops are shown in Fig. 2.9. The fiber/matrix interface completely debonds and the fiber slides partially relative to the matrix upon unloading/reloading at the fatigue peak stress is $\sigma_{max}$ = 450 MPa without considering fibers failure. Upon completely unloading, the fiber/matrix interface counter slip length occupies 70.1% of the interface debonded length, i.e., $y(\sigma_{min})/l_d$ = 70.1%. When the fibers failure is considered, the fibers failure volume fraction reaches 24.5%, which is less than the critical fiber failure volume fraction of 35.7%. Upon completely unloading, the interface counter slip length approaches to 80% of the interface debonded length, i.e., $y_f(\sigma_{min})/l_d$ = 80%. The unloading/reloading interface slip lengths increase when the fibers failure is considered. The residual strain and hysteresis loop area increase, and hysteresis modulus decreases when the fibers failure is considered.

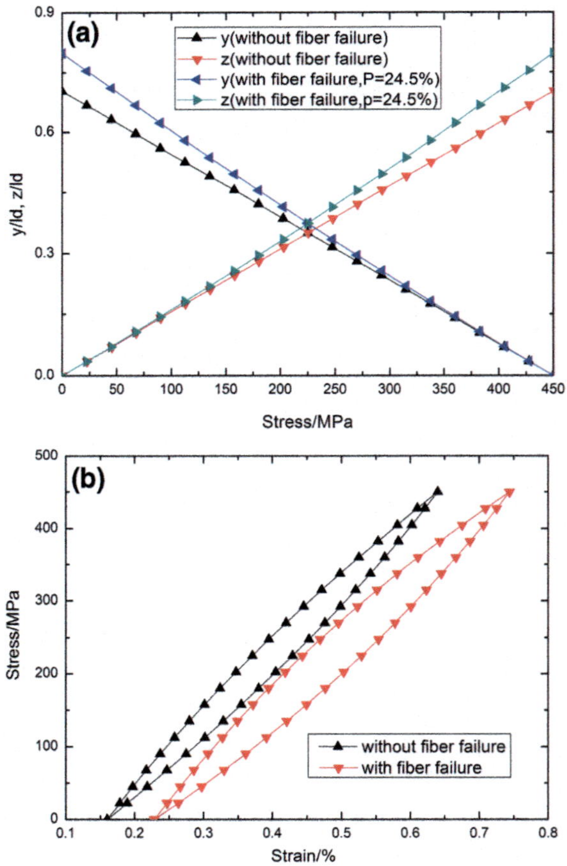


**Fig. 2.9** Effects of fiber failure on **a** unloading/reloading interface slip length; **b** fatigue hysteresis loops

#### 2.2.4.5 Effect of Fatigue Peak Stress

The effects of fatigue peak stress (i.e., $\sigma_{max}$ = 200 and 250 MPa) on the fiber/matrix interface debonding and slip and fatigue stress–strain hysteresis loops are shown in Fig. 2.10. The fibers failure volume fractions at fatigue peak stress of $\sigma_{max}$ = 200 MPa and 250 MPa are 0.38 and 1.08%, respectively. When the fatigue peak stress is $\sigma_{max}$ = 200 MPa, the fiber/matrix interface partially debonds and the interface debonded length occupied 58% of matrix crack spacing; i.e., the fiber/matrix interface debonding ratio of $2l_d/l_c = 0.58$. Upon unloading and reloading, the fiber partially slides relative to the matrix. The completely unloading interface counter slip length occupied 54% of the interface debonded length, i.e., $y(\sigma_{min})/l_d = 0.54$. When the fatigue peak stress is $\sigma_{max}$ = 250 MPa, the fiber/matrix interface partially debonds and the interface debonded length occupied 87% of matrix crack spacing; i.e., the fiber/matrix interface debonding ratio of $2l_d/l_c = 0.87$. Upon unloading and reloading, the fiber partially slides relative to the matrix. The completely unloading interface counter slip length occupied 69% of the interface debonded length, i.e., $y(\sigma_{min})/l_d = 0.69$. The proportion of unloading/reloading interface slip increases with increasing of fatigue peak stress. The residual strain and hysteresis loop area increase, and the hysteresis modulus decreases with increasing of fatigue peak stress.

#### 2.2.4.6 Effect of Applied Cycle Number

The effects of applied cycle number (i.e., $N$ = 1, 5, 109, and 30,040) on the fiber/matrix interface debonding and slip and fatigue stress–strain hysteresis loops at the fatigue peak stress of $\sigma_{max}$ = 280 MPa are shown in Fig. 2.11. When the applied cycle number is $N$ = 1 and 5, the fiber/matrix interface slip corresponds to the interface slip Case II; i.e., the fiber/matrix interface debonding ratio of $2l_d/l_c < 1$ and the fiber/matrix interface sliding ratio of $y/l_d < 1$. When the applied cycle number is $N$ = 109 and 30,040, the fiber/matrix interface slip corresponds to interface slip Case III; i.e., the fiber/matrix interface debonding ratio of $2l_d/l_c = 1$ and the fiber/matrix interface sliding ratio of $y/l_d < 1$. The fatigue stress–strain hysteresis loops at the 1st, 5th, 109th, and 30,040th cycles are shown in Fig. 2.11b. With increasing applied cycles, the fractions of broken fiber increase due to the interface wear and fiber strength degradation. The fiber failure volume fractions corresponding to the 1st, 5th, 109th, and 30040th cycles are 0.018, 0.021, 0.09, and 0.17%. The residual strain increases with increasing applied cycles. The fatigue hysteresis loop area increases with increasing applied cycles and then decreases for different interface slip cases. The fatigue hysteresis modulus decreases with increasing applied cycles due to the fibers failure.

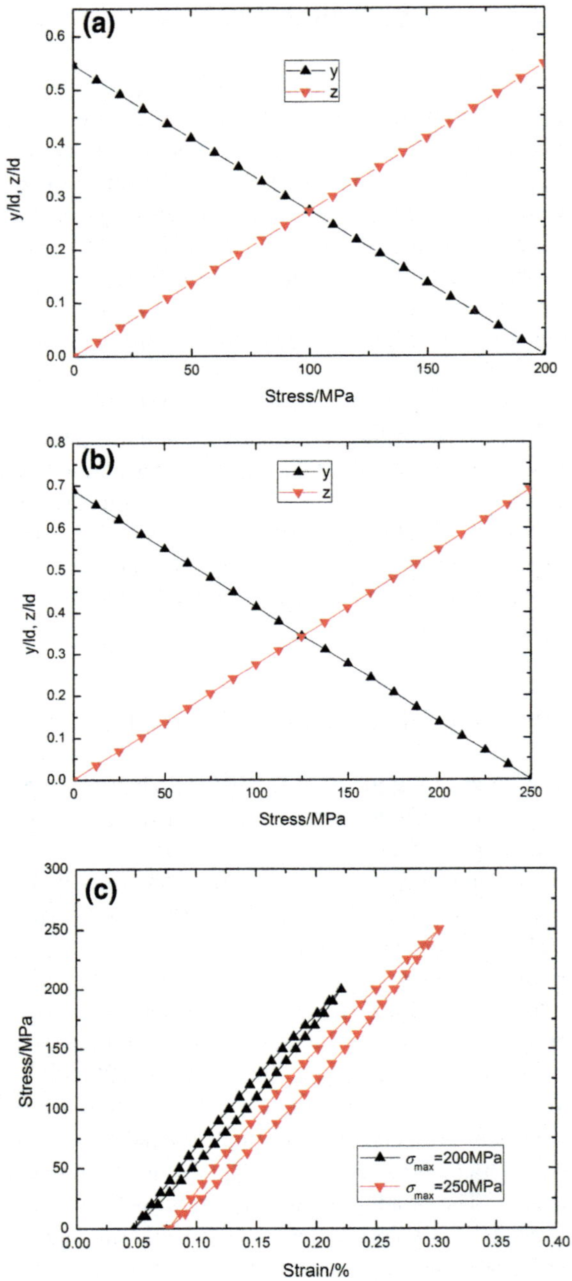


**Fig. 2.10** **a** Unloading/reloading interface slip length at $\sigma_{max}$ = 200 MPa; **b** the unloading/reloading interface slip length at $\sigma_{max}$ = 250 MPa; **c** the effect of fatigue peak stress of $\sigma_{max}$ = 200 and 250 MPa on the fatigue stress–strain hysteresis loops

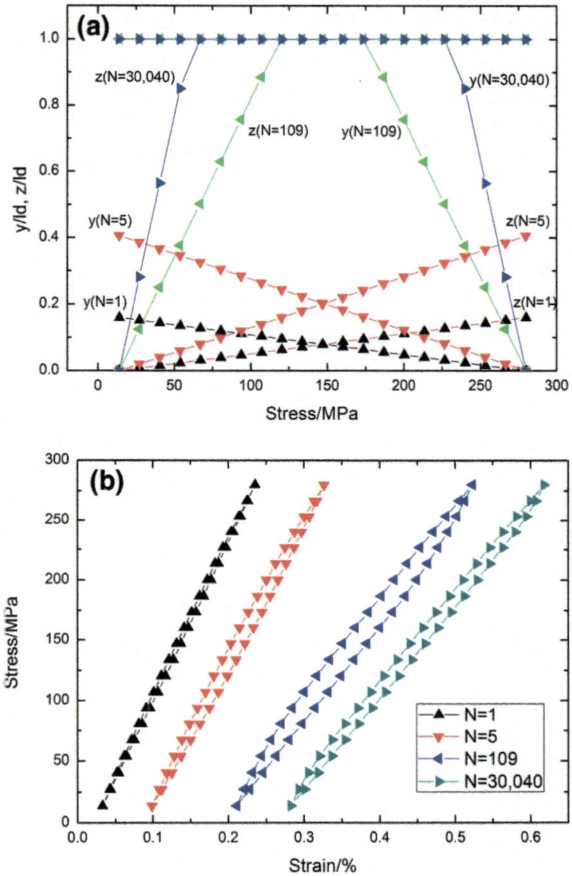


**Fig. 2.11** Effects of applied cycle number on **a** unloading/reloading interface slip; **b** fatigue hysteresis loops under fatigue peak stress of $\sigma_{max} = 280$ MPa

#### 2.2.4.7 Effect of Fibers Volume Fraction

The effect of fiber volume fraction (i.e., $V_f = 34$ and 40%) on the fiber/matrix interface debonding and slip and the fatigue stress–strain hysteresis loops is shown in Fig. 2.12. When the fiber volume fraction is $V_f = 34\%$, the fiber/matrix interface partially debonds, and the fiber/matrix interface debonded length occupies 26% of the entire matrix cracking space; i.e., the fiber/matrix interface debonding ratio of $2l_d/l_c = 0.26$. Upon unloading and reloading, the fiber/matrix interface completely sliding occurs in the debonded region, and upon unloading to the transition stress of $\sigma_{tr\_pu} = 32.7$ MPa, the unloading interface counter slip length approaches to the interface debonded tip, i.e., $y(\sigma_{tr\_pu})/l_d = 1$, and upon reloading to the transition stress of $\sigma_{tr\_pr} = 167.3$ MPa, the reloading interface new slip length approaches to the interface debonding tip, i.e., $z(\sigma_{tr\_pr})/l_d = 1$. When the fiber volume fraction is $V_f = 40\%$, the fiber/matrix interface partially debonds, and the fiber/matrix interface debonded length occupies 12.3% of the entire matrix crack spacing; i.e., the fiber/

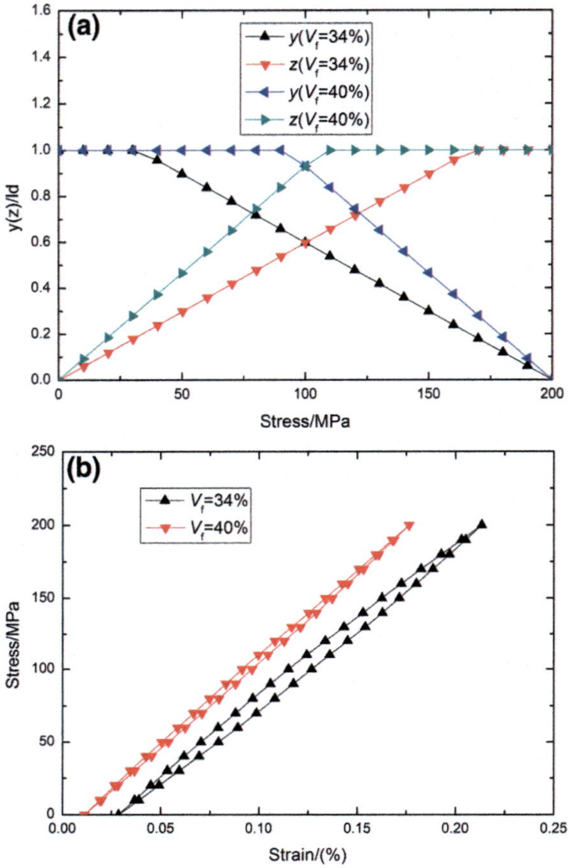


**Fig. 2.12** Effects of fiber volume fraction on **a** unloading/reloading interface slip; **b** fatigue hysteresis loops

matrix interface debonding ratio of $2l_d/l_c = 0.123$, and upon unloading to the transition stress of $\sigma_{tr\_pu} = 92.5$ MPa, the unloading interface counter slip length approaches to the interface debonding tip, i.e., $y(\sigma_{tr\_pu})/l_d = 1$, and upon reloading to the transition stress of $\sigma_{tr\_pr} = 107.5$ MPa, the reloading interface new slip length approaches to the interface debonding tip, i.e., $z(\sigma_{tr\_pr})/l_d = 1$. The unloading transition stress increases with increasing fibers volume fraction, and the reloading transition stress decreases with increasing of fibers volume fraction. Upon unloading and reloading, the fiber/matrix interface sliding ratio increases with increasing fibers volume fraction. The unloading residual strain decreases, the hysteresis loop area decreases, and the hysteresis modulus increases with increasing of fibers volume fraction.

### 2.2.4.8 Effect of Fiber Poisson Contraction

The effect of fiber/matrix interface frictional coefficient (i.e., $\mu = 0.2$ and 0.3) on the fiber/matrix interface debonding and slip and the fatigue stress–strain hysteresis loops is shown in Fig. 2.13. When the fiber/matrix interface frictional coefficient increases, the unloading interface counter slip length and reloading interface new slip length both decrease, and the unloading residual strain decreases, and the hysteresis loop area decreases, and the hysteresis modulus increases.

The effect of fiber Poisson ratio (i.e., $\nu_f = 0.1$ and 0.3) on the fiber/matrix interface debonding and slip and the fatigue stress–strain hysteresis loops is shown in Fig. 2.14. When the fiber Poisson ratio increases, the unloading interface counter slip length and reloading interface new slip length increase, and the unloading residual strain increases, the hysteresis loop area increases, and the hysteresis modulus decreases.

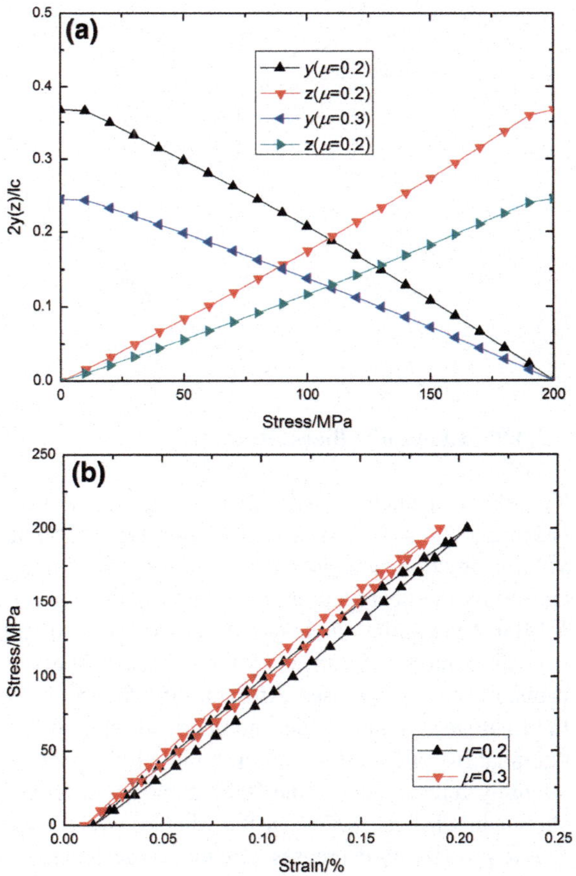


**Fig. 2.13** Effects of fiber/matrix interface frictional coefficient on **a** unloading/reloading interface slip; **b** fatigue hysteresis loops

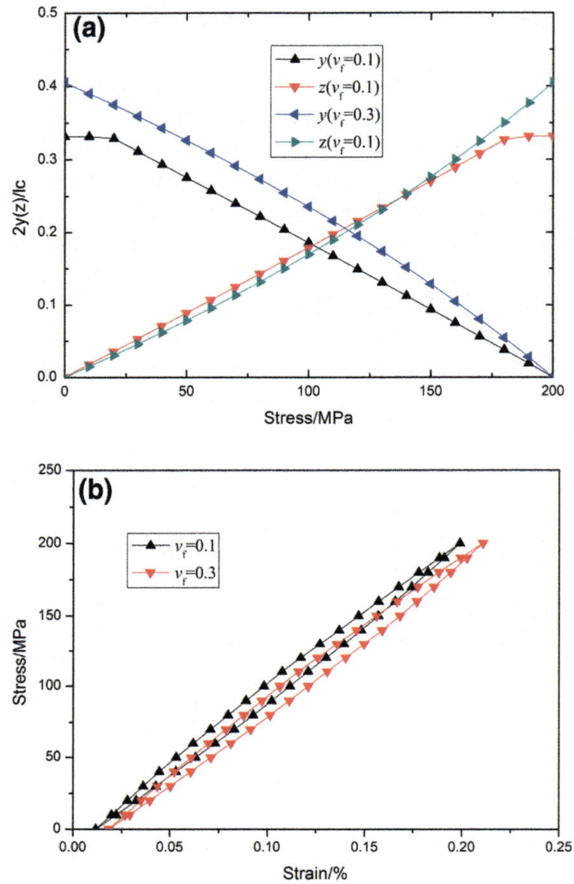


**Fig. 2.14** Effects of fiber Poisson ratio on **a** unloading/reloading interface slip; **b** fatigue hysteresis loops

#### 2.2.4.9 Effect of Fibers Strength

The effect of characteristic fiber strength on the fiber failure under cyclic fatigue loading ($\sigma_{max}$ = 300 MPa) is shown in Fig. 2.15a. It is shown that the fibers failure volume fraction increases rapidly during the initial one hundred cycles, which is caused by the degradation of the fiber/matrix interface shear stress from 15 MPa to 5 MPa at the initial one hundred cycles. When the fiber/matrix interface shear stress decreases, the stress carried by the unbroken fibers is increased, and the fiber failure probability also increases. After the fiber/matrix interface shear stress approaches to the final steady-state value, the fiber failure is mainly caused by the fibers strength degradation, and the fibers failure probability increases linearly. The fatigue stress–strain hysteresis loops at the 100th cycle for different characteristic fiber strength are shown in Fig. 2.15b. When the fibers characteristic strength increases, the fibers failure probability decreases, leading to the decrease of the unloading residual strain and the peak strain, and the area of the fatigue hysteresis loops.

Fig. 2.15 Effect of fiber characteristic strength on **a** fibers failure versus applied cycles; **b** the fatigue hysteresis loops

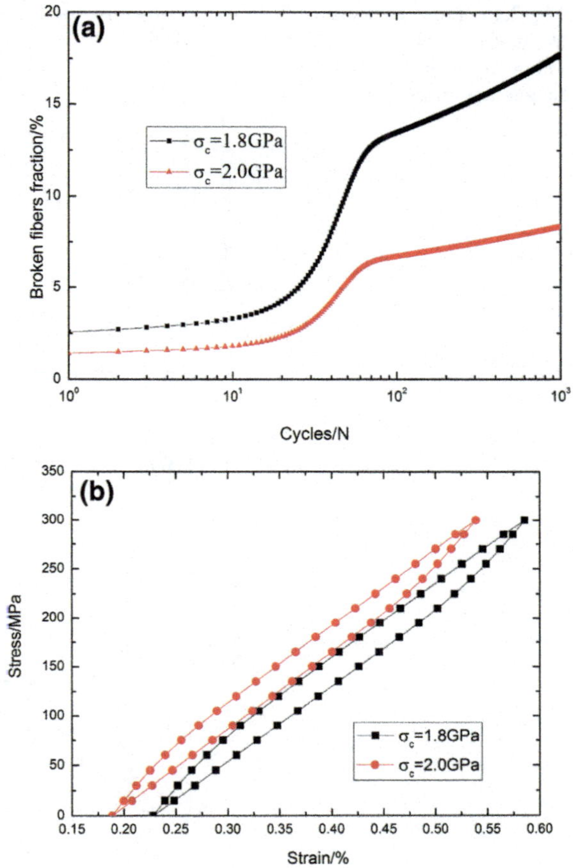


#### 2.2.4.10 Effect of Fibers Weibull Modulus

The effect of fiber Weibull modulus ($m_f$ = 3.6 and 4) on the fibers failure and fatigue stress–strain hysteresis loops at the fatigue peak stress of $\sigma_{max}$ = 300 MPa is shown in Fig. 2.16. When the fibers Weibull modulus increases, the fibers failure probability decrease, and the unloading residual strain and peak strain decreases, and the area of the fatigue hysteresis loops also decreases.

### 2.2.5 *Experimental Comparisons*

The experimental fatigue stress–strain hysteresis loops of unidirectional SiC/CAS, SiC/1723, and C/SiC composites at room and elevated temperatures are predicted using the present analysis.

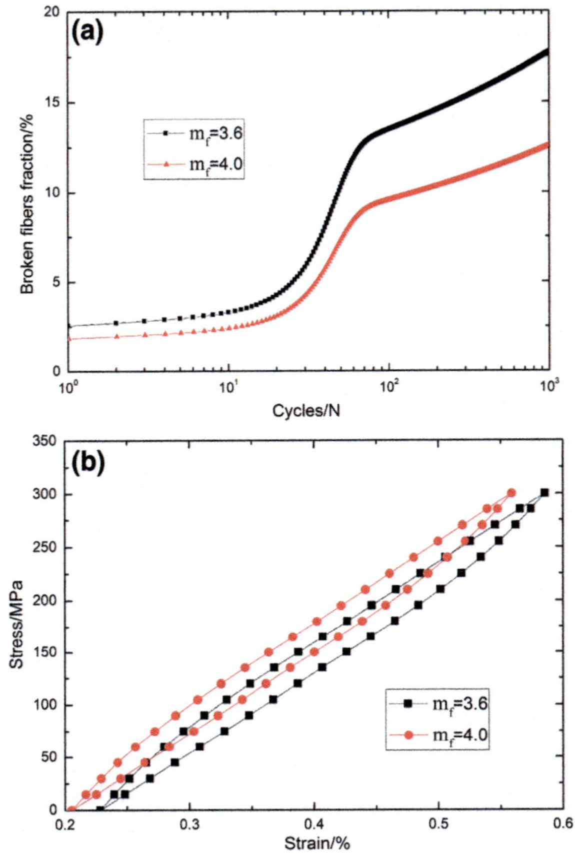


**Fig. 2.16** Effect of fibers Weibull modulus on **a** fiber failure versus cycles; **b** the fatigue hysteresis loops

#### 2.2.5.1 SiC/CAS Composites

Cyclic Loading/Unloading Tensile Hysteresis Loops

Pryce and Smith [20] investigated the cyclic loading/unloading tensile hysteresis behavior of unidirectional SiC/CAS composite at room temperature. The experimental and theoretical stress–strain hysteresis loops and fiber/matrix interface debonding and slip at the fatigue peak stress of $\sigma_{max}$ = 170 MPa are shown in Fig. 2.17. The fiber/matrix interface partially debonds, and the fiber completely slides relative to the matrix in the interface debonded region, corresponding to the interface slip Case I. Upon unloading to the transition stress of $\sigma_{tr\_pu}$ = 62.7 MPa, the unloading interface counter slip length approaches to the interface debonding tip, i.e., $y(\sigma_{tr\_pu}) = l_d$, and upon reloading to the transition stress of $\sigma_{tr\_pr}$ = 107.3 MPa, the reloading interface new slip length approaches to the interface debonding tip, i.e., $z(\sigma_{tr\_pr}) = l_d$.

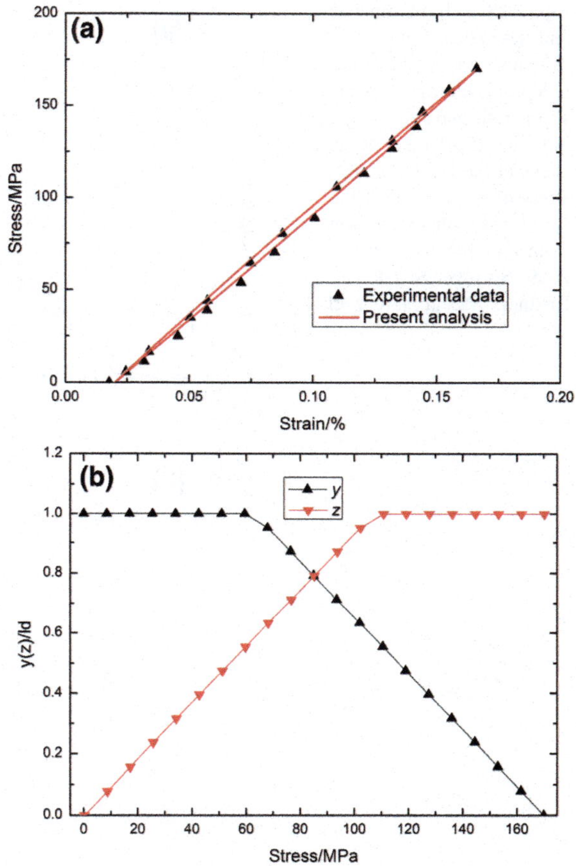


**Fig. 2.17** **a** Experimental and theoretical fatigue hysteresis loops and **b** the unloading interface counter slip length and reloading new interface slip length versus stress at the fatigue peak stress of $\sigma_{max}$ = 170 MPa

The experimental and theoretical stress–strain hysteresis loops and fiber/matrix interface debonding and slip at the fatigue peak stress of $\sigma_{max}$ = 185 MPa are shown in Fig. 2.18. The fiber/matrix interface partially debonding and the fiber completely slides relative to the matrix in the interface debonded region, corresponding to the interface slip Case I. Upon unloading to the transition stress of $\sigma_{tr\_pu}$ = 47.7 MPa, the unloading interface counter slip length approaches to the interface debonding tip, i.e., $y(\sigma_{tr\_pu}) = l_d$; and upon reloading to the transition stress of $\sigma_{tr\_pr}$ = 137.3 MPa, the reloading interface new slip length approaches to the interface debonding tip, i.e., $z(\sigma_{tr\_pr}) = l_d$.

The experimental and theoretical stress–strain hysteresis loops and fiber/matrix interface debonding and slip at the fatigue peak stress of $\sigma_{max}$ = 195 MPa are shown in Fig. 2.19. The fiber/matrix interface partially debonds, and the fiber completely slides relative to the matrix in the interface debonded region, corresponding to the interface slip Case I. Upon unloading to the transition stress of $\sigma_{tr\_pu}$ = 37.7 MPa, the unloading interface counter slip length approaches to the interface debonding tip, i.e., $y(\sigma_{tr\_pu}) = l_d$, and upon reloading to the transition

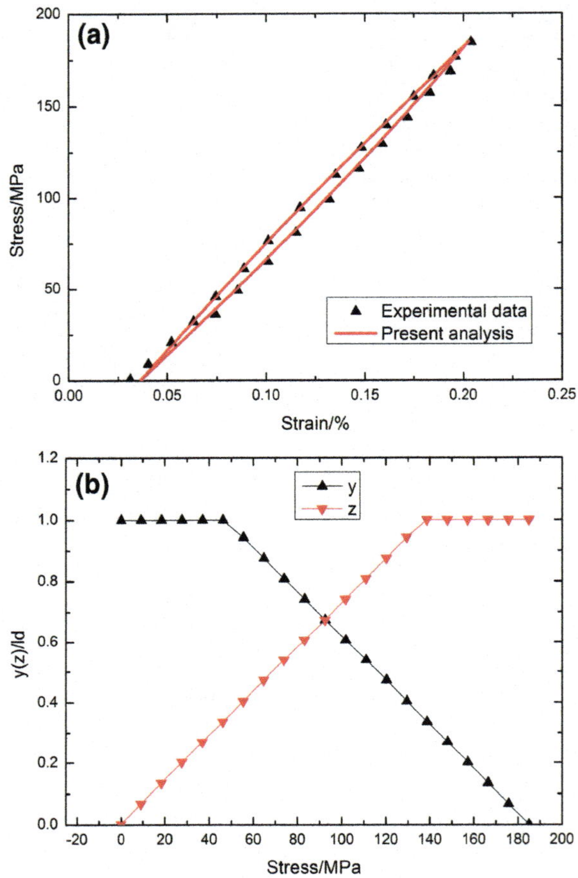


**Fig. 2.18** **a** Experimental and theoretical fatigue hysteresis loops and **b** the unloading interface counter slip length and reloading new interface slip length versus stress at the fatigue peak stress of $\sigma_{max}$ = 185 MPa. Reprinted with permission from Ref. [6]. Copyright 2009, Springer Science Business Media New York

stress of $\sigma_{tr\_pr}$ = 157.3 MPa, the reloading interface new slip length approaches to the interface debonding tip, i.e., $z(\sigma_{tr\_pr}) = l_d$.

The experimental and theoretical stress–strain hysteresis loops and fiber/matrix interface debonding and slip at the fatigue peak stress of $\sigma_{max}$ = 210 MPa are shown in Fig. 2.20. The fiber/matrix interface partially debonds, and the fiber completely slides relative to the matrix in the interface debonded region, corresponding to the interface slip Case I. Upon unloading to the transition stress of $\sigma_{tr\_pu}$ = 22.7 MPa, the unloading interface counter slip length approaches to the interface debonding tip, i.e., $y(\sigma_{tr\_pu}) = l_d$, and upon reloading to the transition stress of $\sigma_{tr\_pr}$ = 187.3 MPa, the reloading interface new slip length approaches to the interface debonding tip, i.e., $z(\sigma_{tr\_pr}) = l_d$.

The experimental and theoretical stress–strain hysteresis loops and fiber/matrix interface debonding and slip at the fatigue peak stress of $\sigma_{max}$ = 242 MPa are shown in Fig. 2.21. The fiber/matrix interface partially debonds, and the fiber partially slides relative to the matrix in the interface debonded region,

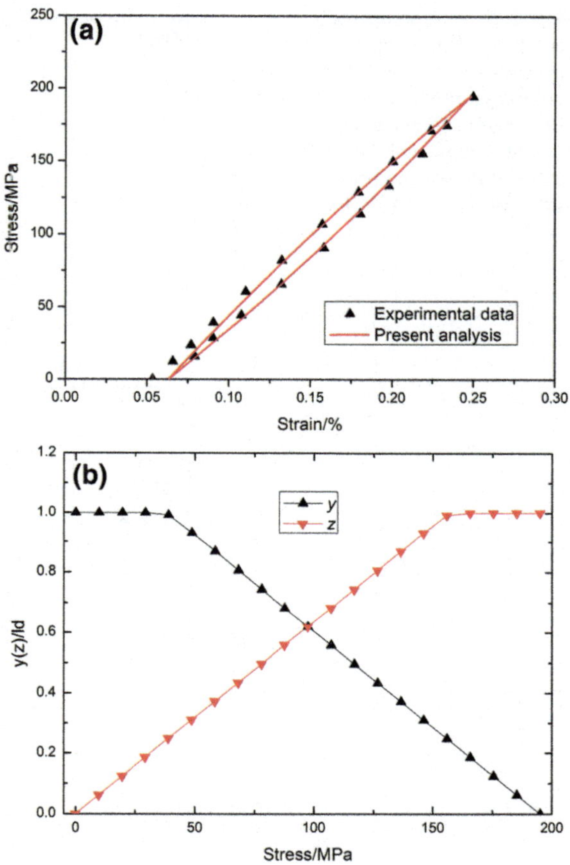


**Fig. 2.19** **a** Experimental and theoretical fatigue hysteresis loops and **b** the unloading interface counter slip length and reloading new interface slip length versus stress at the fatigue peak stress of $\sigma_{max}$ = 195 MPa

corresponding to the interface slip Case II. Upon completely unloading, the interface counter slip length approaches to 96.3% of the fiber/matrix interface debonding length; i.e., the fiber/matrix interface sliding ratio of $y/l_d$ = 0.963.

## Tension–Tension Fatigue Hysteresis Loops

Evans et al. [44] investigated the tensile-tensile fatigue behavior of unidirectional SiC/CAS composite at room temperature. Under cyclic fatigue loading, the fiber/matrix interface wear caused the degradation of the interfacial shear stress. The interface shear stress decreases from 22 MPa at the first applied cycle to 5 MPa at the 100th applied cycle. The theoretical and experimental predicted fatigue hysteresis loops corresponding to $N$ = 1, 5, 109, and 30040 applied cycle are shown in Fig. 2.22.

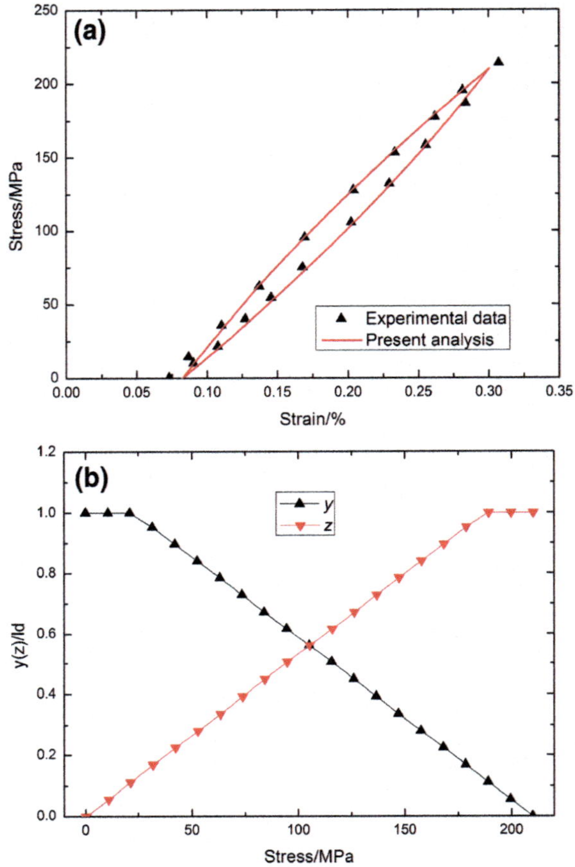


**Fig. 2.20** **a** Experimental and theoretical fatigue hysteresis loops and **b** the unloading interface counter slip length and reloading new interface slip length versus stress at the fatigue peak stress of $\sigma_{\max} = 210$ MPa

Holmes and Cho [16] investigated the tensile–tensile fatigue behavior of unidirectional SiC/CAS-II composite at room temperature. The interface shear stress degraded from 25 MPa in the first cycle to 7 MPa during one hundred cycles. The theoretical and experimental predicted fatigue hysteresis loops corresponding to $N = 3$, 10, 20, and 3200 applied cycle are shown in Fig. 2.23.

#### 2.2.5.2 SiC/1723 Composites

Zawada et al. [45] investigated the tensile–tensile fatigue behavior of unidirectional SiC/1723 composites at room temperature. The interface shear stress degraded from 32 MPa at the first cycle to 27 MPa at the 300th applied cycles. The experimental and theoretical fatigue stress–strain hysteresis loops corresponding to the 10th, 20th, 1010th, and 10,011th applied cycle are shown in Fig. 2.24.

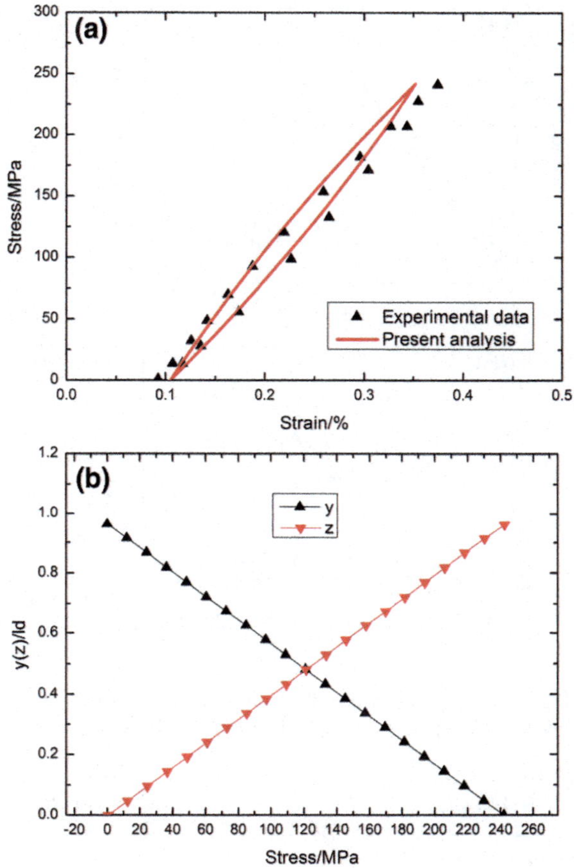


**Fig. 2.21** **a** Experimental and theoretical fatigue hysteresis loops and **b** the unloading interface counter slip length and reloading new interface slip length versus stress at the fatigue peak stress of $\sigma_{max}$ = 242 MPa. Reprinted with permission from Ref. [6]. Copyright 2009, Springer Science Business Media New York

#### 2.2.5.3 C/SiC Composites

Cyclic Loading/Unloading Tensile Hysteresis Loops

The experimental and theoretical fatigue hysteresis loops of unidirectional C/SiC composite at peak stress of $\sigma_{max}$ = 140, 180, and 240 MPa with and without considering fibers failure are shown in Figs. 2.25, 2.26 and 2.27. When the fatigue peak stress is $\sigma_{max}$ = 140 and 180 MPa, the fiber/matrix interface partially debonds and the fiber failure volume fractions are 0.0082 and 1.27%, respectively. The unloading interface counter slip length $y$ increases with stress decreasing until $\sigma_{min}$, i.e., $y(\sigma_{min}) < l_d$. The reloading interface new slip length $z$ increases with stress increasing until $\sigma_{max}$, i.e., $z(\sigma_{max}) < l_d$. At the fatigue peak stress of $\sigma_{max}$ = 180 MPa, the unloading interface counter slip length and reloading interface new slip length considering fibers failure are both larger than those without considering fibers failure. However, the differences at peak stress of $\sigma_{max}$ = 140 MPa are not obvious due to the

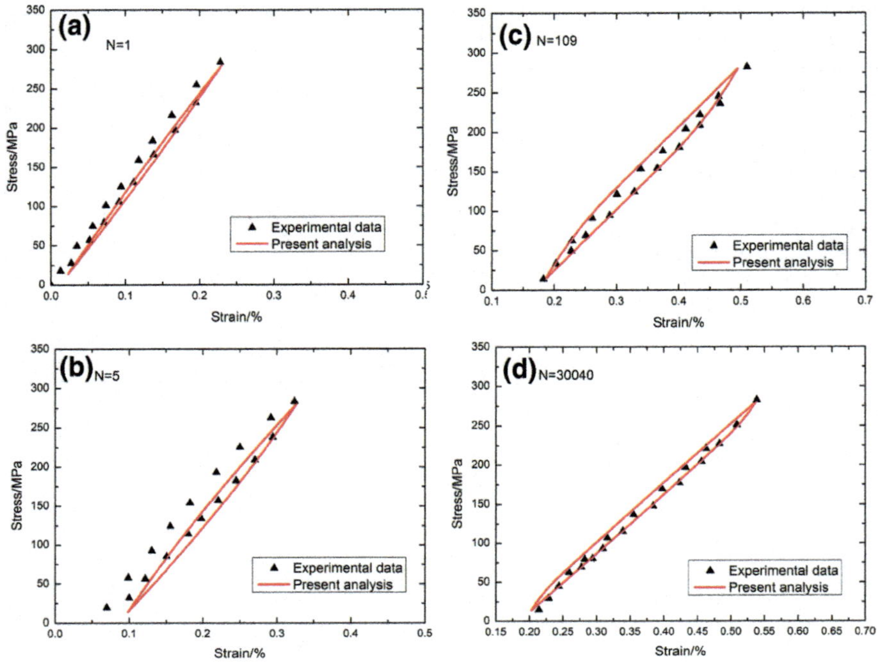


**Fig. 2.22** Experimental and theoretical fatigue stress–strain hysteresis loops of unidirectional SiC/CAS composite corresponding to **a** $N = 1$; **b** $N = 5$; **c** $N = 109$; and **d** $N = 30{,}040$

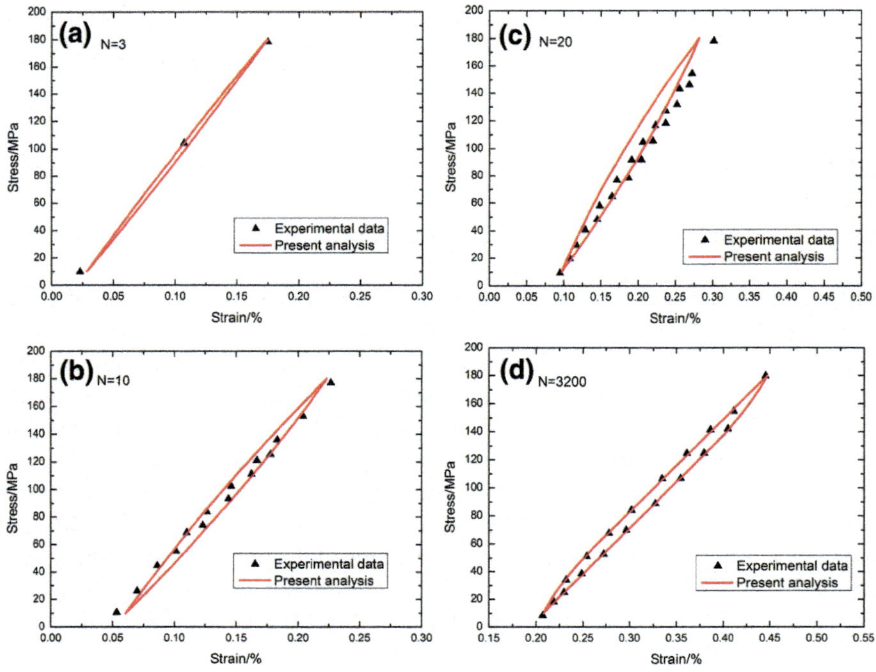


**Fig. 2.23** Experimental and theoretical fatigue stress–strain hysteresis loops of unidirectional SiC/CAS-II corresponding to **a** $N = 3$; **b** $N = 10$; **c** $N = 20$; and **d** $N = 3200$

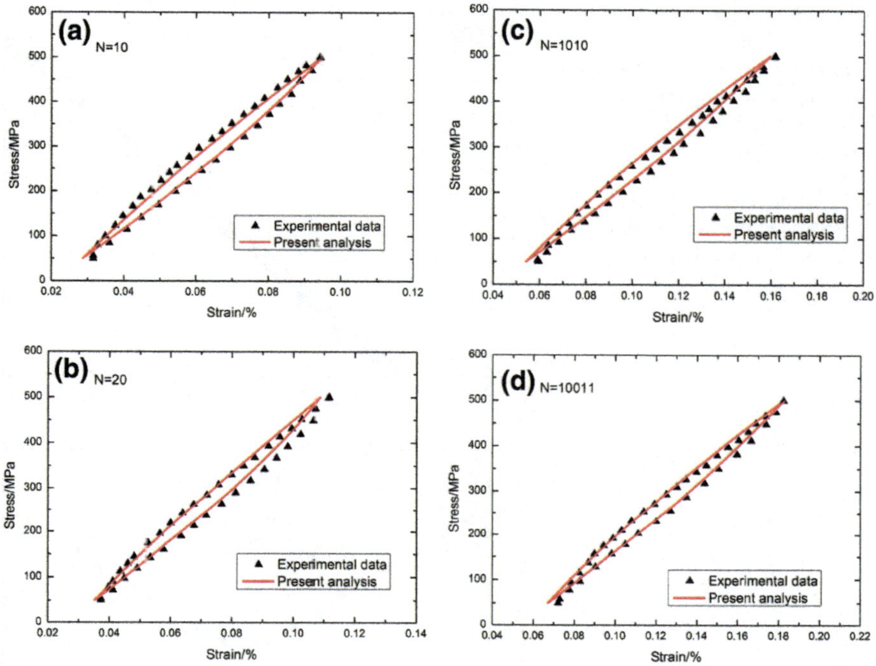


**Fig. 2.24** Experimental and theoretical fatigue stress–strain hysteresis loops of unidirectional SiC/1723 composite corresponding to **a** $N = 10$; **b** $N = 20$; **c** $N = 1010$; and **d** $N = 10{,}011$

low fiber failure rate. The fatigue stress–strain hysteresis loops at $\sigma_{max} = 140$ and 180 MPa correspond to the interface slip Case II.

When the fatigue peak stress is $\sigma_{max} = 240$ MPa, the fiber/matrix interface completely debonds and the fibers failure volume fraction is 8.7%. The unloading interface counter slip length $y$ increases with applied stress decreasing until $\sigma_{min}$, i.e., $y(\sigma_{min}) < l_c/2$. The reloading interface new slip length $z$ increases with applied stress increasing until $\sigma_{max}$, i.e., $z(\sigma_{max}) < l_c/2$. However, the unloading interface counter slip length and reloading interface new slip length considering fibers failure are less than those without considering fibers failure, which is mainly attributed to different interface slip cases with and without considering fibers failure. At the fatigue peak stress of $\sigma_{max} = 240$ MPa, the stress–strain fatigue hysteresis loops considering fibers failure correspond to the interface slip Case III; the fatigue stress–strain hysteresis loops without considering fibers failure correspond to the interface slip Case II.

Cumulative Tension–Tension Fatigue Hysteresis Loops

The unidirectional C/SiC composite is cumulative fatigue loading under peak stresses of $\sigma_{max} = 140$, 180, and 240 MPa each for 1,000,000 applied cycles

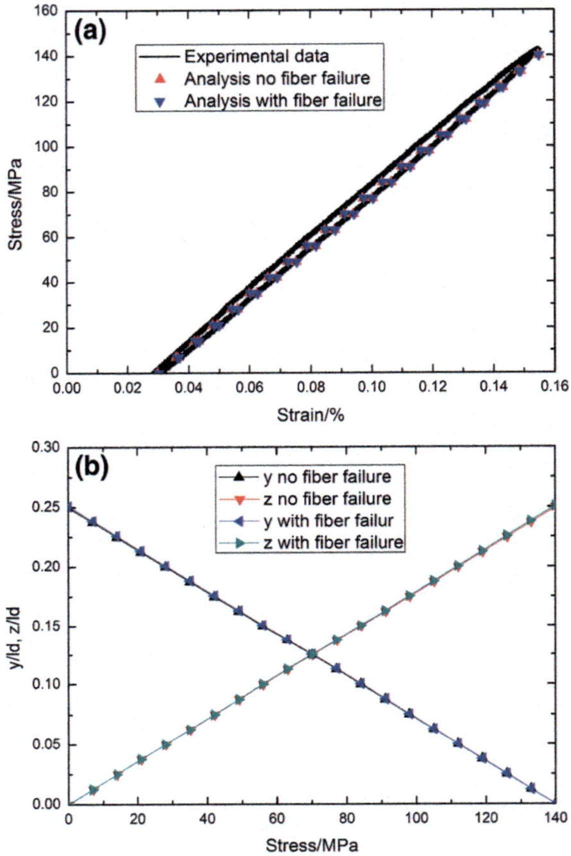


**Fig. 2.25** **a** Experimental and theoretical cyclic tensile stress–strain hysteresis loops; and **b** the unloading interface counter slip length and reloading interface new slip length versus the applied stress of unidirectional C/SiC composite at the fatigue peak stress of $\sigma_{max}$ = 140 MPa

without fatigue failure. The experimental and theoretical fatigue stress–strain hysteresis loops and the fiber/matrix interface debonding and slip at the first cycle for $\sigma_{max}$ = 140, 180, and 240 MPa are shown in Fig. 2.28. During initial fatigue loading at fatigue peak stress of $\sigma_{max}$ = 140 MPa, the fiber failure volume fraction is 0.0082% and the fatigue hysteresis loops correspond to the interface slip Case II. When cycles increasing to $10^6$ cycles, the fatigue peak stress increased to $\sigma_{max}$ = 180 MPa, the fibers failure volume fraction is 1.9% and the fatigue hysteresis loops correspond to the interface slip Case II. After experiencing $10^6$ cycles without fatigue failure, the fatigue peak stress increased to $\sigma_{max}$ = 240 MPa, the fiber failure volume fraction is 10% and the fatigue hysteresis loops correspond to the interface slip Case III.

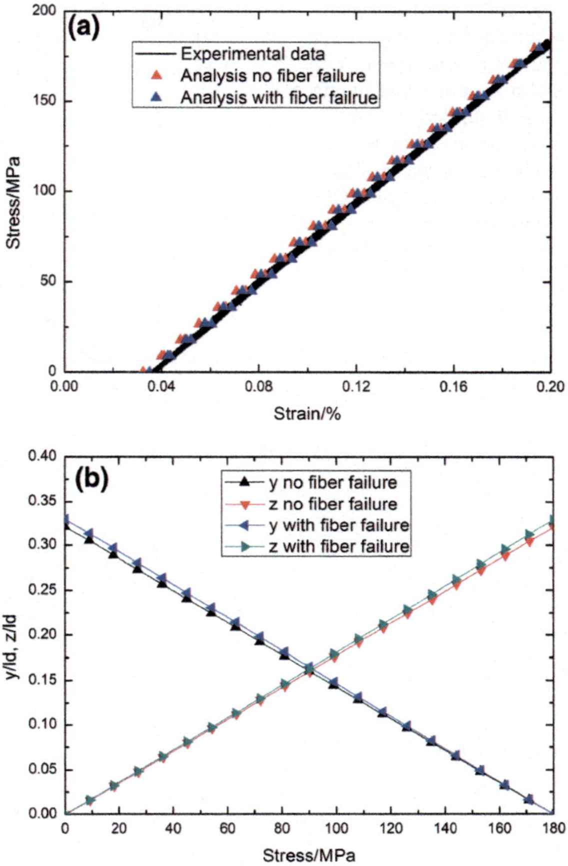


**Fig. 2.26** **a** Experimental and theoretical cyclic tensile stress–strain hysteresis loops and **b** the unloading interface counter slip length and reloading interface new slip length versus the applied stress of unidirectional C/SiC composite at the fatigue peak stress of $\sigma_{max}$ = 180 MPa

## Tension–Tension Fatigue Hysteresis Loops

The experimental and theoretical tension–tension fatigue hysteresis loops of unidirectional C/SiC composites under the fatigue peak stress of $\sigma_{max}$ = 140 MPa at room temperature are shown in Fig. 2.29. Under cyclic fatigue loading, the fiber/matrix interface shear stress decreases from 5.4 MPa at the first cycle to 0.24 MPa at the 816,908th cycle. The fibers failure volume fraction increases from 0.0082% at the first cycle to 0.92% at the 816908th cycle. The fatigue stress–strain hysteresis loops corresponding to $N$ = 1, 15,155, 139,326, 234,783, and 816,908 are predicted using the present analysis.

The experimental and theoretical fatigue hysteresis loops of unidirectional C/SiC composite under fatigue peak stress of $\sigma_{max}$ = 140 MPa at 800 °C in air condition are shown in Fig. 2.30. The fiber/matrix interface shear stress degrades from 5.1 MPa at the first cycle to 0.1 MPa at the 30,415th cycle. The fibers failure volume fraction increases from 0.0082% at the 1st cycle to 1.12% at the 25,000th

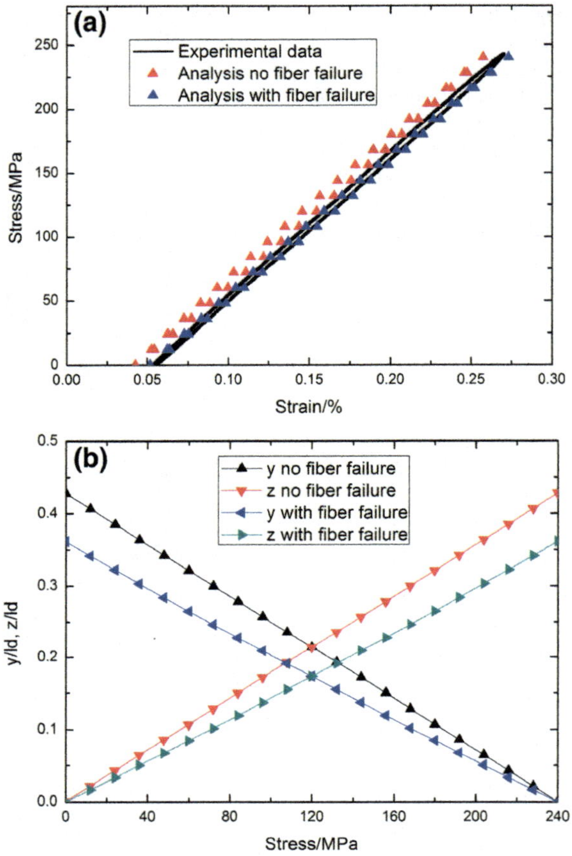


**Fig. 2.27** **a** Experimental and theoretical cyclic tensile stress–strain hysteresis loops and **b** the unloading interface counter slip length and reloading interface new slip length versus the applied stress of unidirectional C/SiC composite at the fatigue peak stress of $\sigma_{max}$ = 240 MPa

cycle. The fatigue stress–strain hysteresis loops corresponding to $N$ = 1, 100, 1000, 10,000, 20,000, and 25,000 are predicted using the present analysis.

## 2.3 Cross-ply and 2D Woven Ceramic-Matrix Composites

Kuo and Chou [46] divided the cross-ply matrix cracking modes into five different types, including:

(1) Matrix cracking mode I, transverse cracking;
(2) Matrix cracking mode II, transverse cracking and matrix cracking with perfect fiber/matrix interface bonding;
(3) Matrix cracking mode III, transverse cracking and matrix cracking with fiber/matrix interface debonding;

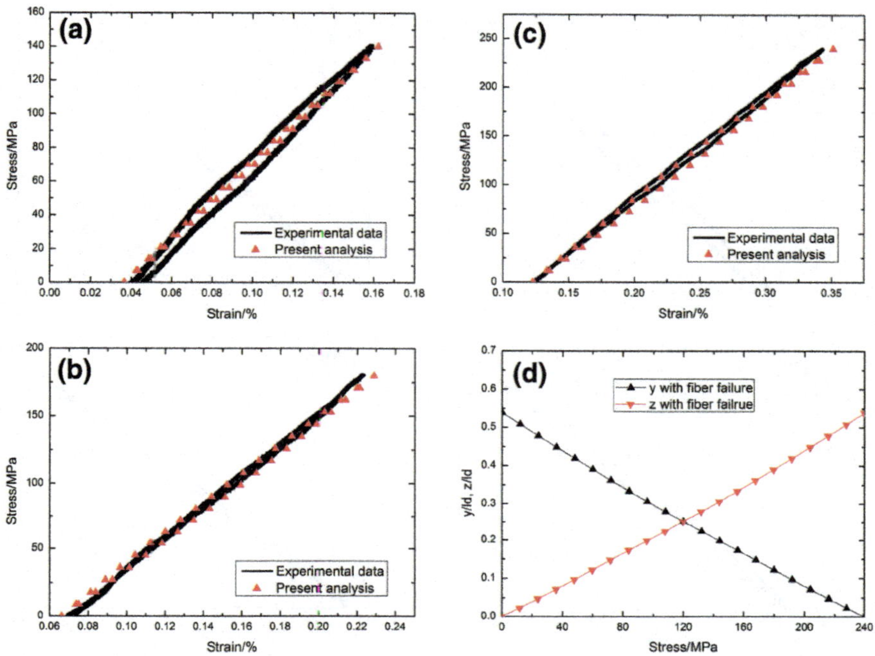


**Fig. 2.28** **a** Experimental and theoretical fatigue hysteresis loops at $\sigma_{max} = 140$ MPa and $N = 1$; **b** the experimental and theoretical fatigue hysteresis loops at $\sigma_{max} = 180$ MPa and $N = 1$; **c** the experimental and theoretical fatigue hysteresis loops at $\sigma_{max} = 240$ MPa and $N = 1$; and **d** the unloading interface counter slip length $y$ and reloading interface new slip length $z$ versus applied stress at $\sigma_{max} = 240$ MPa of unidirectional C/SiC composite

(4) Matrix cracking mode IV, matrix cracking with perfect fiber/matrix interface bonding;
(5) Matrix cracking mode V, matrix cracking with fiber/matrix interface debonding.

Under tensile and cyclic fatigue loading of matrix multicracking in cross-ply CMCs, matrix cracking mode III and mode V also exist [47, 48]. By assuming the mechanical hysteresis behavior of the cross-ply or 2D woven CMCs is mainly attributed to the debonding and sliding in the longitudinal plies or yarns, the evolution of hysteresis loops can be analyzed [49].

In this section, the fatigue hysteresis loops models of cross-ply and 2D woven CMCs considering multiple matrix cracking modes are developed. The Kuo–Chou shear-lag model is used to analyze the microstress field of the damaged CMCs after transverse multicracking, matrix multicracking, and fiber/matrix interface debonding. The fracture mechanics fiber/matrix interface debonding criterion is used to determine the fiber/matrix interface debonding and sliding lengths for different

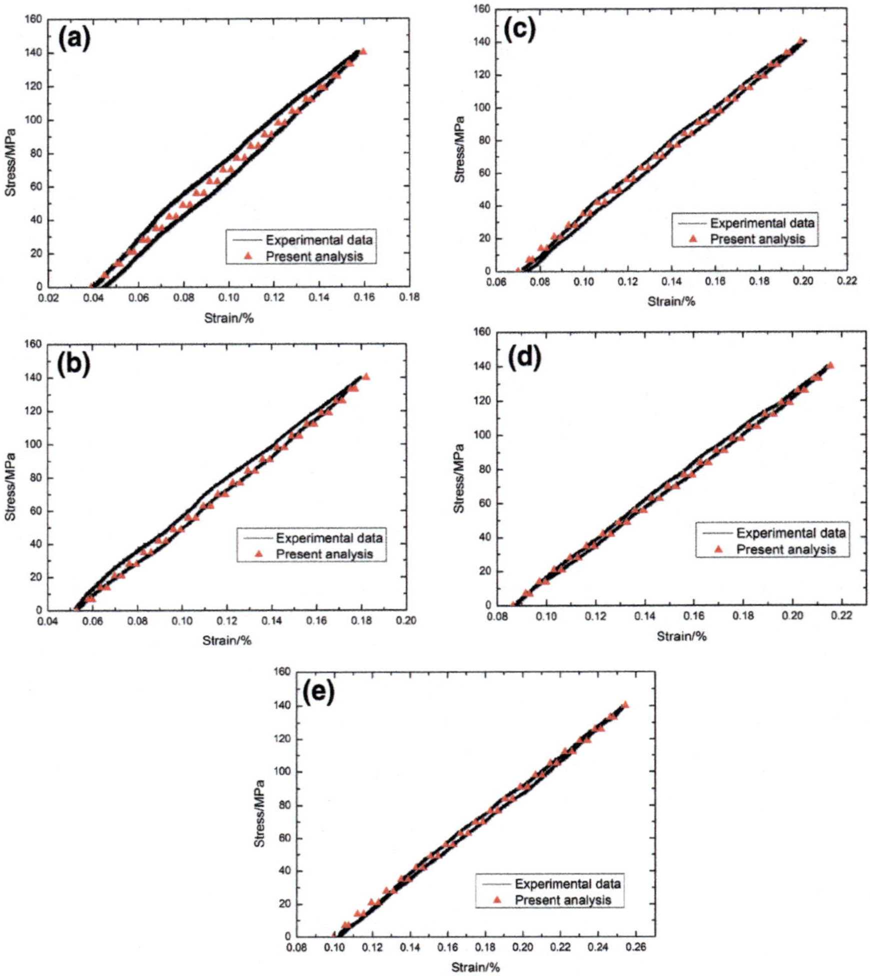


**Fig. 2.29** Fatigue stress–strain hysteresis loops of unidirectional C/SiC composite under the peak stress of $\sigma_{\max} = 140$ MPa at room temperature corresponding to **a** $N = 1$; **b** $N = 15{,}155$; **c** $N = 139{,}326$; **d** $N = 234{,}783$; and **e** $N = 816{,}908$

matrix cracking modes. The hysteresis loops of cross-ply or 2D woven CMCs can be divided into four different cases based on the fiber/matrix interface debonding and sliding condition in the matrix cracking space. The effects of fibers volume fraction, peak stress, matrix crack spacing, fiber/matrix interface shear stress, fiber/matrix interface debonding energy, matrix cracking mode proportion, and fiber/matrix interface wear on the fiber/matrix interface debonding and sliding, and the stress–strain hysteresis loops of CMCs are analyzed. The experimental cyclic loading/unloading and tension–tension fatigue hysteresis loops of cross-ply SiC/CAS and 2D SiC/SiC composites at room and elevated temperatures are predicted.

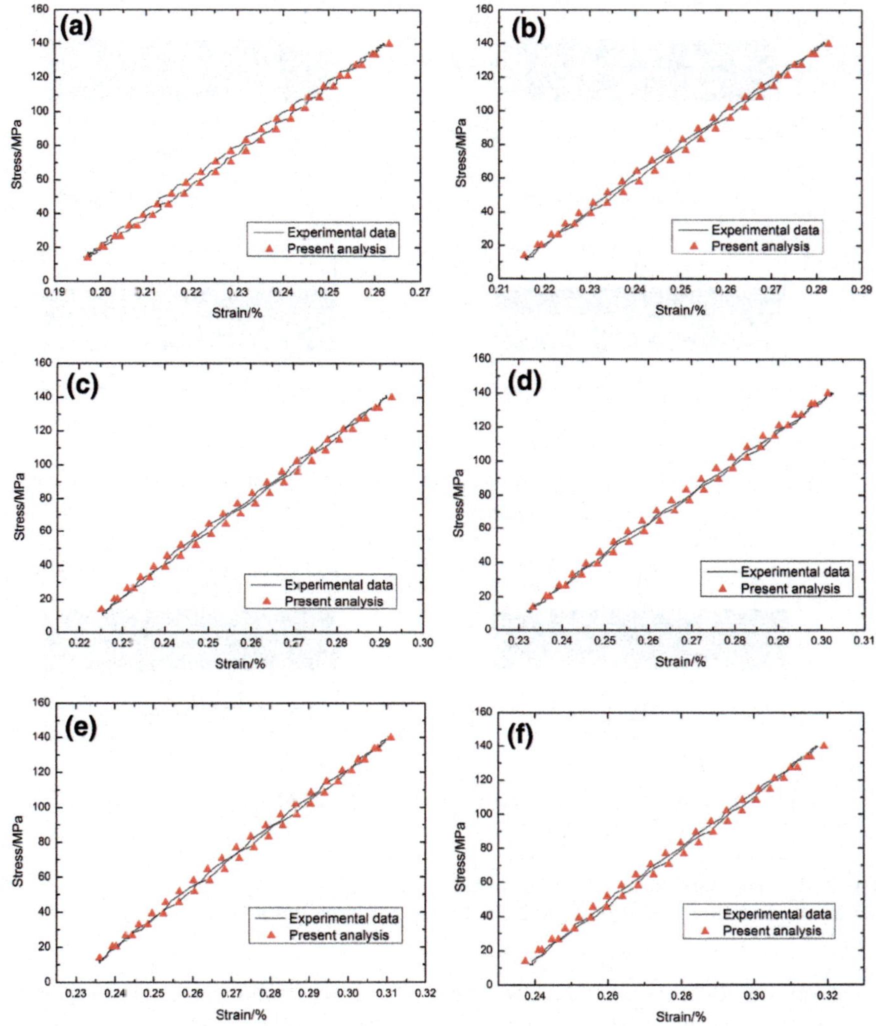


**Fig. 2.30** Fatigue stress–strain hysteresis loops of unidirectional C/SiC composite under the peak stress of $\sigma_{max}$ = 140 MPa at 800 °C in air condition corresponding to **a** $N = 1$; **b** $N = 100$; **c** $N = 1000$; **d** $N = 10{,}000$; **e** $N = 20{,}000$; and **f** $N = 25{,}000$

## *2.3.1 Stress Analysis*

Upon loading to the fatigue peak stress of $\sigma_{max}$, there may exist five different matrix cracking modes inside of cross-ply CMCs [46], as shown in Fig. 2.31. The fiber/matrix interface debonding and sliding in the matrix crack spacing of matrix cracking mode III and mode V are attributed to the stress–strain hysteresis loops of

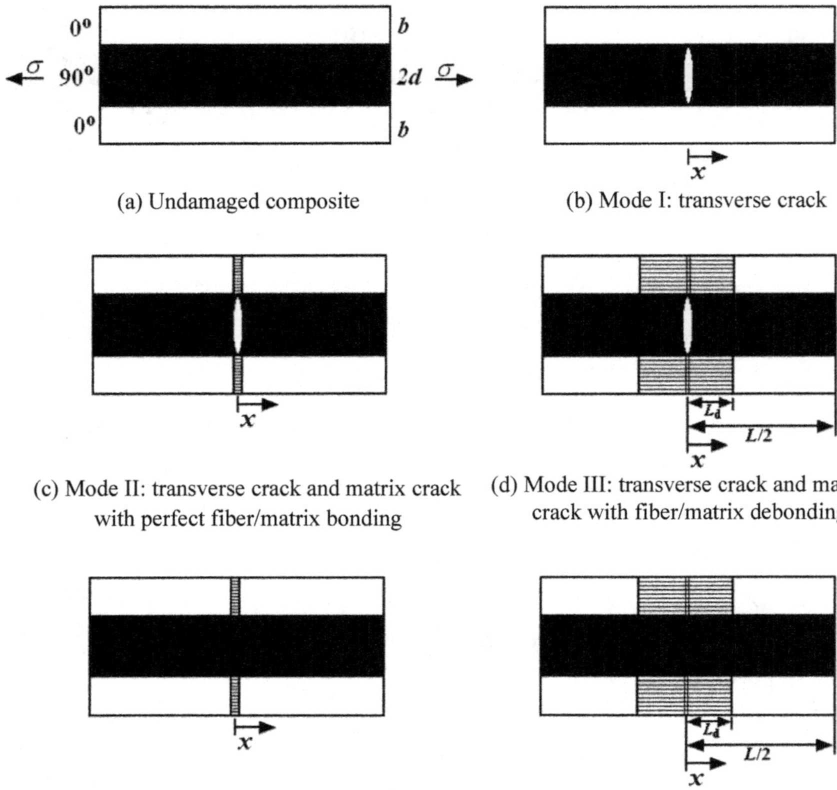


(a) Undamaged composite

(b) Mode I: transverse crack

(c) Mode II: transverse crack and matrix crack with perfect fiber/matrix bonding

(d) Mode III: transverse crack and matrix crack with fiber/matrix debonding

(e) Mode IV: matrix crack with perfect fiber/matrix bonding

(f) Mode V: matrix crack with fiber/matrix debonding

**Fig. 2.31** Undamaged state and five damaged modes in cross-ply or 2D CMCs. Reprinted with permission from Ref. [13]. Copyright 2016, Springer Science Business Media New York

cross-ply or 2D woven CMCs. The fiber and matrix axial stress distributions of matrix cracking mode III and mode V are analyzed using the shear-lag model.

#### 2.3.1.1 Cracking Mode III

The unit cell of matrix cracking mode III is shown in Fig. 2.31d, containing the transverse cracking and matrix cracking and the fiber/matrix interface debonding. The axial stress distributions of fiber, matrix, and 90° ply can be described using the following equations.

$$\sigma_{\mathrm{f}}(x)=\begin{cases}\frac{1}{V_{\mathrm{f}}}\frac{b+d}{b}\sigma-\frac{2\tau_{\mathrm{i}}}{r_{\mathrm{f}}}x, & x\in(0,l_{\mathrm{d}})\\ \sigma_{\mathrm{fo}}+\left(\frac{1}{V_{\mathrm{f}}}\frac{b+d}{b}\sigma-\frac{2\tau_{\mathrm{i}}}{r_{\mathrm{f}}}l_{\mathrm{d}}-\sigma_{\mathrm{fo}}\right)\exp\left(-\rho\frac{x-l_{\mathrm{d}}}{r_{\mathrm{f}}}\right), & x\in\left(l_{\mathrm{d}},\frac{l_{\mathrm{c}}}{2}\right)\end{cases}\tag{2.33}$$

$$\sigma_{\mathrm{m}}(x)=\begin{cases}2\tau_{\mathrm{i}}\frac{V_{\mathrm{f}}}{V_{\mathrm{m}}}\frac{x}{r_{\mathrm{f}}}-\frac{d}{b}\frac{\sigma_{\mathrm{t}}(x)}{V_{\mathrm{m}}}, & x\in(0,l_{\mathrm{d}})\\ \sigma_{\mathrm{mo}}+\frac{d}{b}\frac{\sigma_{\mathrm{to}}-\sigma_{\mathrm{t}}(x)}{V_{\mathrm{m}}}-\left(\sigma_{\mathrm{mo}}+\frac{d}{b}\frac{\sigma_{\mathrm{to}}}{V_{\mathrm{m}}}-2\tau_{\mathrm{i}}\frac{V_{\mathrm{f}}}{V_{\mathrm{m}}}\frac{l_{\mathrm{d}}}{r_{\mathrm{f}}}\right)\exp\left(-\rho\frac{x-l_{\mathrm{d}}}{r_{\mathrm{f}}}\right), & x\in\left(l_{\mathrm{d}},\frac{l_{\mathrm{c}}}{2}\right)\end{cases}\tag{2.34}$$

$$\sigma_{\mathrm{t}}(x)=\sigma_{\mathrm{to}}(1-\exp(-\lambda x)),\quad x\in(0,l_{\mathrm{c}}/2)\tag{2.35}$$

where $\lambda$ denotes the shear-lag parameter [50].

#### 2.3.1.2 Cracking Mode V

The unit cell of matrix cracking mode V is shown in Fig. 2.31f, containing the matrix cracking and fiber/matrix interface debonding. The axial stress distribution of fiber, matrix, and 90° ply can be described using the following equations.

$$\sigma_{\mathrm{f}}(x)=\begin{cases}\frac{1}{V_{\mathrm{f}}}\left(\frac{b+d}{b}\sigma-\frac{d}{b}\sigma_{\mathrm{to}}\right)-\frac{2\tau_{\mathrm{i}}}{r_{\mathrm{f}}}x, & x\in(0,l_{\mathrm{d}})\\ \sigma_{\mathrm{fo}}+\left[\frac{1}{V_{\mathrm{f}}}\left(\frac{b+d}{b}\sigma-\frac{d}{b}\sigma_{\mathrm{to}}\right)-\frac{2\tau_{\mathrm{i}}}{r_{\mathrm{f}}}l_{\mathrm{d}}-\sigma_{\mathrm{fo}}\right]\exp\left(-\rho\frac{x-l_{\mathrm{d}}}{r_{\mathrm{f}}}\right), & x\in(l_{\mathrm{d}},l_{\mathrm{c}}/2)\end{cases}\tag{2.36}$$

$$\sigma_{\mathrm{m}}(x)=\begin{cases}2\tau_{\mathrm{i}}\frac{V_{\mathrm{f}}}{V_{\mathrm{m}}}\frac{x}{r_{\mathrm{f}}}, x\in(0,l_{\mathrm{d}})\\ \sigma_{\mathrm{mo}}-\left(\sigma_{\mathrm{mo}}-2\tau_{\mathrm{i}}\frac{V_{\mathrm{f}}}{V_{\mathrm{m}}}\frac{l_{\mathrm{d}}}{r_{\mathrm{f}}}\right)\exp\left(-\rho\frac{x-l_{\mathrm{d}}}{r_{\mathrm{f}}}\right), x\in(l_{\mathrm{d}},l_{\mathrm{c}}/2)\end{cases}\tag{2.37}$$

$$\sigma_{\mathrm{t}}(x)=\sigma_{\mathrm{to}}, x\in(0,l_{\mathrm{c}}/2)\tag{2.38}$$

### 2.3.2 *Hysteresis Theories*

Based on the debonding and sliding condition in the matrix crack spacing, the fatigue stress–strain hysteresis loops of cracking mode III and mode V under cyclic loading can be divided into four different cases, as following:

(1) Case I, the fiber/matrix interface partially debonding ($l_{\mathrm{d}} < l_{\mathrm{c}}/2$), and the fiber sliding completely relative to the matrix in the interface debonded region ($y/l_{\mathrm{d}} = z/l_{\mathrm{d}} = 1$).
(2) Case II, the fiber/matrix interface partially debonding ($l_{\mathrm{d}} < l_{\mathrm{c}}/2$), and the fiber sliding partially relative to the matrix in the interface debonded region ($y/l_{\mathrm{d}} = z/l_{\mathrm{d}} < 1$).

(3) Case III, the fiber/matrix interface completely debonding ($l_d = l_c/2$), and the fiber sliding partially relative to the matrix in the interface debonded region ($y/l_d = z/l_d < 1$).
(4) Case IV, the fiber/matrix interface completely debonding ($l_d = l_c/2$), and the fiber sliding completely relative to the matrix in the interface debonded region ($y/l_d = z/l_d = 1$).

#### 2.3.2.1 Cracking Mode III

When the fiber/matrix interface partially debonds, the unloading and reloading stress–strain relationships can be described using the following equation.

$$\varepsilon_c = \frac{b+d}{b}\frac{\sigma}{V_f E_f} + 4\frac{\tau_i}{E_f}\frac{y^2}{r_f L} - 2\frac{\tau_i}{E_f}\frac{(2y-l_d)(2y-l_c+l_d)}{r_f l_c} - (\alpha_c - \alpha_f)\Delta T \quad (2.39)$$

$$\begin{aligned}\varepsilon_c = &\frac{b+d}{b}\frac{\sigma}{V_f E_f} - 4\frac{\tau_i}{E_f}\frac{z^2}{r_f l_c} + \frac{4\tau_i}{E_f}\frac{(y-2z)^2}{r_f l_c} \\ &+ 2\frac{\tau_i}{E_f}\frac{(l_d-2y+2z)(l_d+2y-2z-l_c)}{r_f l_c} - (\alpha_c - \alpha_f)\Delta T\end{aligned} \quad (2.40)$$

When the fiber/matrix interface completely debonds, the unloading and reloading stress–strain relationships can be described using the following equations.

$$\varepsilon_c = \frac{b+d}{b}\frac{\sigma}{V_f E_f} + 4\frac{\tau_i}{E_f}\frac{y^2}{r_f l_c} - 2\frac{\tau_i}{E_f}\frac{(2y-l_c/2)^2}{r_f l_c} - (\alpha_c - \alpha_f)\Delta T \quad (2.41)$$

$$\begin{aligned}\varepsilon_c = &\frac{b+d}{b}\frac{\sigma}{V_f E_f} - 4\frac{\tau_i}{E_f}\frac{z^2}{r_f l_c} + 4\frac{\tau_i}{E_f}\frac{(y-2z)^2}{r_f l_c} - 2\frac{\tau_i}{E_f}\frac{(l_c/2-2y+2z)^2}{r_f l_c} \\ &- (\alpha_c - \alpha_f)\Delta T\end{aligned} \quad (2.42)$$

#### 2.3.2.2 Cracking Mode V

When the fiber/matrix interface partially debonds, the unloading and reloading stress–strain relationships can be described using the following equations.

$$\begin{aligned}\varepsilon_c = &\frac{b+d}{b}\frac{1}{V_f E_f}\left(\sigma - \frac{d}{b+d}\sigma_{to}\right) + 4\frac{\tau_i}{E_f}\frac{y^2}{r_f l_c} - 2\frac{\tau_i}{E_f}\frac{(2y-l_d)(2y+l_d-l_c)}{r_f l_c} \\ &- (\alpha_c - \alpha_f)\Delta T\end{aligned} \quad (2.43)$$

$$\varepsilon_c = \frac{b+d}{b}\frac{1}{V_f E_f}\left(\sigma - \frac{d}{b+d}\sigma_{to}\right) - 4\frac{\tau_i}{E_f}\frac{z^2}{r_f l_c} + \frac{4\tau_i}{E_f}\frac{(y-2z)^2}{r_f l_c} + 2\frac{\tau_i}{E_f}\frac{(l_d - 2y + 2z)(l_d + 2y - 2z - l_c)}{r_f l_c} - (\alpha_c - \alpha_f)\Delta T \tag{2.44}$$

When the fiber/matrix interface completely debonds, the unloading and reloading stress–strain relationships can be described using the following equations.

$$\varepsilon_c = \frac{b+d}{b}\frac{1}{V_f E_f}\left(\sigma - \frac{d}{b+d}\sigma_{to}\right) + 4\frac{\tau_i}{E_f}\frac{y^2}{r_f l_c} - 2\frac{\tau_i}{E_f}\frac{(2y - l_c/2)^2}{r_f l_c} - (\alpha_c - \alpha_f)\Delta T \tag{2.45}$$

$$\varepsilon_c = \frac{b+d}{b}\frac{1}{V_f E_f}\left(\sigma - \frac{d}{b+d}\sigma_{to}\right) - 4\frac{\tau_i}{E_f}\frac{z^2}{r_f l_c} + 4\frac{\tau_i}{E_f}\frac{(y-2z)^2}{r_f l_c} - 2\frac{\tau_i}{E_f}\frac{(l_c/2 - 2y + 2z)^2}{r_f l_c} - (\alpha_c - \alpha_f)\Delta T \tag{2.46}$$

### 2.3.3 Results and Discussions

Considering matrix cracking mode on the fatigue stress–strain hysteresis loops of 2D CMCs, the unloading and reloading strains can be descried using the following equations.

$$\varepsilon_{c\_unloading} = \eta\varepsilon_{III\_unloading} + (1-\eta)\varepsilon_{V\_unloading} \tag{2.47}$$

$$\varepsilon_{c\_reloading} = \eta\varepsilon_{III\_reloading} + (1-\eta)\varepsilon_{V\_reloading} \tag{2.48}$$

where $\eta$ denotes the matrix cracking proportion.

Under cyclic fatigue loading, the effects of fiber volume fraction, peak stress, matrix crack spacing, fiber/matrix interface shear stress, fiber/matrix interface debonded energy, matrix cracking mode proportion, and fiber/matrix interface wear on the fiber/matrix interface debonding and sliding and the fatigue stress–strain hysteresis loops of 2D SiC/SiC composite are analyzed.

#### 2.3.3.1 Effect of Fiber Volume Fraction

The effect of fiber volume fraction (i.e., $V_f$ = 30 and 40%) on the fiber/matrix interface debonding and sliding and the stress–strain hysteresis loops of matrix cracking mode III and V of 2D SiC/SiC composite is shown in Fig. 2.32.

For the matrix cracking mode III, when the fiber volume fraction is $V_f$ = 30%, the fiber/matrix interface debonding ratio is $2l_d/l_c = 1$, and upon unloading to the

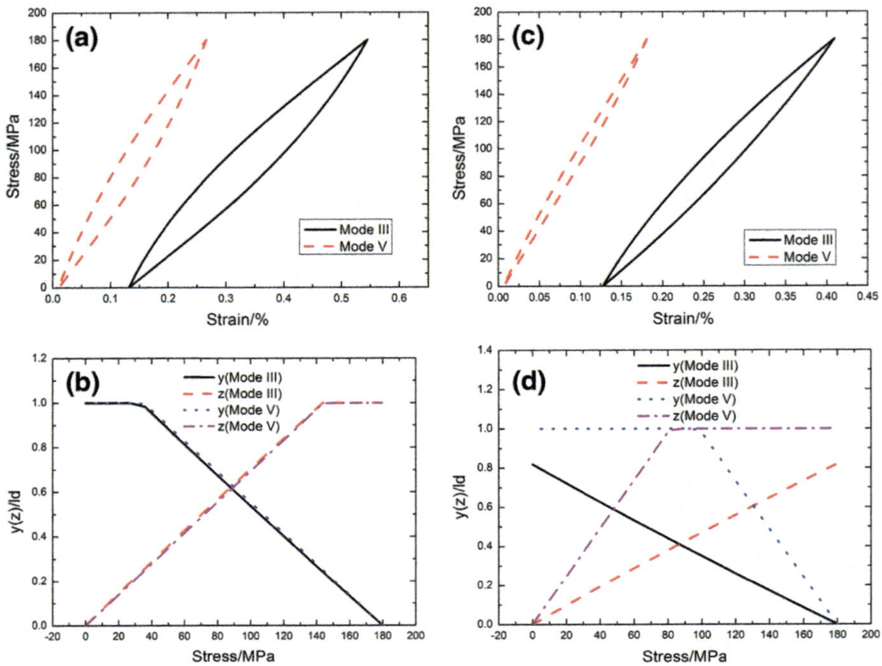


**Fig. 2.32** **a** Fatigue stress–strain hysteresis loops of matrix cracking mode III and mode V when the fiber volume fraction is $V_f = 30\%$; **b** the fiber/matrix interface sliding ratio of matrix cracking mode III and mode V when the fiber volume fraction is $V_f = 30\%$; **c** the fatigue stress–strain hysteresis loops of matrix cracking mode III and mode V when the fiber volume fraction is $V_f = 40\%$; and **d** the fiber/matrix interface sliding ratio of matrix cracking mode III and mode V when the fiber volume fraction is $V_f = 40\%$

transition stress of $\sigma_{tr\_fu} = 36$ MPa, the unloading interface counter slip length approaches to the fiber/matrix interface debonding tip; i.e., the fiber/matrix interface sliding ratio of $y(\sigma_{tr\_fu})/l_d = 1$, and upon reloading to the transition stress of $\sigma_{tr\_fr} = 144$ MPa, the reloading interface new slip length approaches to the fiber/matrix interface debonding tip; i.e., the fiber/matrix interface sliding ratio of $z(\sigma_{tr\_fr})/l_d = 1$, and the fatigue stress–strain hysteresis loops correspond to the interface slip Case IV; i.e., the fiber/matrix interface debonding ratio of $2l_d/l_c = 1$, and the fiber/matrix interface sliding ratio of $y(\sigma_{tr\_fu})/l_d = 1$. When the fiber volume fraction is $V_f = 40\%$, the fiber/matrix interface partially debonds; i.e., the fiber/matrix interface debonding ratio of $2l_d/l_c < 1$, and upon unloading to the valley stress, the interface counter slip length occupies 81.8% of the interface debonded length; i.e., the fiber/matrix interface sliding ratio of $y(\sigma_{min})/l_d = 0.818$, and upon reloading to the peak stress, the interface new slip length occupies 81.8% of the interface debonded length; i.e., the fiber/matrix interface sliding ratio of $z(\sigma_{min})/l_d = 0.818$, and the fatigue stress–strain hysteresis loops correspond to the interface slip Case II.

For the matrix cracking mode V, when the fiber volume fraction is $V_f = 30\%$, the fiber/matrix interface debonding ratio is $2l_d/l_c < 1$, and upon unloading to the transition stress of $\sigma_{tr\_pu} = 27$ MPa, the unloading interface counter slip length approaches to the fiber/matrix interface debonding tip; i.e., the fiber/matrix interface sliding ratio of $y(\sigma_{tr\_pu})/l_d = 1$, and upon reloading to the transition stress of $\sigma_{tr\_pr} = 153$ MPa, the reloading interface new slip length approaches to the fiber/matrix interface debonding tip; i.e., the fiber/matrix interface sliding ratio of $z(\sigma_{tr\_pr})/l_d = 1$, and the fatigue stress–strain hysteresis loops correspond to the interface slip Case I; i.e., the fiber/matrix interface debonding ratio of $2l_d/l_c < 1$, and the fiber/matrix interface sliding ratio of $y(\sigma_{tr\_pu})/l_d = 1$. When the fiber volume fraction is $V_f = 40\%$, the fiber/matrix interface debonding ratio is $2l_d/l_c < 1$, and upon unloading to the transition stress of $\sigma_{tr\_pu} = 90$ MPa, the unloading interface counter slip length approaches to the fiber/matrix interface debonding tip; i.e., the fiber/matrix interface sliding ratio of $y(\sigma_{tr\_pu})/l_d = 1$, and upon reloading to the transition stress of $\sigma_{tr\_pr} = 90$ MPa, the reloading interface new slip length approaches to the fiber/matrix interface debonding tip; i.e., the fiber/matrix interface sliding ratio of $z(\sigma_{tr\_pr})/l_d = 1$, and the fatigue stress–strain hysteresis loops correspond to the interface slip Case I; i.e., the fiber/matrix interface debonding ratio of $2l_d/l_c < 1$, and the fiber/matrix interface sliding ratio of $y(\sigma_{tr\_pu})/l_d = 1$.

#### 2.3.3.2 Effect of Peak Stress

The effect of fatigue peak stress (i.e., $\sigma_{max} = 180$ and 200 MPa) on the fiber/matrix interface debonding and sliding and the stress–strain hysteresis loops of matrix cracking mode III and V of 2D SiC/SiC composite is shown in Fig. 2.33.

For the matrix cracking mode III, when the fatigue peak stress is $\sigma_{max} = 180$ MPa, the fiber/matrix interface debonding ratio is $2l_d/l_c < 1$, and upon unloading, the fiber/matrix interface counter sliding length at the valley stress occupies 83.2% of the interface debonding length; i.e., the fiber/matrix interface sliding ratio of $y(\sigma_{min})/l_d = 0.832$, and upon reloading, the fiber/matrix interface new slip length at the peak stress occupies 83.2% of the interface debonding length; i.e., the fiber/matrix interface sliding ratio of $z(\sigma_{max})/l_d = 0.832$, and the fatigue stress–strain hysteresis loops correspond to the interface slip Case II. When the fatigue peak stress is $\sigma_{max} = 200$ MPa, the fiber/matrix interface debonding ratio is $2l_d/l_c < 1$, and upon unloading to the valley stress, the fiber/matrix interface counter sliding length occupies 92% of the interface debonding length; i.e., the fiber/matrix interface sliding ratio of $y(\sigma_{min})/l_d = 0.92$, and upon reloading to the peak stress, the fiber/matrix interface new slip length occupies 92% of the interface debonding length; i.e., the fiber matrix interface sliding ratio of $z(\sigma_{max})/l_d = 0.92$.

For the matrix cracking mode V, when the fatigue peak stress is $\sigma_{max} = 180$ MPa, the fiber/matrix interface debonding ratio is $2l_d/l_c < 1$, and upon unloading to the transition stress of $\sigma_{tr\_pu} = 54$ MPa, the fiber/matrix interface counter slip length approaches to the interface debonding tip, i.e., $y(\sigma_{tr\_pu})/l_d = 1$, and upon reloading to the transition stress of $\sigma_{tr\_pr} = 126$ MPa, the fiber/matrix

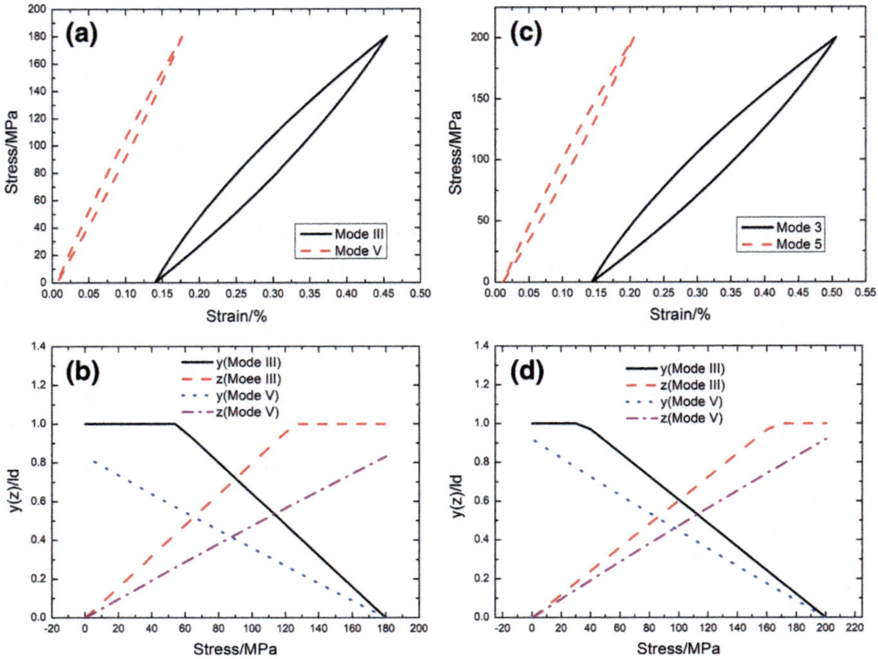


**Fig. 2.33** **a** Fatigue stress–strain hysteresis loops of matrix cracking mode III and mode V when the fatigue peak stress is $\sigma_{max}$ = 180 MPa; **b** the fiber/matrix interface sliding ratio of matrix cracking mode III and mode V when the fatigue peak stress is $\sigma_{max}$ = 180 MPa; **c** the fatigue stress–strain hysteresis loops of matrix cracking mode III and mode V when the fatigue peak stress is $\sigma_{max}$ = 200 MPa; and **d** the fiber/matrix interface sliding ratio of matrix cracking mode III and mode V when the fatigue peak stress is $\sigma_{max}$ = 200 MPa

interface new slip length approaches to the interface debonding tip, i.e., $z(\sigma_{tr\_pr})/l_d = 1$, and the fatigue stress–strain hysteresis loops correspond to the interface slip Case I. When the fatigue peak stress is $\sigma_{max}$ = 200 MPa, the fiber/matrix interface debonding ratio is $2l_d/l_c < 1$, and upon unloading to the transition stress of $\sigma_{tr\_pu}$ = 30 MPa, the fiber/matrix interface counter slip length approaches to the interface debonding tip, i.e., $y(\sigma_{tr\_pu})/l_d = 1$, and upon reloading to the transition stress of $\sigma_{tr\_pr}$ = 170 MPa, the fiber/matrix interface new slip length approaches to the interface debonding tip, i.e., $z(\sigma_{tr\_pr})/l_d = 1$, and the fatigue stress–strain hysteresis loops correspond to the interface slip Case I.

#### 2.3.3.3 Effect of Matrix Crack Spacing

The effect of matrix crack spacing (i.e., $l_c = 20r_f$ and $30r_f$) on the fiber/matrix interface debonding and sliding and the stress–strain hysteresis loops of matrix cracking mode III and V of 2D SiC/SiC composite is shown in Fig. 2.34.

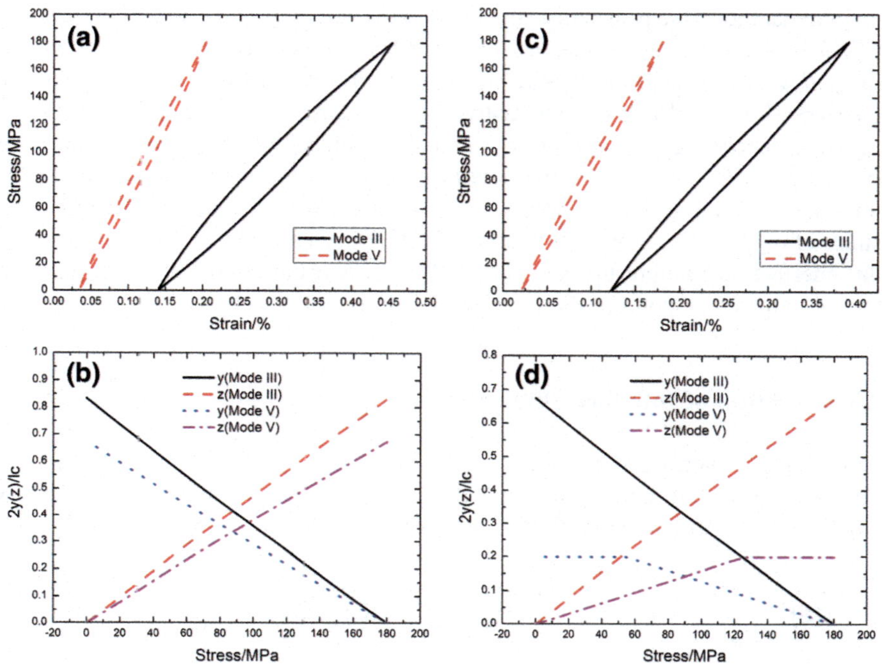


**Fig. 2.34** **a** Fatigue stress–strain hysteresis loops of matrix cracking mode III and mode V when the matrix crack spacing is $l_c = 20r_f$; **b** the fiber/matrix interface sliding ratio of matrix cracking mode III and mode V when the fatigue peak stress is $l_c = 20r_f$; **c** the fatigue stress–strain hysteresis loops of matrix cracking mode III and mode V when the fatigue peak stress is $l_c = 30r_f$; and **d** the fiber/matrix interface sliding ratio of matrix cracking mode III and mode V when the fatigue peak stress is $l_c = 30r_f$

For the matrix cracking mode III, when the matrix crack spacing is $l_c = 20r_f$, the fiber/matrix interface debonding ratio is $2l_d/l_c = 1$, and upon unloading to the valley stress, the fiber/matrix interface counter sliding length occupies 83.2% of the matrix crack spacing; i.e., the fiber/matrix interface sliding ratio of $2y(\sigma_{min})/l_c = 0.832$, and upon reloading to the peak stress, the fiber/matrix interface new slip length occupies 83.2% of the matrix crack spacing; i.e., the fiber/matrix interface sliding ratio of $2z(\sigma_{max})/l_c = 0.832$, and the fatigue stress–strain hysteresis loops correspond to the interface slip Case III. When the matrix crack spacing is $l_c = 30r_f$, the fiber/matrix interface debonding ratio is $2l_d/l_c < 1$, and upon unloading to the valley stress, the fiber/matrix interface counter slip length occupies 67.3% of the matrix cracking space; i.e., the fiber/matrix interface sliding ratio of $2y(\sigma_{min})/l_c = 0.673$, and upon reloading, the fiber/matrix interface new slip length occupies 67.3% of the matrix crack spacing; i.e., the fiber/matrix interface sliding ratio of $2z(\sigma_{max})/l_c = 0.673$.

For the matrix cracking mode V, when the matrix crack spacing is $l_c = 20r_f$, the fiber/matrix interface completely debonds, and upon unloading to the valley stress, the fiber/matrix interface counter slip length occupies 67.3% of the matrix crack spacing; i.e., the fiber/matrix interface sliding ratio of $2y(\sigma_{min})/l_c = 67.3\%$, and upon reloading to the peak stress, the fiber/matrix interface new slip length occupies 67.3% of the matrix crack spacing; i.e., the fiber/matrix interface sliding ratio of $2z(\sigma_{max})/l_c$ = 67.3%, and the fatigue stress–strain hysteresis loops correspond to the interface slip Case III. When the matrix crack spacing is $l_c = 30r_f$, the fiber/matrix interface debonding ratio is $2l_d/l_c < 1$, and upon unloading to the transition stress of $\sigma_{tr\_pu}$ = 54 MPa, the fiber/matrix interface counter slip length approaches to the interface debonding tip, i.e., $y(\sigma_{tr\_pu})/l_d = 1$, and upon reloading to the transition stress of $\sigma_{tr\_pr}$ = 126 MPa, the fiber/matrix interface new slip length approaches to the interface debonding tip, i.e., $z(\sigma_{tr\_pr})/l_d = 1$, and the fatigue stress–strain hysteresis loops correspond to the interface slip Case I.

#### 2.3.3.4 Effect of Interface Shear Stress

The effect of interface shear stress (i.e., $\tau_i$ = 15 and 20 MPa) on the fiber/matrix interface debonding and sliding and the stress–strain hysteresis loops of matrix cracking mode III and V of 2D SiC/SiC composite is shown in Fig. 2.35.

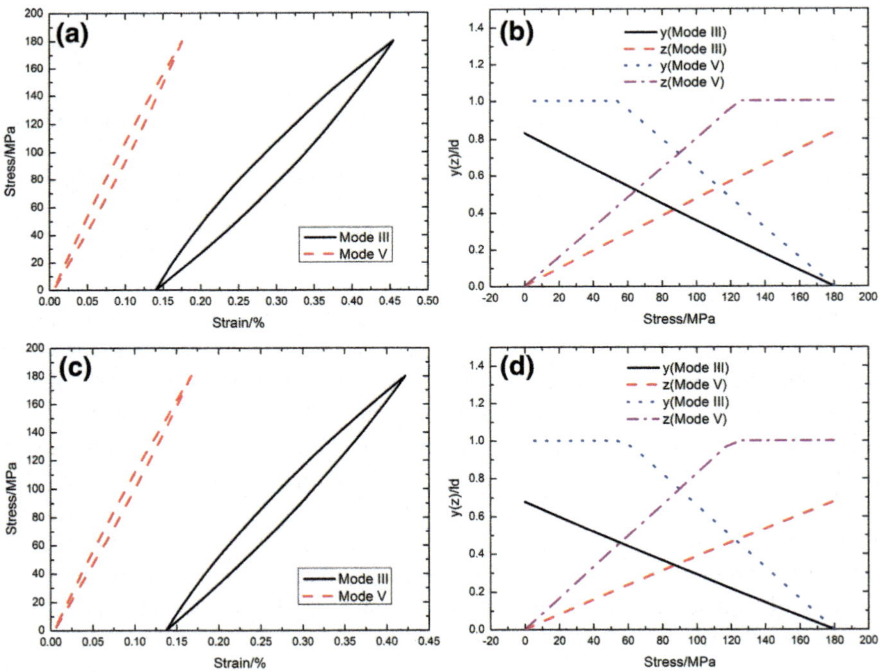


**Fig. 2.35** **a** Fatigue stress–strain hysteresis loops of matrix cracking mode III and mode V when the interface shear stress is $\tau_i$ = 15 MPa; **b** the fiber/matrix interface sliding ratio of matrix cracking mode III and mode V when the interface shear stress is $\tau_i$ = 15 MPa; **c** the fatigue stress–strain hysteresis loops of matrix cracking mode III and mode V when the interface shear stress is $\tau_i$ = 20 MPa; and **d** the fiber/matrix interface sliding ratio of matrix cracking mode III and mode V when the interface shear stress is $\tau_i$ = 20 MPa

For the matrix cracking mode III, when the fiber/matrix interface shear stress is $\tau_i = 15$ MPa, the fiber/matrix interface debonding ratio is $2l_d/l_c < 1$, and upon unloading to the valley stress, the fiber/matrix interface counter slip length occupies 83.2% of the interface debonded length; i.e., the fiber/matrix interface sliding ratio of $y(\sigma_{min})/l_d = 0.832$, and upon reloading to the peak stress, the fiber/matrix interface new slip length occupies 83.2% of the interface debonded length; i.e., the fiber/matrix interface sliding ratio of $z(\sigma_{max})/l_d = 0.832$, and the fatigue stress–strain hysteresis loops correspond to the interface slip Case II. When the interface shear stress is $\tau_i = 20$ MPa, the fiber/matrix interface debonding ratio is $2l_d/l_c < 1$, and upon unloading to the valley stress, the interface counter slip length occupies 67.8% of the interface debonded length, i.e., $y(\sigma_{min})/l_d = 0.678$, and upon reloading to the peak stress, the interface new slip length occupies 67.8% of the interface debonded length; i.e., $z(\sigma_{max})/l_d = 0.678$, and the fatigue stress–strain hysteresis loops correspond to the interface slip Case II.

For the matrix cracking mode V, when the fiber/matrix interface shear stress is $\tau_i = 15$ MPa, the fiber/matrix interface debonding ratio is $2l_d/l_c < 1$, and upon unloading to the transition stress of $\sigma_{tr\_pu} = 54$ MPa, the fiber/matrix interface counter slip length occupies the entire interface debonded length; i.e., the fiber/matrix interface sliding ratio of $y(\sigma_{tr\_pu})/l_d = 1$, and upon reloading to the transition stress of $\sigma_{tr\_pr} = 126$ MPa, the fiber/matrix interface new slip length occupies the entire interface debonded length; i.e., the fiber/matrix interface sliding ratio of $z(\sigma_{tr\_pr})/l_d = 1$, and the fatigue stress–strain hysteresis loops correspond to the interface slip Case I. When the interface shear stress is $\tau_i = 20$ MPa, the fiber/matrix interface debonding ratio is $2l_d/l_c < 1$, and upon unloading to the transition stress of $\sigma_{tr\_pu} = 57$ MPa, the fiber/matrix interface counter slip length occupies the entire interface debonded length; i.e., the fiber/matrix interface sliding ratio of $y(\sigma_{tr\_pu})/l_d = 1$, and upon reloading to the transition stress of $\sigma_{tr\_pr} = 123$ MPa, the fiber/matrix interface new slip length occupies the entire interface debonded length; i.e., the fiber/matrix interface sliding ratio of $z(\sigma_{tr\_pr})/l_d = 1$, and the fatigue stress–strain hysteresis loops correspond to the interface slip Case I.

#### 2.3.3.5 Effect of Interface Debonded Energy

The effect of interface debonded energy (i.e., $\zeta_d = 0.5$ and $1.5$ J/m$^2$) on the fiber/matrix interface debonding and sliding and the stress–strain hysteresis loops of matrix cracking mode III and V of 2D SiC/SiC composite is shown in Fig. 2.36.

For the matrix cracking mode III, when the interface debonded energy is $\zeta_d = 0.5$ J/m$^2$, the fiber/matrix interface debonding ratio is $2l_d/l_c < 1$, and upon unloading to the valley stress, the fiber/matrix interface counter slip length occupies 74.1% of the interface debonded length; i.e., the fiber/matrix interface sliding ratio of $y(\sigma_{min})/l_d = 0.741$, and upon reloading to the peak stress, the fiber/matrix interface new slip length occupies 74.1% of the interface debonded length; i.e., the fiber/matrix interface sliding ratio of $z(\sigma_{max})/l_d = 0.741$, and the fatigue stress–strain hysteresis loops correspond to the interface slip Case II. When the fiber/

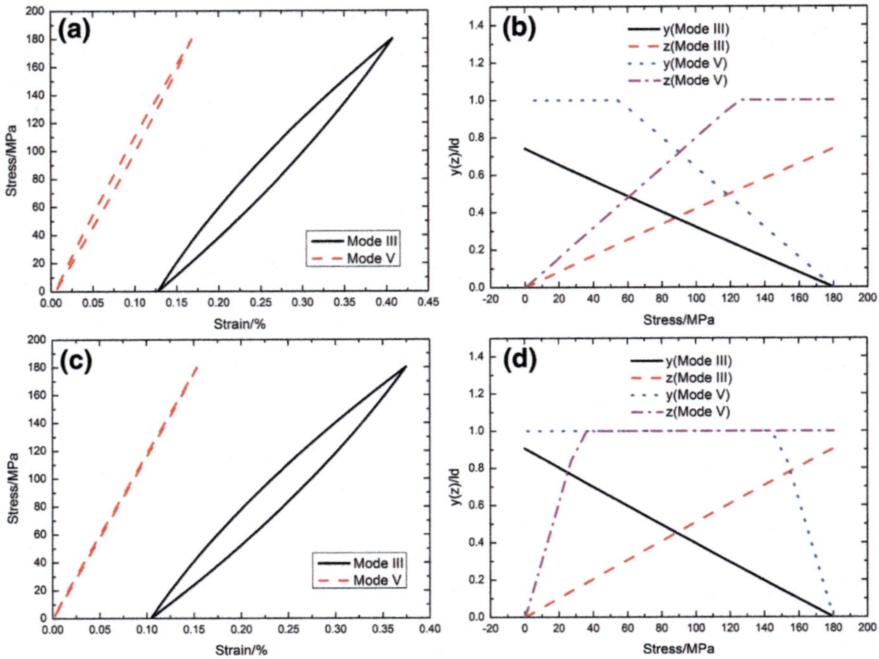


**Fig. 2.36** **a** Fatigue stress–strain hysteresis loops of matrix cracking mode III and mode V when the interface debonded energy is $\zeta_d = 0.5$ J/m$^2$; **b** the fiber/matrix interface sliding ratio of matrix cracking mode III and mode V when the interface debonded energy is $\zeta_d = 0.5$ J/m$^2$; **c** the fatigue stress–strain hysteresis loops of matrix cracking mode III and mode V when the interface debonded energy is $\zeta_d = 1.5$ J/m$^2$; and **d** the fiber/matrix interface sliding ratio of matrix cracking mode III and mode V when the interface debonded energy is $\zeta_d = 1.5$ J/m$^2$

matrix interface debonded energy is $\zeta_d = 1.5$ J/m$^2$, the fiber/matrix interface partially debonds, and upon unloading to the valley stress, the fiber/matrix interface counter slip length occupies 90.6% of the interface debonded length; i.e., the fiber/matrix interface sliding ratio of $y(\sigma_{min})/l_d = 0.906$, and upon reloading to the peak stress, the fiber/matrix interface new slip length occupies 90.6% of the interface debonded length; i.e., the fiber/matrix interface sliding ratio of $z(\sigma_{max})/l_d = 0.906$, and the fatigue stress–strain hysteresis loops correspond to the interface slip Case II.

For the matrix cracking mode V, when the fiber/matrix interface debonded energy is $\zeta_d = 0.5$ J/m$^2$, the fiber/matrix interface debonding ratio is $2l_d/l_c < 1$, and upon unloading to the transition stress of $\sigma_{tr\_pu} = 54$ MPa, the interface counter slip length occupies the entire interface debonded length; i.e., the fiber/matrix interface sliding ratio of $y(\sigma_{tr\_pu})/l_d = 1$, and upon reloading to the transition stress of $\sigma_{tr\_pr} = 126$ MPa, the interface new slip length occupies the entire interface debonded length; i.e., the fiber/matrix interface sliding ratio of $z(\sigma_{tr\_pr})/l_d = 1$, and the fatigue stress–strain hysteresis loops correspond to the interface slip Case I. When the fiber/matrix interface debonded energy is $\zeta_d = 1.5$ J/m$^2$, the fiber/matrix interface debonding ratio is $2l_d/l_c < 1$, and upon unloading to the transition stress of

$\sigma_{tr\_pu}$ = 144 MPa, the interface counter slip length occupies the entire interface debonded length; i.e., the fiber/matrix interface sliding ratio of $y(\sigma_{tr\_pu})/l_d = 1$, and upon reloading to the transition stress of $\sigma_{tr\_pr}$ = 36 MPa, the interface new slip length occupies the entire interface debonded length; i.e., the fiber/matrix interface sliding ratio of $z(\sigma_{tr\_pr})/l_d = 1$, and the fatigue stress–strain hysteresis loops correspond to the interface slip Case I.

#### 2.3.3.6 Effect of Matrix Cracking Mode Proportion

The effect of matrix cracking mode proportion (i.e., $\eta = 0$ to 1) on the fatigue stress–strain hysteresis loops of 2D SiC/SiC composite is shown in Fig. 2.37. When the matrix cracking mode proportion increases, the fatigue stress–strain hysteresis loop area and strain increase.

#### 2.3.3.7 Effect of Interface Wear

The effect of fiber/matrix interface wear on the fatigue stress–strain hysteresis loops and fiber/matrix interface debonding and sliding of matrix cracking mode III is shown in Fig. 2.38.

When the fiber/matrix interface shear stress is $\tau_i$ = 35 MPa, the fiber/matrix interface debonding ratio is $2l_d/l_c < 1$, and upon unloading to the transition stress of $\sigma_{tr\_pu}$ = 50 MPa, the interface counter slip length occupies the entire interface debonded length; i.e., the fiber/matrix interface sliding ratio of $y(\sigma_{tr\_pu})/l_d = 1$, and upon reloading to the transition stress of $\sigma_{tr\_pr}$ = 150 MPa, the interface new slip length occupies the entire interface debonded length; i.e., the fiber/matrix interface sliding ratio of $z(\sigma_{tr\_pr})/l_d = 1$, and the fatigue stress–strain hysteresis loops correspond to the interface slip Case I.

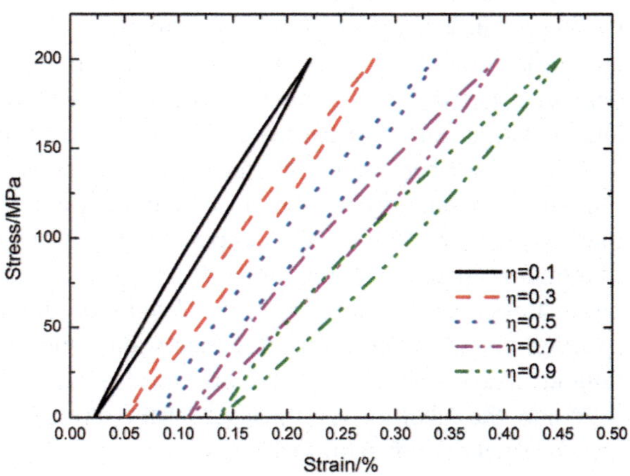


**Fig. 2.37** Effect of matrix cracking mode proportion on the fatigue stress–strain hysteresis loops of 2D SiC/SiC composite

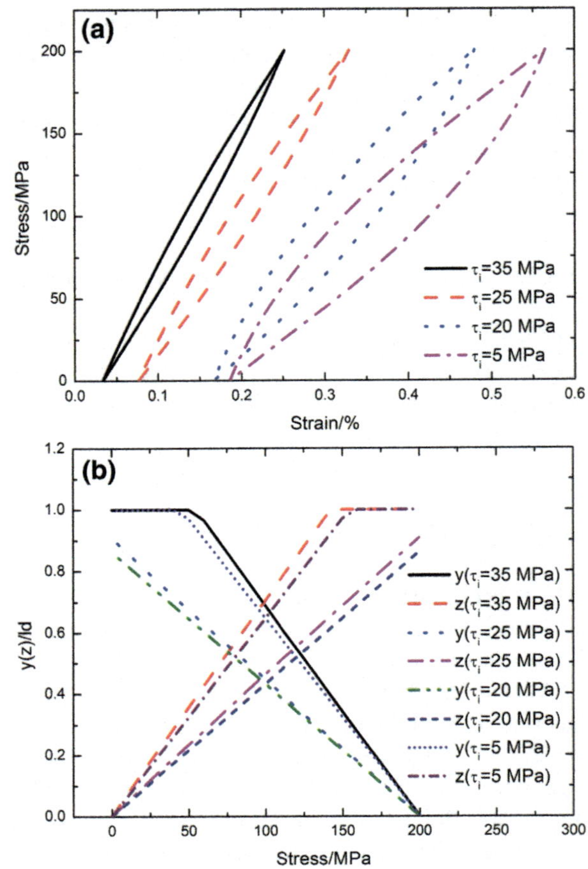


**Fig. 2.38** Effect of fiber/matrix interface wear on **a** the fatigue stress–strain hysteresis loops and **b** the fiber/matrix interface debonding and sliding of matrix cracking mode III

When the fiber/matrix interface shear stress is $\tau_i = 25$ MPa, the fiber/matrix interface debonding ratio is $2l_d/l_c < 1$, and upon unloading to the valley stress, the interface counter slip length occupies 90.7% of the interface debonded length; i.e., the fiber/matrix interface sliding ratio of $y(\sigma_{min})/l_d = 0.907$, and upon reloading to the peak stress, the interface new slip length occupies 90.7% of the interface debonded length; i.e., the fiber/matrix interface sliding ratio of $z(\sigma_{max})/l_d = 0.907$, and the fatigue stress–strain hysteresis loops correspond to the interface slip Case II.

When the fiber/matrix interface shear stress is $\tau_i = 20$ MPa, the fiber/matrix interface debonding ratio is $2l_d/l_c = 1$, and upon unloading to the valley stress, the interface counter slip length occupies 86.2% of the matrix crack spacing; i.e., the fiber/matrix interface sliding ratio of $2y(\sigma_{min})/l_c = 0.862$, and upon reloading to the peak stress, the interface new slip length occupies 86.2% of the matrix crack spacing; i.e., the fiber/matrix interface sliding ratio of $2z(\sigma_{max})/l_c = 0.862$, and the fatigue stress–strain hysteresis loops correspond to the interface slip Case III.

When the fiber/matrix interface shear stress is $\tau_i = 5$ MPa, the fiber/matrix interface debonding ratio is $2l_d/l_c = 1$, upon unloading to the transition stress of

$\sigma_{tr\_fu}$ = 40 MPa, the interface counter slip length occupies the entire matrix crack spacing; i.e., the fiber/matrix interface sliding ratio of $2y(\sigma_{tr\_fu})/l_c = 1$, and upon reloading to the transition stress of $\sigma_{tr\_fr}$ = 160 MPa, the interface new slip length occupies the entire matrix crack spacing; i.e., the fiber/matrix interface sliding ratio of $2z(\sigma_{tr\_fr})/l_c = 1$, and the fatigue stress–strain hysteresis loops correspond to the interface slip Case IV.

The effect of fiber/matrix interface wear on the fatigue stress–strain hysteresis loops and fiber/matrix interface debonding and sliding of matrix cracking mode V is shown in Fig. 2.39.

When the fiber/matrix interface shear stress is $\tau_i$ = 35 MPa, the fiber/matrix interface debonding ratio is $2l_d/l_c < 1$, and upon unloading to the transition stress of $\sigma_{tr\_pu}$ = 50 MPa, the interface counter slip length occupies the entire interface debonded length; i.e., the fiber/matrix interface sliding ratio of $y(\sigma_{tr\_pu})/l_d = 1$, and upon reloading to the transition stress of $\sigma_{tr\_pr}$ = 150 MPa, the interface new slip length occupies the entire interface debonded length; i.e., the fiber/matrix interface

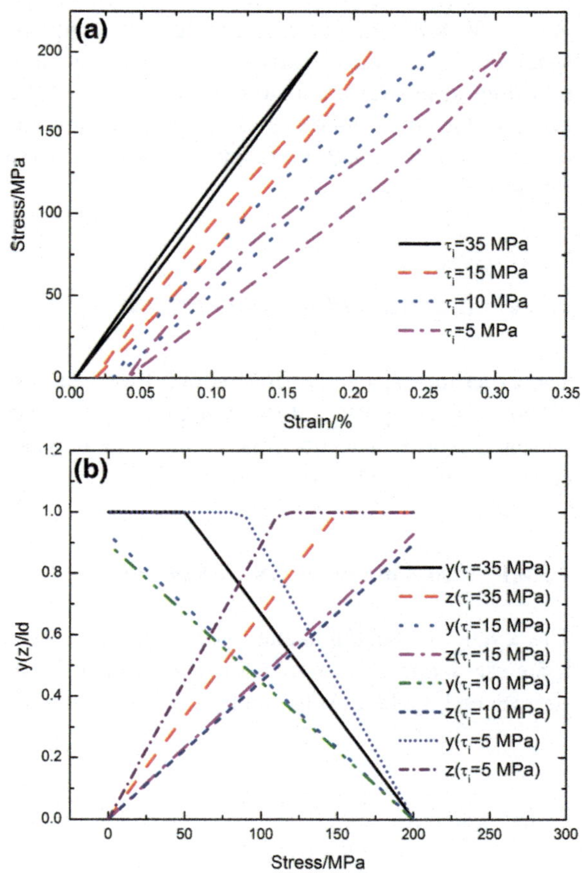


**Fig. 2.39** Effect of fiber/matrix interface wear on **a** the fatigue stress–strain hysteresis loops and **b** the fiber/matrix interface debonding and sliding of matrix cracking mode V

sliding ratio of $z(\sigma_{tr\_pr})/l_d = 1$, and the fatigue stress–strain hysteresis loops correspond to the interface slip Case I.

When the fiber/matrix interface shear stress is $\tau_i = 15$ MPa, the fiber/matrix interface debonding ratio is $2l_d/l_c < 1$, and upon unloading to the valley stress, the interface counter slip length occupies 92.8% of the interface debonded length; i.e., the fiber/matrix interface sliding ratio of $y(\sigma_{min})/l_d = 0.928$, and upon reloading to the peak stress, the interface new slip length occupies 92.8% of the interface debonded length; i.e., the fiber/matrix interface sliding ratio of $z(\sigma_{max})/l_d = 0.928$, and the fatigue stress–strain hysteresis loops correspond to the interface slip Case II.

When the fiber/matrix interface shear stress is $\tau_i = 10$ MPa, the fiber/matrix interface debonding ratio is $2l_d/l_c = 1$, and upon unloading to the valley stress, the interface counter slip length occupies 89.5% of the matrix crack spacing; i.e., the fiber/matrix interface sliding ratio of $2y(\sigma_{min})/l_c = 0.895$, and upon reloading to the peak stress, the interface new slip length occupies 89.5% of the matrix crack spacing; i.e., the fiber/matrix interface sliding ratio of $2z(\sigma_{max})/l_c = 0.895$, and the fatigue stress–strain hysteresis loops correspond to the interface slip Case III.

When the fiber/matrix interface shear stress is $\tau_i = 5$ MPa, the fiber/matrix interface debonding ratio is $2l_d/l_c = 1$, upon unloading to the transition stress of $\sigma_{tr\_fu} = 80$ MPa, the interface counter slip length occupies the entire matrix crack spacing; i.e., the fiber/matrix interface sliding ratio of $2y(\sigma_{tr\_fu})/l_c = 1$, and upon reloading to the transition stress of $\sigma_{tr\_fr} = 120$ MPa, the interface new slip length occupies the entire matrix crack spacing; i.e., the fiber/matrix interface sliding ratio of $2z(\sigma_{tr\_fr})/l_c = 1$, and the fatigue stress–strain hysteresis loops correspond to the interface slip Case IV.

## 2.3.4 Experimental Comparisons

The experimental fatigue stress–strain hysteresis loops of cross-ply SiC/CAS and 2D SiC/SiC composites at room and elevated temperatures are predicted using the present analysis. The fiber/matrix interface debonding and sliding in the matrix crack spacing are also analyzed.

### 2.3.4.1 Cross-ply SiC/CAS Composite

The cyclic loading/unloading tensile and tension–tension fatigue stress–strain hysteresis loops of cross-ply SiC/CAS composite corresponding to different peak stresses and applied cycle numbers are predicted using the present analysis.

## Cyclic Loading/Unloading Tensile Hysteresis Loops at Room Temperature

Wang et al. [51] investigated the cyclic loading/unloading tensile hysteresis loops of cross-ply $[0_3/90/0_3]$-SiC/CAS composite. The experimental and theoretical tensile stress–strain hysteresis loops and fiber/matrix interface debonding and sliding at the peak stress of $\sigma_{max}$ = 165 MPa are shown in Fig. 2.40. The fiber/matrix interface debonding is $2l_d/l_c < 1$ for the matrix cracking mode III and mode V, and the fiber/matrix interface sliding ratio is $y(\sigma_{min})/l_d < 1$ for matrix cracking mode III and $y(\sigma_{tr\_pu})/l_d = 1$ for the matrix cracking mode V.

The experimental and theoretical tensile stress–strain hysteresis loops and fiber/matrix interface debonding and sliding at the peak stress of $\sigma_{max}$ = 185 MPa are shown in Fig. 2.41. The fiber/matrix interface debonding is $2l_d/l_c = 1$ for the matrix cracking mode III, and $2l_d/l_c < 1$ for the matrix cracking mode V, and the fiber/matrix interface sliding ratio is $y(\sigma_{min})/l_d < 1$ for the matrix cracking mode III, and $y(\sigma_{tr\_pu})/l_d = 1$ for the matrix cracking mode V.

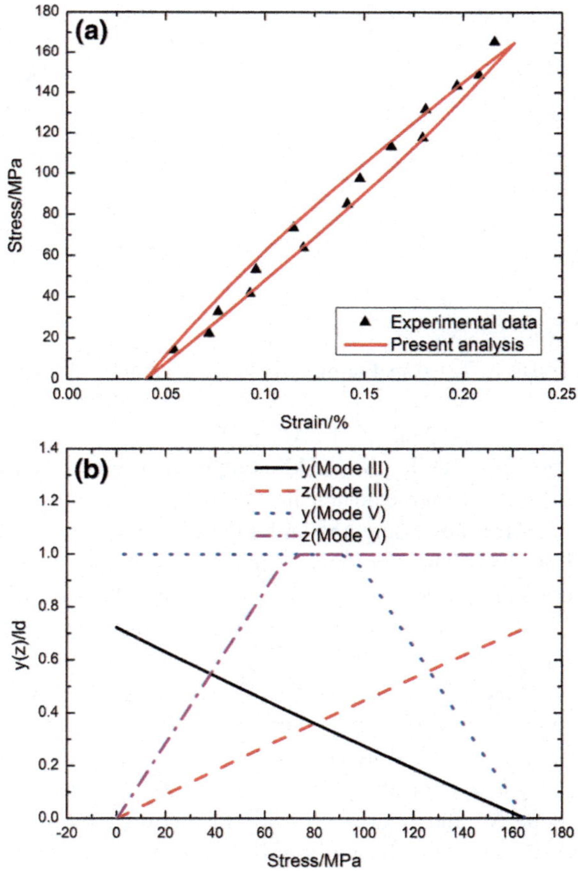


**Fig. 2.40** **a** Experimental and theoretical fatigue stress–strain hysteresis loops; **b** the fiber/matrix interface sliding of matrix cracking mode III and V at the peak stress of $\sigma_{max}$ = 165 MPa

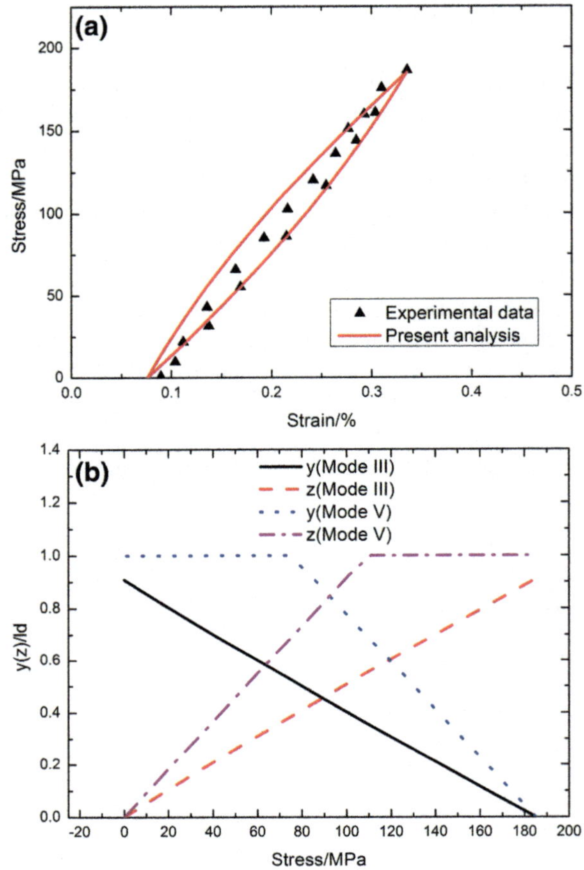


**Fig. 2.41** **a** Experimental and theoretical fatigue stress–strain hysteresis loops; **b** the fiber/matrix interface sliding of matrix cracking mode III and V at the peak stress of $\sigma_{max}$ = 185 MPa

## Tension–Tension Fatigue Hysteresis Loops at Room Temperature

Opalski and Mall [52] investigated the tension–tension fatigue hysteresis loops of cross-ply $[0/90]_{2s}$ SiC/CAS composite at room temperature. The experimental and theoretical predicted tensile fatigue stress–strain hysteresis loops and fiber/matrix interface debonding and sliding at the peak stress of $\sigma_{max}$ = 140 MPa are shown in Fig. 2.42. The fiber/matrix interface debonding ratio is $2l_d/l_c < 1$ for the matrix cracking mode III and mode V, and fiber/matrix interface sliding ratio of $y(\sigma_{min})/l_d < 1$ for the matrix cracking mode III, and $y(\sigma_{tr\_pu})/l_d = 1$ for the matrix cracking mode V.

The experimental and theoretical predicted tensile fatigue stress–strain hysteresis loops and fiber/matrix interface debonding and sliding at the peak stress of $\sigma_{max}$ = 180 MPa are shown in Fig. 2.43. The fiber/matrix interface debonding ratio is $2l_d/l_c = 1$ for the matrix cracking mode III, and $2l_d/l_c < 1$ for the matrix cracking mode V, and fiber/matrix interface sliding ratio of $y(\sigma_{tr\_fu})/l_d = 1$ for the matrix cracking mode III, and $y(\sigma_{tr\_pu})/l_d = 1$ for the matrix cracking mode V.

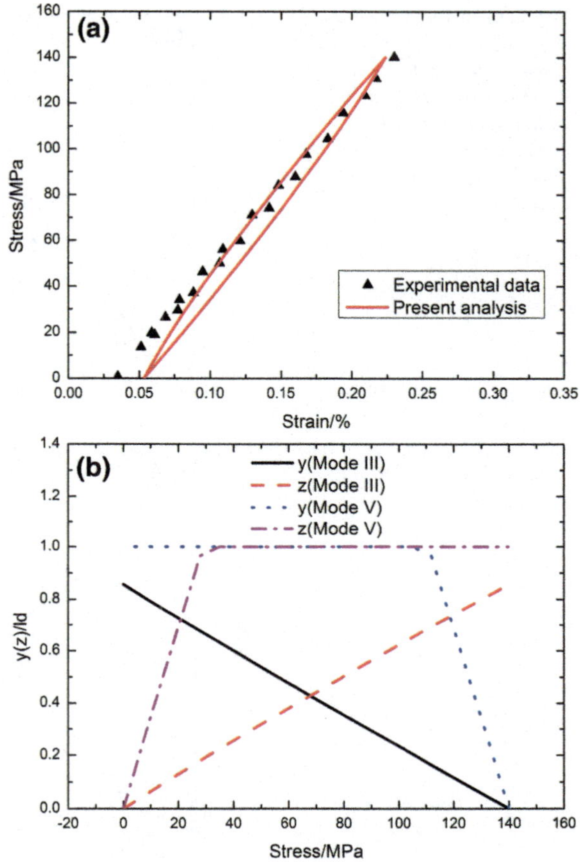


**Fig. 2.42** **a** Experimental and theoretical fatigue stress–strain hysteresis loops and **b** the fiber/matrix interface sliding ratio of matrix cracking mode III and mode V at the peak stress of $\sigma_{max}$ = 140 MPa

#### 2.3.4.2 2D SiC/SiC Composite

The cyclic loading/unloading tensile and tension–tension fatigue stress–strain hysteresis loops of 2D SiC/SiC composite corresponding to different peak stresses and applied cycle numbers are predicted using the present analysis.

##### Cyclic Loading/Unloading Tensile Hysteresis Loops at Room Temperature

Li et al. [53] investigated the cyclic loading/unloading tensile fatigue hysteresis loops of 2D SiC/SiC composite at room temperature. The experimental and theoretical tensile fatigue stress–strain hysteresis loops at the peak stresses of $\sigma_{max}$ = 152, 166, 181, 195, 210, 224, 239, and 253 MPa are shown in Fig. 2.44. When $\sigma_{max}$ = 152 MPa, the fiber/matrix interface debonding ratio is $2l_d/l_c < 1$ for the matrix cracking mode III and mode V, and the fiber/matrix interface sliding ratio is $y(\sigma_{tr\_pu} = 53\ \text{MPa})/l_d = 1$ for the matrix cracking mode III, and $y(\sigma_{tr\_pu} = 114\ \text{MPa})/$

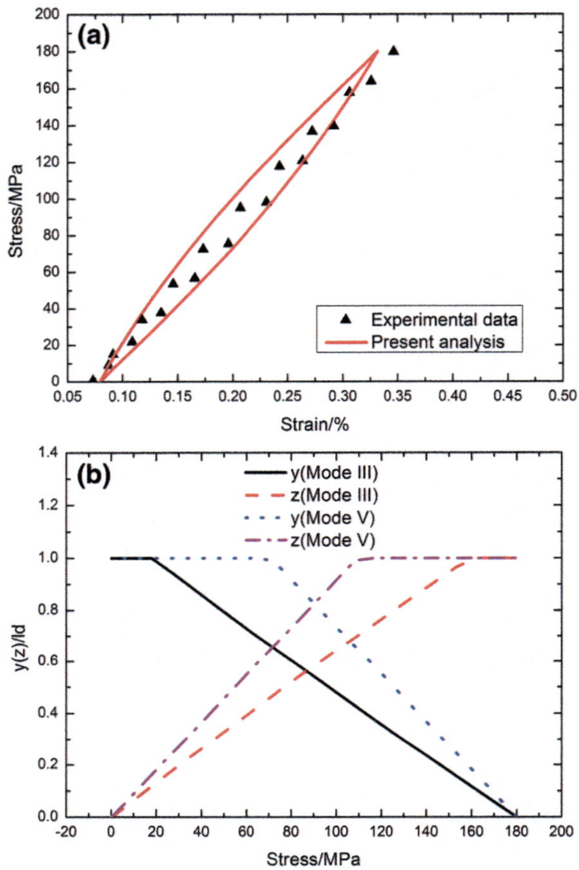


**Fig. 2.43** **a** Experimental and theoretical fatigue stress–strain hysteresis loops and **b** the fiber/matrix interface sliding ratio of matrix cracking mode III and mode V at the peak stress of $\sigma_{\max} = 180$ MPa

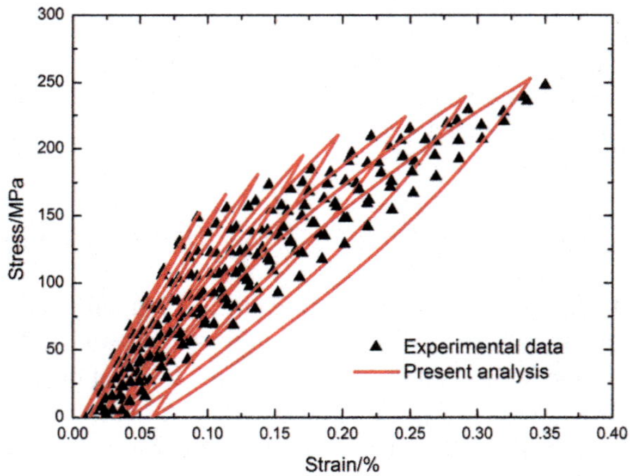


**Fig. 2.44** Experimental and theoretical hysteresis loops of 2D SiC/SiC composite at room temperature

$l_d = 1$ for the matrix cracking mode V. When $\sigma_{max} = 166$ MPa, the fiber/matrix interface debonding ratio is $2l_d/l_c < 1$ for the matrix cracking mode III and mode V, and the fiber/matrix interface sliding ratio is $y(\sigma_{tr\_pu} = 41\ \text{MPa})/l_d = 1$ for the matrix cracking mode III, and $y(\sigma_{tr\_pu} = 100\ \text{MPa})/l_d = 1$ for the matrix cracking mode V. When $\sigma_{max} = 181$ MPa, the fiber/matrix interface debonding ratio is $2l_d/l_c < 1$ for the matrix cracking mode III and mode V, and the fiber/matrix interface sliding ratio is $y(\sigma_{tr\_pu} = 27\ \text{MPa})/l_d = 1$ for the matrix cracking mode III, and $y(\sigma_{tr\_pu} = 81\ \text{MPa})/l_d = 1$ for the matrix cracking mode V. When $\sigma_{max} = 195$ MPa, the fiber/matrix interface debonding ratio is $2l_d/l_c < 1$ for the matrix cracking mode III and mode V, and the fiber/matrix interface sliding ratio is $y(\sigma_{tr\_pu} = 10\ \text{MPa})/l_d = 1$ for the matrix cracking mode III, and $y(\sigma_{tr\_pu} = 68\ \text{MPa})/l_d = 1$ for the matrix cracking mode V. When $\sigma_{max} = 210$ MPa, the fiber/matrix interface debonding ratio is $2l_d/l_c < 1$ for the matrix cracking mode III and mode V, and the fiber/matrix interface sliding ratio is $y(\sigma_{min})/l_d = 1$ for the matrix cracking mode III, and $y(\sigma_{tr\_pu} = 52\ \text{MPa})/l_d = 1$ for the matrix cracking mode V. When $\sigma_{max} = 224$ MPa, the fiber/matrix interface debonding ratio is $2l_d/l_c = 1$ for the matrix cracking mode III, and $2l_d/l_c < 1$ for the matrix cracking mode V, and the fiber/matrix interface sliding ratio is $y(\sigma_{min})/l_d = 0.95$ for the matrix cracking mode III, and $y(\sigma_{tr\_pu} = 34\ \text{MPa})/l_d = 1$ for the matrix cracking mode V. When $\sigma_{max} = 239$ MPa, the fiber/matrix interface debonding ratio is $2l_d/l_c = 1$ for the matrix cracking mode III, and $2l_d/l_c < 1$ for the matrix cracking mode V, and the fiber/matrix interface sliding ratio is $y(\sigma_{min})/l_d = 0.9$ for the matrix cracking mode III, and $y(\sigma_{tr\_pu} = 24\ \text{MPa})/l_d = 1$ for the matrix cracking mode V. When $\sigma_{max} = 253$ MPa, the fiber/matrix interface debonding ratio is $2l_d/l_c = 1$ for the matrix cracking mode III, and $2l_d/l_c < 1$ for the matrix cracking mode V, and the fiber/matrix interface sliding ratio is $y(\sigma_{min})/l_d = 0.86$ for the matrix cracking mode III, and $y(\sigma_{tr\_pu} = 13\ \text{MPa})/l_d = 1$ for the matrix cracking mode V.

### Tension–Tension Fatigue Hysteresis Loops at 1000 °C in Air Condition

Michael [54] investigated the tension–tension cyclic fatigue hysteresis loops of 2D SiC/SiC composite at 1000 °C in air condition. Under the fatigue peak stress of $\sigma_{max} = 80$ MPa, the experimental and theoretical fatigue stress–strain hysteresis loops at the applied cycle number of $N = 2$, 1000, 10,000, 20,000, and 30,000 are shown in Fig. 2.45.

When $N = 2$, for the matrix cracking mode III, the fiber/matrix interface debonding ratio is $2l_d/l_c < 1$, and upon unloading to the valley stress, the interface counter slip length occupies 52.5% of interface debonded length; i.e., the fiber/matrix interface sliding ratio of $y(\sigma_{min})/l_d = 0.525$, and upon reloading to the peak stress, the interface new slip length occupies 52.5% of interface debonded length; i.e., the fiber/matrix interface sliding ratio of $z(\sigma_{max})/l_d = 0.525$. For the matrix cracking mode V, the fiber/matrix interface debonding ratio is $2l_d/l_c < 1$, and upon unloading to the valley stress, the interface counter slip length occupies 64.5% of interface debonded length; i.e., the fiber/matrix interface sliding ratio of $y(\sigma_{min})/$

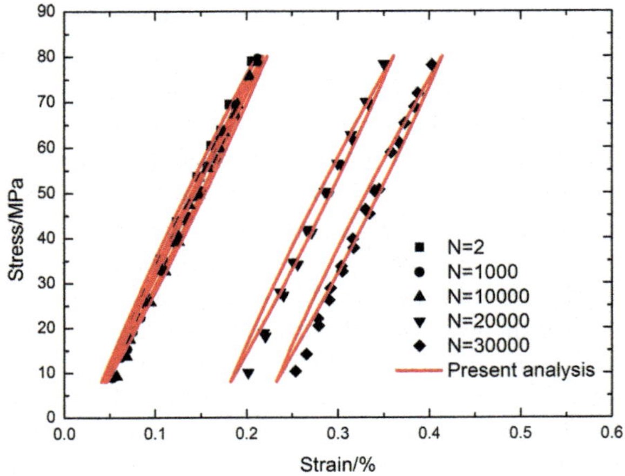


**Fig. 2.45** Experimental and theoretical fatigue hysteresis loops of 2D SiC/SiC composite under $\sigma_{max}$ = 80 MPa at 1000 °C in air condition

$l_d$ = 0.645, and upon reloading to the peak stress, the interface new slip length occupies 64.5% of interface debonded length; i.e., the fiber/matrix interface sliding ratio of $z(\sigma_{max})/l_d = 0.645$.

When $N = 1000$, for the matrix cracking mode III, the fiber/matrix interface debonding ratio is $2l_d/l_c < 1$, and upon unloading to the valley stress, the interface counter slip length occupies 52.3% of interface debonded length; i.e., the fiber/matrix interface sliding ratio of $y(\sigma_{min})/l_d = 0.523$, and upon reloading to the peak stress, the interface new slip length occupies 52.3% of interface debonded length; i.e., the fiber/matrix interface sliding ratio of $z(\sigma_{max})/l_d = 0.523$. For the matrix cracking mode V, the fiber/matrix interface debonding ratio is $2l_d/l_c < 1$, and upon unloading to the valley stress, the interface counter slip length occupies 63.7% of interface debonded length; i.e., the fiber/matrix interface sliding ratio of $y(\sigma_{min})/l_d = 0.637$, and upon reloading to the peak stress, the interface new slip length occupies 63.7% of interface debonded length; i.e., the fiber/matrix interface sliding ratio of $z(\sigma_{max})/l_d = 0.637$.

When $N = 10{,}000$, for the matrix cracking mode III, the fiber/matrix interface debonding ratio is $2l_d/l_c < 1$, and upon unloading to the valley stress, the interface counter slip length occupies 52.2% of interface debonded length; i.e., the fiber/matrix interface sliding ratio of $y(\sigma_{min})/l_d = 0.522$, and upon reloading to the peak stress, the interface new slip length occupies 52.2% of interface debonded length; i.e., the fiber/matrix interface sliding ratio of $z(\sigma_{max})/l_d = 0.522$. For the matrix cracking mode V, the fiber/matrix interface debonding ratio is $2l_d/l_c < 1$, and upon unloading to the valley stress, the interface counter slip length occupies 63.4% of interface debonded length; i.e., the fiber/matrix interface sliding ratio of $y(\sigma_{min})/l_d = 0.634$, and upon reloading to the peak stress, the interface new slip length occupies 63.4% of interface debonded length; i.e., the fiber/matrix interface sliding ratio of $z(\sigma_{max})/l_d = 0.634$.

When $N = 20{,}000$, for the matrix cracking mode III, the fiber/matrix interface debonding ratio is $2l_d/l_c < 1$, and upon unloading to the valley stress, the interface counter slip length occupies 52.1% of interface debonded length; i.e., the fiber/matrix interface sliding ratio of $y(\sigma_{min})/l_d = 0.521$, and upon reloading to the peak stress, the interface new slip length occupies 52.1% of interface debonded length; i.e., the fiber/matrix interface sliding ratio of $z(\sigma_{max})/l_d = 0.521$. For the matrix cracking mode V, the fiber/matrix interface debonding ratio is $2l_d/l_c < 1$, and upon unloading to the valley stress, the interface counter slip length occupies 63% of interface debonded length; i.e., the fiber/matrix interface sliding ratio of $y(\sigma_{min})/l_d = 0.63$, and upon reloading to the peak stress, the interface new slip length occupies 63% of interface debonded length; i.e., the fiber/matrix interface sliding ratio of $z(\sigma_{max})/l_d = 0.63$.

When $N = 20{,}000$, for the matrix cracking mode III, the fiber/matrix interface debonding ratio is $2l_d/l_c < 1$, and upon unloading to the valley stress, the interface counter slip length occupies 52.1% of interface debonded length; i.e., the fiber/matrix interface sliding ratio of $y(\sigma_{min})/l_d = 0.521$, and upon reloading to the peak stress, the interface new slip length occupies 52.1% of interface debonded length; i.e., the fiber/matrix interface sliding ratio of $z(\sigma_{max})/l_d = 0.521$. For the matrix cracking mode V, the fiber/matrix interface debonding ratio is $2l_d/l_c < 1$, and upon unloading to the valley stress, the interface counter slip length occupies 63% of interface debonded length; i.e., the fiber/matrix interface sliding ratio of $y(\sigma_{min})/l_d = 0.63$, and upon reloading to the peak stress, the interface new slip length occupies 63% of interface debonded length; i.e., the fiber/matrix interface sliding ratio of $z(\sigma_{max})/l_d = 0.63$.

When $N = 30{,}000$, for the matrix cracking mode III, the fiber/matrix interface debonding ratio is $2l_d/l_c < 1$, and upon unloading to the valley stress, the interface counter slip length occupies 52% of interface debonded length; i.e., the fiber/matrix interface sliding ratio of $y(\sigma_{min})/l_d = 0.52$, and upon reloading to the peak stress, the interface new slip length occupies 52% of interface debonded length; i.e., the fiber/matrix interface sliding ratio of $z(\sigma_{max})/l_d = 0.52$. For the matrix cracking mode V, the fiber/matrix interface debonding ratio is $2l_d/l_c < 1$, and upon unloading to the valley stress, the interface counter slip length occupies 62.7% of interface debonded length; i.e., the fiber/matrix interface sliding ratio of $y(\sigma_{min})/l_d = 0.627$, and upon reloading to the peak stress, the interface new slip length occupies 62.7% of interface debonded length; i.e., the fiber/matrix interface sliding ratio of $z(\sigma_{max})/l_d = 0.627$.

### Tension–Tension Fatigue Hysteresis Loops at 1000 °C in Steam Condition

Michael [54] investigated the tension–tension cyclic fatigue hysteresis loops of 2D SiC/SiC composite at 1000 °C in steam condition. Under the fatigue peak stress of $\sigma_{max} = 60$ MPa, the experimental and theoretical fatigue stress–strain hysteresis loops of 2D SiC/SiC composite corresponding to different cycle numbers of $N = 2$, 10,000, 100,000, 150,000, and 190,000 are shown in Fig. 2.46.

When $N = 2$, for the matrix cracking mode III, the fiber/matrix interface debonding ratio is $2l_d/l_c < 1$, and upon unloading to the valley stress, the interface

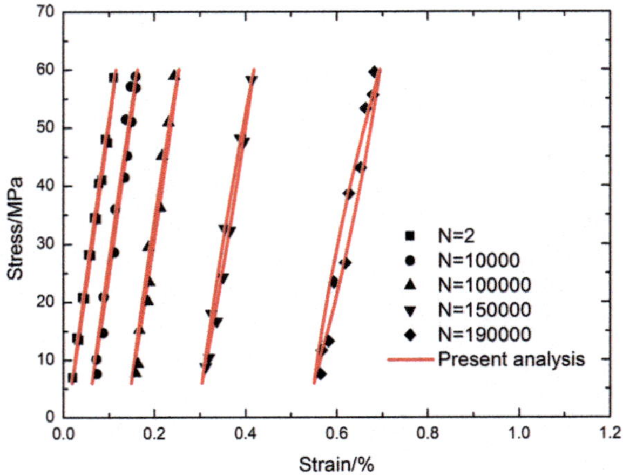


**Fig. 2.46** Experimental and theoretical fatigue hysteresis loops of 2D SiC/SiC composite under $\sigma_{max}$ = 60 MPa at 1000 °C in steam condition

counter slip length occupies 56.6% of interface debonded length; i.e., the fiber/matrix interface sliding ratio of $y(\sigma_{min})/l_d = 0.566$, and upon reloading to the peak stress, the interface new slip length occupies 56.6% of interface debonded length; i.e., the fiber/matrix interface sliding ratio of $z(\sigma_{max})/l_d = 0.566$. For the matrix cracking mode V, the fiber/matrix interface debonding ratio is $2l_d/l_c < 1$, and upon unloading to the valley stress, the interface counter slip length occupies 79% of interface debonded length; i.e., the fiber/matrix interface sliding ratio of $y(\sigma_{min})/l_d = 0.79$, and upon reloading to the peak stress, the interface new slip length occupies 79% of interface debonded length; i.e., the fiber/matrix interface sliding ratio of $z(\sigma_{max})/l_d = 0.79$.

When $N$ = 10,000, for the matrix cracking mode III, the fiber/matrix interface debonding ratio is $2l_d/l_c < 1$, and upon unloading to the valley stress, the interface counter slip length occupies 55.8% of interface debonded length; i.e., the fiber/matrix interface sliding ratio of $y(\sigma_{min})/l_d = 0.558$, and upon reloading to the peak stress, the interface new slip length occupies 55.8% of interface debonded length; i.e., the fiber/matrix interface sliding ratio of $z(\sigma_{max})/l_d = 0.558$. For the matrix cracking mode V, the fiber/matrix interface debonding ratio is $2l_d/l_c < 1$, and upon unloading to the valley stress, the interface counter slip length occupies 75.6% of interface debonded length; i.e., the fiber/matrix interface sliding ratio of $y(\sigma_{min})/l_d = 0.756$, and upon reloading to the peak stress, the interface new slip length occupies 75.6% of interface debonded length; i.e., the fiber/matrix interface sliding ratio of $z(\sigma_{max})/l_d = 0.756$.

When $N$ = 100,000, for the matrix cracking mode III, the fiber/matrix interface debonding ratio is $2l_d/l_c < 1$, and upon unloading to the valley stress, the interface counter slip length occupies 55.5% of interface debonded length; i.e., the fiber/matrix interface sliding ratio of $y(\sigma_{min})/l_d = 0.555$, and upon reloading to the peak stress, the interface new slip length occupies 55.5% of interface debonded length; i.e., the fiber/matrix interface sliding ratio of $z(\sigma_{max})/l_d = 0.555$. For the matrix cracking mode V, the fiber/matrix interface debonding ratio is $2l_d/l_c < 1$, and upon unloading to the valley stress, the interface counter slip length occupies 74.4% of interface debonded length; i.e., the fiber/matrix interface sliding ratio of $y(\sigma_{min})/l_d = 0.744$, and upon reloading to the peak stress, the interface new slip length occupies 74.4% of interface debonded length; i.e., the fiber/matrix interface sliding ratio of $z(\sigma_{max})/l_d = 0.744$.

When $N$ = 150,000, for the matrix cracking mode III, the fiber/matrix interface debonding ratio is $2l_d/l_c < 1$, and upon unloading to the valley stress, the interface counter slip length occupies 55.1% of interface debonded length; i.e., the fiber/matrix interface sliding ratio of $y(\sigma_{min})/l_d = 0.551$, and upon reloading to the peak stress, the interface new slip length occupies 55.1% of interface debonded length; i.e., the fiber/matrix interface sliding ratio of $z(\sigma_{max})/l_d = 0.551$. For the matrix cracking mode V, the fiber/matrix interface debonding ratio is $2l_d/l_c < 1$, and upon unloading to the valley stress, the interface counter slip length occupies 72.7% of interface debonded length; i.e., the fiber/matrix interface sliding ratio of $y(\sigma_{min})/l_d = 0.727$, and upon reloading to the peak stress, the interface new slip length occupies 72.7% of interface debonded length; i.e., the fiber/matrix interface sliding ratio of $z(\sigma_{max})/l_d = 0.727$.

When $N$ = 190,000, for the matrix cracking mode III, the fiber/matrix interface debonding ratio is $2l_d/l_c < 1$, and upon unloading to the valley stress, the interface counter slip length occupies 60.6% of interface debonded length; i.e., the fiber/matrix interface sliding ratio of $y(\sigma_{min})/l_d = 0.606$, and upon reloading to the peak stress, the interface new slip length occupies 60.6% of interface debonded length; i.e., the fiber/matrix interface sliding ratio of $z(\sigma_{max})/l_d = 0.606$. For the matrix cracking mode V, the fiber/matrix interface debonding ratio is $2l_d/l_c < 1$, and upon unloading to the valley stress, the interface counter slip length occupies 71.6% of interface debonded length; i.e., the fiber/matrix interface sliding ratio of $y(\sigma_{min})/l_d = 0.716$, and upon reloading to the peak stress, the interface new slip length occupies 71.6% of interface debonded length; i.e., the fiber/matrix interface sliding ratio of $z(\sigma_{max})/l_d = 0.716$.

Under the fatigue peak stress of $\sigma_{max}$ = 100 MPa, the experimental and theoretical fatigue stress–strain hysteresis loops of 2D SiC/SiC composite corresponding to different cycle numbers of $N$ = 2, 500, 3000, and 10,000 are shown in Fig. 2.47.

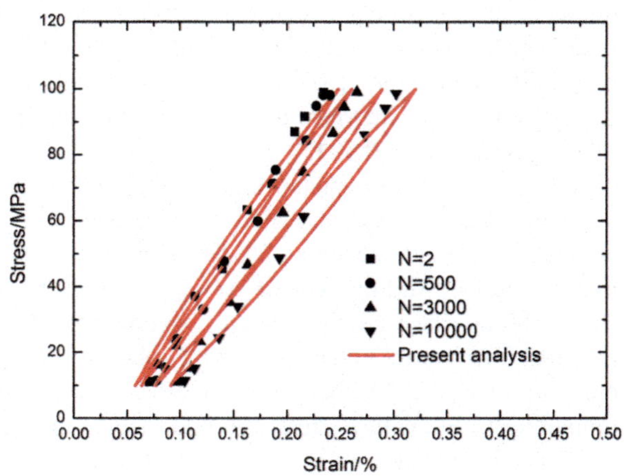


**Fig. 2.47** Experimental and theoretical fatigue hysteresis loops of 2D SiC/SiC composite under $\sigma_{max}$ = 100 MPa at 1000 °C in steam condition

When $N = 2$, for the matrix cracking mode III, the fiber/matrix interface debonding ratio is $2l_d/l_c < 1$, and upon unloading to the valley stress, the interface counter slip length occupies 51.2% of interface debonded length; i.e., the fiber/matrix interface sliding ratio of $y(\sigma_{min})/l_d = 0.512$, and upon reloading to the peak stress, the interface new slip length occupies 51.2% of interface debonded length; i.e., the fiber/matrix interface sliding ratio of $z(\sigma_{max})/l_d = 0.512$. For the matrix cracking mode V, the fiber/matrix interface debonding ratio is $2l_d/l_c < 1$, and upon unloading to the valley stress, the interface counter slip length occupies 60.6% of interface debonded length; i.e., the fiber/matrix interface sliding ratio of $y(\sigma_{min})/l_d = 0.606$, and upon reloading to the peak stress, the interface new slip length occupies 60.6% of interface debonded length; i.e., the fiber/matrix interface sliding ratio of $z(\sigma_{max})/l_d = 0.606$.

When $N = 500$, for the matrix cracking mode III, the fiber/matrix interface debonding ratio is $2l_d/l_c < 1$, and upon unloading to the valley stress, the interface counter slip length occupies 51.1% of interface debonded length; i.e., the fiber/matrix interface sliding ratio of $y(\sigma_{min})/l_d = 0.511$, and upon reloading to the peak stress, the interface new slip length occupies 51.1% of interface debonded length; i.e., the fiber/matrix interface sliding ratio of $z(\sigma_{max})/l_d = 0.511$. For the matrix cracking mode V, the fiber/matrix interface debonding ratio is $2l_d/l_c < 1$, and upon unloading to the valley stress, the interface counter slip length occupies 60.1% of interface debonded length; i.e., the fiber/matrix interface sliding ratio of $y(\sigma_{min})/l_d = 0.601$, and upon reloading to the peak stress, the interface new slip length occupies 60.1% of interface debonded length; i.e., the fiber/matrix interface sliding ratio of $z(\sigma_{max})/l_d = 0.601$.

When $N = 3000$, for the matrix cracking mode III, the fiber/matrix interface debonding ratio is $2l_d/l_c < 1$, and upon unloading to the valley stress, the interface counter slip length occupies 50.8% of interface debonded length; i.e., the fiber/matrix interface sliding ratio of $y(\sigma_{min})/l_d = 0.508$, and upon reloading to the peak stress, the interface new slip length occupies 50.8% of interface debonded length; i.e., the fiber/matrix interface sliding ratio of $z(\sigma_{max})/l_d = 0.508$. For the matrix cracking mode V, the fiber/matrix interface debonding ratio is $2l_d/l_c < 1$, and upon unloading to the valley stress, the interface counter slip length occupies 59.4% of interface debonded length; i.e., the fiber/matrix interface sliding ratio of $y(\sigma_{min})/l_d = 0.594$, and upon reloading to the peak stress, the interface new slip length occupies 59.4% of interface debonded length; i.e., the fiber/matrix interface sliding ratio of $z(\sigma_{max})/l_d = 0.594$.

When $N = 10{,}000$, for the matrix cracking mode III, the fiber/matrix interface debonding ratio is $2l_d/l_c < 1$, and upon unloading to the valley stress, the interface counter slip length occupies 50.7% of interface debonded length; i.e., the fiber/matrix interface sliding ratio of $y(\sigma_{min})/l_d = 0.507$, and upon reloading to the peak stress, the interface new slip length occupies 50.7% of interface debonded length; i.e., the fiber/matrix interface sliding ratio of $z(\sigma_{max})/l_d = 0.507$. For the matrix cracking mode V, the fiber/matrix interface debonding ratio is $2l_d/l_c < 1$, and upon unloading to the valley stress, the interface counter slip length occupies 58.9% of interface debonded length; i.e., the fiber/matrix interface sliding ratio of $y(\sigma_{min})/$

$l_d = 0.589$, and upon reloading to the peak stress, the interface new slip length occupies 58.9% of interface debonded length; i.e., the fiber/matrix interface sliding ratio of $z(\sigma_{max})/l_d = 0.589$.

### Tension–Tension Fatigue Hysteresis Loops at 1200 °C in Air Condition

Jacob [55] investigated the tension–tension cyclic fatigue hysteresis loops of 2D SiC/SiC composite at 1200 °C in air condition. Under the fatigue peak stress of $\sigma_{max} = 140$ MPa and the loading frequency of 0.1 Hz, the experimental fatigue stress–strain hysteresis loops of 2D SiC/SiC composite corresponding to the cycle number of $N = 1000$, 10,000, and 30,000 are shown in Fig. 2.48.

When $N = 1000$, for the matrix cracking mode III, the fiber/matrix interface debonding ratio is $2l_d/l_c < 1$, and upon unloading to the valley stress, the interface counter slip length occupies 52.9% of interface debonded length; i.e., the fiber/matrix interface sliding ratio of $y(\sigma_{min})/l_d = 0.529$, and upon reloading to the peak stress, the interface new slip length occupies 52.9% of interface debonded length; i.e., the fiber/matrix interface sliding ratio of $z(\sigma_{max})/l_d = 0.529$. For the matrix cracking mode V, the fiber/matrix interface debonding ratio is $2l_d/l_c < 1$, and upon unloading to the valley stress, the interface counter slip length occupies 67% of interface debonded length; i.e., the fiber/matrix interface sliding ratio of $y(\sigma_{min})/l_d = 0.67$, and upon reloading to the peak stress, the interface new slip length occupies 67% of interface debonded length; i.e., the fiber/matrix interface sliding ratio of $z(\sigma_{max})/l_d = 0.67$.

When $N = 10{,}000$, for the matrix cracking mode III, the fiber/matrix interface debonding ratio is $2l_d/l_c = 1$, and upon unloading to the transition stress of $\sigma_{tr\_fu} = 89.6$ MPa, the interface counter slip length occupies the entire matrix crack spacing; i.e., the fiber/matrix interface sliding ratio of $y(\sigma_{tr\_fu})/l_d = 1$, and upon reloading to the transition stress of $\sigma_{tr\_fr} = 64.4$ MPa, the interface new slip length

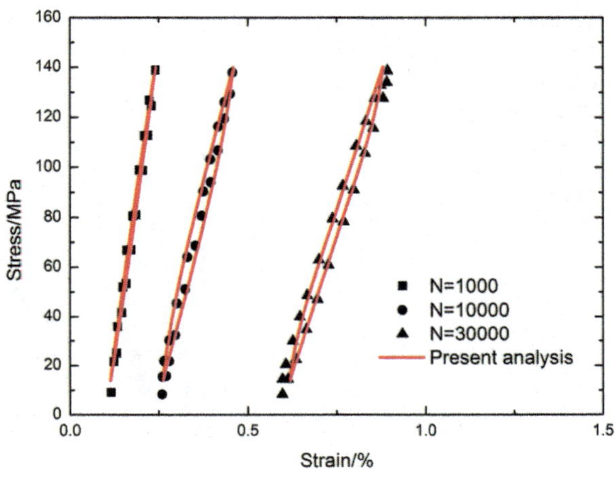


**Fig. 2.48** Experimental and theoretical fatigue hysteresis loops of 2D SiC/SiC composite under $\sigma_{max} = 140$ MPa and $f = 0.1$ Hz at 1200 °C in air condition

occupies the entire matrix crack spacing; i.e., the fiber/matrix interface sliding ratio of $z(\sigma_{tr\_fr})/l_d = 1$. For the matrix cracking mode V, the fiber/matrix interface debonding ratio is $2l_d/l_c < 1$, and upon unloading to the valley stress, the interface counter slip length occupies 64.8% of interface debonded length; i.e., the fiber/matrix interface sliding ratio of $y(\sigma_{min})/l_d = 0.648$, and upon reloading to the peak stress, the interface new slip length occupies 64.8% of interface debonded length; i.e., the fiber/matrix interface sliding ratio of $z(\sigma_{max})/l_d = 0.648$.

When $N = 30{,}000$, for the matrix cracking mode III, the fiber/matrix interface debonding ratio is $2l_d/l_c = 1$, and upon unloading to the transition stress of $\sigma_{tr\_fu} = 108.5$ MPa, the interface counter slip length occupies the entire matrix crack spacing; i.e., the fiber/matrix interface sliding ratio of $y(\sigma_{tr\_fu})/l_d = 1$, and upon reloading to the transition stress of $\sigma_{tr\_fr} = 45.5$ MPa, the interface new slip length occupies the entire matrix crack spacing; i.e., the fiber/matrix interface sliding ratio of $z(\sigma_{tr\_fr})/l_d = 1$. For the matrix cracking mode V, the fiber/matrix interface debonding ratio is $2l_d/l_c = 1$, and upon unloading to the transition stress of $\sigma_{tr\_fu} = 95.9$ MPa, the interface counter slip length occupies the entire matrix crack spacing; i.e., the fiber/matrix interface sliding ratio of $y(\sigma_{tr\_fu})/l_d = 1$, and upon reloading to the transition stress of $\sigma_{tr\_fr} = 58.1$ MPa, the interface new slip length occupies the entire matrix crack spacing; i.e., the fiber/matrix interface sliding ratio of $z(\sigma_{tr\_fr})/l_d = 1$.

Under the fatigue peak stress of $\sigma_{max} = 140$ MPa and the loading frequency of 1 Hz, the experimental and theoretical fatigue stress–strain hysteresis loops of 2D SiC/SiC composite corresponding to the cycle number of $N = 10{,}000$, 30,000, and 60,000 are shown in Fig. 2.49.

When $N = 10{,}000$, for the matrix cracking mode III, the fiber/matrix interface debonding ratio is $2l_d/l_c < 1$, and upon unloading to the valley stress, the interface counter slip length occupies 52.6% of interface debonded length; i.e., the fiber/matrix interface sliding ratio of $y(\sigma_{min})/l_d = 0.526$, and upon reloading to the peak stress, the interface new slip length occupies 52.6% of interface debonded length;

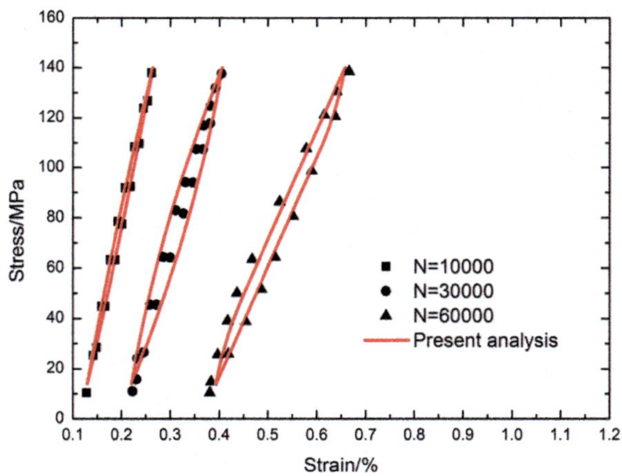


**Fig. 2.49** Experimental and theoretical fatigue hysteresis loops of 2D SiC/SiC composite under $\sigma_{max} = 140$ MPa and $f = 1$ Hz at 1200 °C in air condition

i.e., the fiber/matrix interface sliding ratio of $z(\sigma_{max})/l_d = 0.526$. For the matrix cracking mode V, the fiber/matrix interface debonding ratio is $2l_d/l_c < 1$, and upon unloading to the valley stress, the interface counter slip length occupies 65.9% of interface debonded length; i.e., the fiber/matrix interface sliding ratio of $y(\sigma_{min})/l_d = 0.659$, and upon reloading to the peak stress, the interface new slip length occupies 65.9% of interface debonded length; i.e., the fiber/matrix interface sliding ratio of $z(\sigma_{max})/l_d = 0.659$.

When $N = 30{,}000$, for the matrix cracking mode III, the fiber/matrix interface debonding ratio is $2l_d/l_c = 1$, and upon unloading to the valley stress, the interface counter slip length occupies 98.3% of interface debonded length; i.e., the fiber/matrix interface sliding ratio of $y(\sigma_{min})/l_d = 0.983$, and upon reloading to the peak stress, the interface new slip length occupies 98.3% of interface debonded length; i.e., the fiber/matrix interface sliding ratio of $z(\sigma_{max})/l_d = 0.983$. For the matrix cracking mode V, the fiber/matrix interface debonding ratio is $2l_d/l_c = 1$, and upon unloading to the valley stress, the interface counter slip length occupies 76.7% of interface debonded length; i.e., the fiber/matrix interface sliding ratio of $y(\sigma_{min})/l_d = 0.767$, and upon reloading to the peak stress, the interface new slip length occupies 76.7% of interface debonded length; i.e., the fiber/matrix interface sliding ratio of $z(\sigma_{max})/l_d = 0.767$.

When $N = 60{,}000$, for the matrix cracking mode III, the fiber/matrix interface debonding ratio is $2l_d/l_c = 1$, and upon unloading to the transition stress of $\sigma_{tr\_fu} = 95.9$ MPa, the interface counter slip length occupies the entire matrix crack spacing; i.e., the fiber/matrix interface sliding ratio of $y(\sigma_{tr\_fu})/l_d = 1$, and upon reloading to the transition stress of $\sigma_{tr\_fr} = 58.1$ MPa, the interface new slip length occupies the entire matrix crack spacing; i.e., the fiber/matrix interface sliding ratio of $z(\sigma_{tr\_fr})/l_d = 1$. For the matrix cracking mode V, the fiber/matrix interface debonding ratio is $2l_d/l_c = 1$, and upon unloading to the transition stress of $\sigma_{tr\_fu} = 102.2$ MPa, the interface counter slip length occupies the entire matrix crack spacing; i.e., the fiber/matrix interface sliding ratio of $y(\sigma_{tr\_fu})/l_d = 1$, and upon reloading to the transition stress of $\sigma_{tr\_fr} = 51.8$ MPa, the interface new slip length occupies the entire matrix crack spacing; i.e., the fiber/matrix interface sliding ratio of $z(\sigma_{tr\_fr})/l_d = 1$.

#### Tension–Tension Fatigue Hysteresis Loops at 1200 °C in Steam Condition

Jacob [55] investigated the tension–tension cyclic fatigue hysteresis loops of 2D SiC/SiC composite at 1200 °C in steam condition. Under the fatigue peak stress of $\sigma_{max} = 140$ MPa and $f = 0.1$ Hz, the experimental and theoretical fatigue stress–strain hysteresis loops corresponding to the cycle number of $N = 100$, 1000, and 10,000 are shown in Fig. 2.50.

When $N = 100$, for the matrix cracking mode III, the fiber/matrix interface debonding ratio is $2l_d/l_c < 1$, and upon unloading to the valley stress, the interface counter slip length occupies 54.6% of interface debonded length; i.e., the fiber/matrix interface sliding ratio of $y(\sigma_{min})/l_d = 0.546$, and upon reloading to the peak

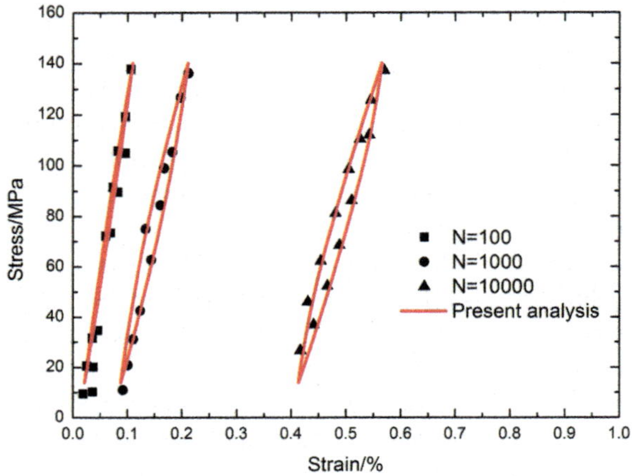


**Fig. 2.50** Experimental and theoretical fatigue hysteresis loops of 2D SiC/SiC composite under $\sigma_{max}$ = 140 MPa and $f$ = 0.1 Hz at 1200 °C in steam condition

stress, the interface new slip length occupies 54.6% of interface debonded length; i.e., the fiber/matrix interface sliding ratio of $z(\sigma_{max})/l_d = 0.546$. For the matrix cracking mode V, the fiber/matrix interface debonding ratio is $2l_d/l_c < 1$, and upon unloading to the valley stress, the interface counter slip length occupies 72.8% of interface debonded length; i.e., the fiber/matrix interface sliding ratio of $y(\sigma_{min})/l_d = 0.728$, and upon reloading to the peak stress, the interface new slip length occupies 72.8% of interface debonded length; i.e., the fiber/matrix interface sliding ratio of $z(\sigma_{max})/l_d = 0.728$.

When $N = 1000$, for the matrix cracking mode III, the fiber/matrix interface debonding ratio is $2l_d/l_c = 1$, and upon unloading to the transition stress of $\sigma_{tr\_fu}$ = 39.2 MPa, the interface counter slip length occupies the entire matrix crack spacing; i.e., the fiber/matrix interface sliding ratio of $y(\sigma_{tr\_fu})/l_d = 1$, and upon reloading to the transition stress of $\sigma_{tr\_fr}$ = 114.8 MPa, the interface new slip length occupies the entire matrix crack spacing; i.e., the fiber/matrix interface sliding ratio of $z(\sigma_{tr\_fr})/l_d = 1$. For the matrix cracking mode V, the fiber/matrix interface debonding ratio is $2l_d/l_c < 1$, and upon unloading to the valley stress, the interface counter slip length occupies 71.3% of interface debonded length; i.e., the fiber/matrix interface sliding ratio of $y(\sigma_{min})/l_d = 0.713$, and upon reloading to the peak stress, the interface new slip length occupies 71.3% of interface debonded length; i.e., the fiber/matrix interface sliding ratio of $z(\sigma_{max})/l_d = 0.713$.

When $N = 10{,}000$, for the matrix cracking mode III, the fiber/matrix interface debonding ratio is $2l_d/l_c = 1$, and upon unloading to the transition stress of $\sigma_{tr\_fu}$ = 89.6 MPa, the interface counter slip length occupies the entire matrix crack spacing; i.e., the fiber/matrix interface sliding ratio of $y(\sigma_{tr\_fu})/l_d = 1$, and upon reloading to the transition stress of $\sigma_{tr\_fr}$ = 64.4 MPa, the interface new slip length occupies the entire matrix crack spacing; i.e., the fiber/matrix interface sliding ratio of $z(\sigma_{tr\_fr})/l_d = 1$. For the matrix cracking mode V, the fiber/matrix interface debonding ratio is $2l_d/l_c = 1$, and upon unloading to the transition stress of $\sigma_{tr\_fu}$ = 14 MPa, the interface counter slip length occupies the entire matrix crack spacing; i.e., the fiber/matrix interface sliding ratio of $y(\sigma_{tr\_fu})/l_d = 1$, and upon reloading to the transition stress of $\sigma_{tr\_fr} = 140$ MPa, the interface new slip length occupies the entire matrix crack spacing; i.e., the fiber/matrix interface sliding ratio of $z(\sigma_{tr\_fr})/l_d = 1$.

Under the fatigue peak stress of $\sigma_{max} = 140$ MPa and $f = 1$ Hz, the experimental and theoretical fatigue stress–strain hysteresis loops corresponding to the cycle number of $N = 1000$, 10,000, and 30,000 are shown in Fig. 2.51.

When $N = 1000$, for the matrix cracking mode III, the fiber/matrix interface debonding ratio is $2l_d/l_c < 1$, and upon unloading to the valley stress, the interface counter slip length occupies 53.9% of interface debonded length; i.e., the fiber/matrix interface sliding ratio of $y(\sigma_{min})/l_d = 0.539$, and upon reloading to the peak stress, the interface new slip length occupies 53.9% of interface debonded length; i.e., the fiber/matrix interface sliding ratio of $z(\sigma_{max})/l_d = 0.539$. For the matrix cracking mode V, the fiber/matrix interface debonding ratio is $2l_d/l_c < 1$, and upon unloading to the valley stress, the interface counter slip length occupies 70.2% of interface debonded length; i.e., the fiber/matrix interface sliding ratio of $y(\sigma_{min})/l_d = 0.702$, and upon reloading to the peak stress, the interface new slip length occupies 70.2% of interface debonded length; i.e., the fiber/matrix interface sliding ratio of $z(\sigma_{max})/l_d = 0.702$.

When $N = 10{,}000$, for the matrix cracking mode III, the fiber/matrix interface debonding ratio is $2l_d/l_c = 1$, and upon unloading to the transition stress of $\sigma_{tr\_fu} = 77$ MPa, the interface counter slip length occupies the entire matrix crack spacing; i.e., the fiber/matrix interface sliding ratio of $y(\sigma_{tr\_fu})/l_d = 1$, and upon reloading to the transition stress of $\sigma_{tr\_fr} = 77$ MPa, the interface new slip length occupies the entire matrix crack spacing; i.e., the fiber/matrix interface sliding ratio of $z(\sigma_{tr\_fr})/l_d = 1$. For the matrix cracking mode V, the fiber/matrix interface debonding ratio is $2l_d/l_c < 1$, and upon unloading to the valley stress, the interface counter slip length occupies 69.3% of interface debonded length; i.e., the fiber/matrix interface sliding ratio of $y(\sigma_{min})/l_d = 0.693$, and upon reloading to the peak

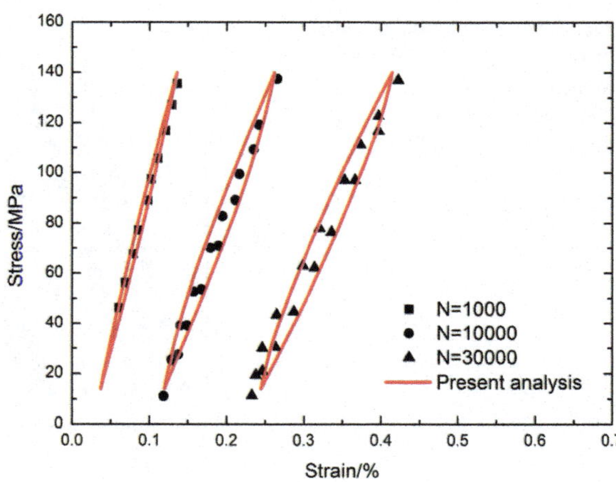


**Fig. 2.51** Experimental and theoretical fatigue hysteresis loops of 2D SiC/SiC composite under $\sigma_{max} = 140$ MPa and $f = 1$ Hz at 1200 °C in steam condition

stress, the interface new slip length occupies 69.3% of interface debonded length; i.e., the fiber/matrix interface sliding ratio of $z(\sigma_{max})/l_d = 0.693$.

When $N = 30{,}000$, for the matrix cracking mode III, the fiber/matrix interface debonding ratio is $2l_d/l_c = 1$, and upon unloading to the transition stress of $\sigma_{tr\_fu} = 108.5$ MPa, the interface counter slip length occupies the entire matrix crack spacing; i.e., the fiber/matrix interface sliding ratio of $y(\sigma_{tr\_fu})/l_d = 1$, and upon reloading to the transition stress of $\sigma_{tr\_fr} = 45.5$ MPa, the interface new slip length occupies the entire matrix crack spacing; i.e., the fiber/matrix interface sliding ratio of $z(\sigma_{tr\_fr})/l_d = 1$. For the matrix cracking mode V, the fiber/matrix interface debonding ratio is $2l_d/l_c = 1$, and upon unloading to the transition stress of $\sigma_{tr\_fu} = 14$ MPa, the interface counter slip length occupies the entire matrix crack spacing; i.e., the fiber/matrix interface sliding ratio of $y(\sigma_{tr\_fu})/l_d = 1$, and upon reloading to the transition stress of $\sigma_{tr\_fr} = 140$ MPa, the interface new slip length occupies the entire matrix crack spacing; i.e., the fiber/matrix interface sliding ratio of $z(\sigma_{tr\_fr})/l_d = 1$.

### Tension–Tension Fatigue Hysteresis Loops at 1300 °C in Air Condition

Zhu et al. [56] investigated the tension–tension cyclic fatigue hysteresis loop of 2D SiC/SiC composite at 1300 °C in air condition. Under the fatigue peak stress of $\sigma_{max} = 90$ MPa, the experimental and theoretical fatigue stress–strain hysteresis loops corresponding to the applied cycle number of $N = 6000$, 24,000, 90,000, 650,000, 1,200,000, and 2,800,000 are shown in Fig. 2.52.

When $N = 6000$, for the matrix cracking mode III, the fiber/matrix interface debonding ratio is $2l_d/l_c < 1$, and upon unloading to the valley stress, the interface counter slip length occupies 60% of interface debonded length; i.e., the fiber/matrix interface sliding ratio of $y(\sigma_{min})/l_d = 0.6$, and upon reloading to the peak stress, the interface new slip length occupies 60% of interface debonded length; i.e., the fiber/

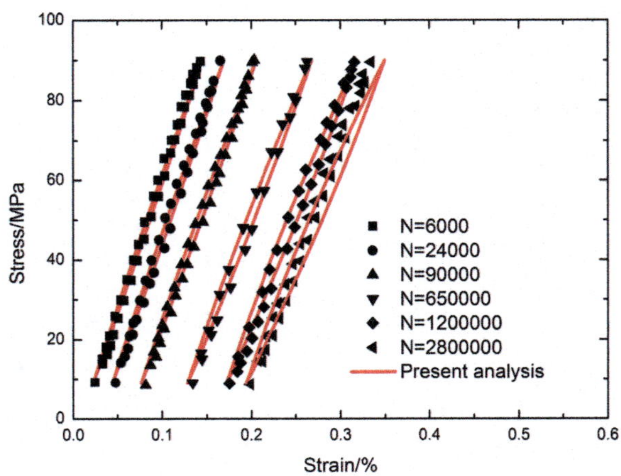


**Fig. 2.52** Experimental and theoretical fatigue hysteresis loops of 2D SiC/SiC composite under $\sigma_{max} = 90$ MPa at 1300 °C in air condition

matrix interface sliding ratio of $z(\sigma_{max})/l_d = 0.6$. For the matrix cracking mode V, the fiber/matrix interface debonding ratio is $2l_d/l_c < 1$, and upon unloading to the transition stress of $\sigma_{tr\_pu} = 9$ MPa, the interface counter slip length occupies the entire matrix crack spacing; i.e., the fiber/matrix interface sliding ratio of $y(\sigma_{tr\_pu})/l_d = 1$, and upon reloading to the transition stress of $\sigma_{tr\_pr} = 90$ MPa, the interface new slip length occupies the entire matrix crack spacing; i.e., the fiber/matrix interface sliding ratio of $z(\sigma_{tr\_pr})/l_d = 1$.

When $N = 90{,}000$, for the matrix cracking mode III, the fiber/matrix interface debonding ratio is $2l_d/l_c < 1$, and upon unloading to the valley stress, the interface counter slip length occupies 59.5% of interface debonded length; i.e., the fiber/matrix interface sliding ratio of $y(\sigma_{min})/l_d = 0.595$, and upon reloading to the peak stress, the interface new slip length occupies 59.5% of interface debonded length; i.e., the fiber/matrix interface sliding ratio of $z(\sigma_{max})/l_d = 0.595$. For the matrix cracking mode V, the fiber/matrix interface debonding ratio is $2l_d/l_c < 1$, and upon unloading to the valley stress, the interface counter slip length occupies 99.5% of interface debonded length; i.e., the fiber/matrix interface sliding ratio of $y(\sigma_{min})/l_d = 0.995$, and upon reloading to the peak stress, the interface new slip length occupies 99.5% of interface debonded length; i.e., the fiber/matrix interface sliding ratio of $z(\sigma_{max})/l_d = 0.995$.

When $N = 650{,}000$, for the matrix cracking mode III, the fiber/matrix interface debonding ratio is $2l_d/l_c < 1$, and upon unloading to the valley stress, the interface counter slip length occupies 59.1% of interface debonded length; i.e., the fiber/matrix interface sliding ratio of $y(\sigma_{min})/l_d = 0.591$, and upon reloading to the peak stress, the interface new slip length occupies 59.1% of interface debonded length; i.e., the fiber/matrix interface sliding ratio of $z(\sigma_{max})/l_d = 0.591$. For the matrix cracking mode V, the fiber/matrix interface debonding ratio is $2l_d/l_c < 1$, and upon unloading to the valley stress, the interface counter slip length occupies 97% of interface debonded length; i.e., the fiber/matrix interface sliding ratio of $y(\sigma_{min})/l_d = 0.97$, and upon reloading to the peak stress, the interface new slip length occupies 97% of interface debonded length; i.e., the fiber/matrix interface sliding ratio of $z(\sigma_{max})/l_d = 0.97$.

When $N = 1{,}200{,}000$, for the matrix cracking mode III, the fiber/matrix interface debonding ratio is $2l_d/l_c = 1$, and upon unloading to the valley stress, the interface counter slip length occupies 70.4% of interface debonded length; i.e., the fiber/matrix interface sliding ratio of $y(\sigma_{min})/l_d = 0.704$, and upon reloading to the peak stress, the interface new slip length occupies 70.4% of interface debonded length; i.e., the fiber/matrix interface sliding ratio of $z(\sigma_{max})/l_d = 0.704$. For the matrix cracking mode V, the fiber/matrix interface debonding ratio is $2l_d/l_c < 1$, and upon unloading to the valley stress, the interface counter slip length occupies 96.2% of interface debonded length; i.e., the fiber/matrix interface sliding ratio of $y(\sigma_{min})/l_d = 0.962$, and upon reloading to the peak stress, the interface new slip length occupies 96.2% of interface debonded length; i.e., the fiber/matrix interface sliding ratio of $z(\sigma_{max})/l_d = 0.962$.

When $N = 2{,}800{,}000$, for the matrix cracking mode III, the fiber/matrix interface debonding ratio is $2l_d/l_c = 1$, and upon unloading to the valley stress, the interface counter slip length occupies 92.3% of interface debonded length; i.e., the fiber/matrix interface sliding ratio of $y(\sigma_{min})/l_d = 0.923$, and upon reloading to the peak stress, the interface new slip length occupies 92.3% of interface debonded length; i.e., the fiber/matrix interface sliding ratio of $z(\sigma_{max})/l_d = 0.923$. For the matrix cracking mode V, the fiber/matrix interface debonding ratio is $2l_d/l_c < 1$, and upon unloading to the valley stress, the interface counter slip length occupies 95.4% of interface debonded length; i.e., the fiber/matrix interface sliding ratio of $y(\sigma_{min})/l_d = 0.954$, and upon reloading to the peak stress, the interface new slip length occupies 95.4% of interface debonded length; i.e., the fiber/matrix interface sliding ratio of $z(\sigma_{max})/l_d = 0.954$.

Under the fatigue peak stress of $\sigma_{max}$ = 120 MPa, the experimental and theoretical fatigue stress–strain hysteresis loops of 2D SiC/SiC composite corresponding to the applied cycle numbers of $N$ = 100, 6000, 18,000, and 36,000 are shown in Fig. 2.53.

When $N$ = 100, for the matrix cracking mode III, the fiber/matrix interface debonding ratio is $2l_d/l_c < 1$, and upon unloading to the valley stress, the interface counter slip length occupies 55.9% of interface debonded length; i.e., the fiber/matrix interface sliding ratio of $y(\sigma_{min})/l_d = 0.559$, and upon reloading to the peak stress, the interface new slip length occupies 55.9% of interface debonded length; i.e., the fiber/matrix interface sliding ratio of $z(\sigma_{max})/l_d = 0.559$. For the matrix cracking mode V, the fiber/matrix interface debonding ratio is $2l_d/l_c < 1$, and upon unloading to the valley stress, the interface counter slip length occupies 80.7% of interface debonded length; i.e., the fiber/matrix interface sliding ratio of $y(\sigma_{min})/l_d = 0.807$, and upon reloading to the peak stress, the interface new slip length occupies 80.7% of interface debonded length; i.e., the fiber/matrix interface sliding ratio of $z(\sigma_{max})/l_d = 0.807$.

When $N$ = 6000, for the matrix cracking mode III, the fiber/matrix interface debonding ratio is $2l_d/l_c < 1$, and upon unloading to the valley stress, the interface counter slip length occupies 55.2% of interface debonded length; i.e., the fiber/

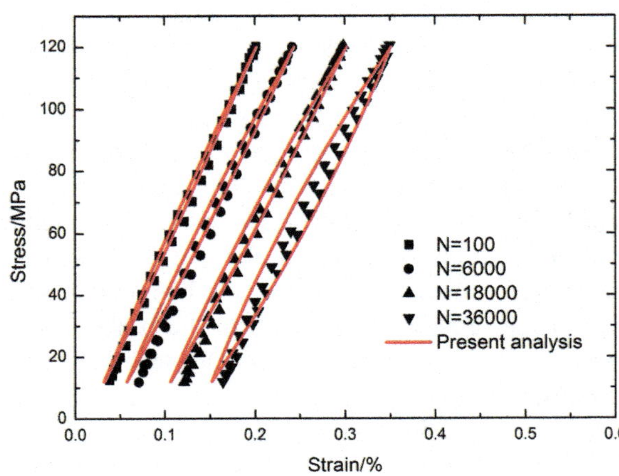


**Fig. 2.53** Experimental and theoretical fatigue hysteresis loops of 2D SiC/SiC composite under $\sigma_{max}$ = 120 MPa at 1300 °C in air condition

matrix interface sliding ratio of $y(\sigma_{min})/l_d = 0.552$, and upon reloading to the peak stress, the interface new slip length occupies 55.2% of interface debonded length; i.e., the fiber/matrix interface sliding ratio of $z(\sigma_{max})/l_d = 0.552$. For the matrix cracking mode V, the fiber/matrix interface debonding ratio is $2l_d/l_c < 1$, and upon unloading to the valley stress, the interface counter slip length occupies 77.2% of interface debonded length; i.e., the fiber/matrix interface sliding ratio of $y(\sigma_{min})/l_d = 0.772$, and upon reloading to the peak stress, the interface new slip length occupies 77.2% of interface debonded length; i.e., the fiber/matrix interface sliding ratio of $z(\sigma_{max})/l_d = 0.772$.

When $N = 18{,}000$, for the matrix cracking mode III, the fiber/matrix interface debonding ratio is $2l_d/l_c < 1$, and upon unloading to the valley stress, the interface counter slip length occupies 55% of interface debonded length; i.e., the fiber/matrix interface sliding ratio of $y(\sigma_{min})/l_d = 0.55$, and upon reloading to the peak stress, the interface new slip length occupies 55% of interface debonded length; i.e., the fiber/matrix interface sliding ratio of $z(\sigma_{max})/l_d = 0.55$. For the matrix cracking mode V, the fiber/matrix interface debonding ratio is $2l_d/l_c < 1$, and upon unloading to the valley stress, the interface counter slip length occupies 76.4% of interface debonded length; i.e., the fiber/matrix interface sliding ratio of $y(\sigma_{min})/l_d = 0.764$, and upon reloading to the peak stress, the interface new slip length occupies 76.4% of interface debonded length; i.e., the fiber/matrix interface sliding ratio of $z(\sigma_{max})/l_d = 0.764$.

When $N = 36{,}000$, for the matrix cracking mode III, the fiber/matrix interface debonding ratio is $2l_d/l_c = 1$, and upon unloading to the valley stress, the interface counter slip length occupies 61.5% of interface debonded length; i.e., the fiber/matrix interface sliding ratio of $y(\sigma_{min})/l_d = 0.615$, and upon reloading to the peak stress, the interface new slip length occupies 61.5% of interface debonded length; i.e., the fiber/matrix interface sliding ratio of $z(\sigma_{max})/l_d = 0.615$. For the matrix cracking mode V, the fiber/matrix interface debonding ratio is $2l_d/l_c < 1$, and upon unloading to the valley stress, the interface counter slip length occupies 75.6% of interface debonded length; i.e., the fiber/matrix interface sliding ratio of $y(\sigma_{min})/l_d = 0.756$, and upon reloading to the peak stress, the interface new slip length occupies 75.6% of interface debonded length; i.e., the fiber/matrix interface sliding ratio of $z(\sigma_{max})/l_d = 0.756$.

## 2.4 5D Woven Ceramic-Matrix Composites

Based on the microstructure analysis, the 2.5D woven CMCs are divided into four elements of warp yarns, weft yarns, matrix outside of yarns, and open porosity [57]. By assuming the mechanical fatigue hysteresis loops of 2.5D woven CMCs is mainly controlled by fiber/matrix interface slip in the warp yarns, the interface counter slip length and interface new slip length are determined. The fatigue stress–strain hysteresis loops of 2.5D C/SiC composite are predicted.

### 2.4.1 Hysteresis Theories

When matrix multicracking and fiber/matrix interface debonding occur in the warp yarns upon loading to fatigue peak stress, the fatigue stress–strain hysteresis loops develop due to the interface frictional sliding in the warp yarns upon unloading and subsequent reloading. The fatigue stress–strain hysteresis loops can be divided into three cases, as following:

(1) Case I, the fiber/matrix interface debonding ratio of $2l_d/l_c < 1$, and the fiber/matrix interface sliding ratio of $y(\sigma_{min})/l_d < 1$;
(2) Case II, the fiber/matrix interface debonding ratio of $2l_d/l_c = 1$, and the fiber/matrix interface sliding ratio of $y(\sigma_{min})/l_d < 1$;
(3) Case III, the fiber/matrix interface debonding ratio of $2l_d/l_c = 1$, and the fiber/matrix interface sliding ratio of $y(\sigma_{min})/l_d = 1$.

#### 2.4.1.1 Case 1

Upon unloading, the fiber axial stress distribution can be described using the following equation.

$$\begin{cases} \sigma_f(x) = \frac{\sigma}{V_{f0}} + \frac{2\tau_i}{r_f}x, x \in (0, y) \\ \sigma_f(x) = \frac{\sigma}{V_{f0}} + \frac{2\tau_i}{r_f}(2y - x), x \in (y, l_d) \\ \sigma_f(x) = \frac{E_f}{E_c}\sigma - \frac{(1-V_{f0})E_{m-mini}}{V_{f0}E_{innier0}}\sigma_{th}, x \in (l_d, l_c/2) \end{cases} \quad (2.49)$$

where

$$y = \frac{r_f}{4\tau_i}\frac{1 - V_{f0}}{V_{f0}}\frac{E_{m-mini}}{E_{inner0}}(\sigma_{max} - \sigma) \quad (2.50)$$

Upon reloading, the fiber axial stress distribution can be described using the following equation.

$$\begin{cases} \sigma_f(x) = \frac{\sigma}{V_{f0}} - \frac{2\tau_i}{r_f}x, \quad x \in (0, z) \\ \sigma_f(x) = \frac{\sigma}{V_{f0}} - \frac{2\tau_i}{r_f}(2z - x), \quad x \in (z, y) \\ \sigma_f(x) = \frac{\sigma}{V_{f0}} + \frac{2\tau_i}{r_f}2(y - z) - \frac{2\tau_i}{r_f}x, \quad x \in (y, l_d) \\ \sigma_f(x) = \frac{E_f}{E_c}\sigma - \frac{(1-V_{f0})E_{m-mini}}{V_{f0}E_{inner0}}\sigma_{th}, \quad x \in (l_d, l_c/2) \end{cases} \quad (2.51)$$

where

$$z = y(\sigma_{min}) - \frac{r_f}{4\tau_i}\frac{1 - V_{f0}}{V_{f0}}\frac{E_{m-mini}}{E_{inner0}}(\sigma_{max} - \sigma) \quad (2.52)$$

#### 2.4.1.2 Case II

Upon unloading, the fiber axial stress distribution can be described using the following equation.

$$\begin{cases} \sigma_{\mathrm{f}}(x) = \frac{\sigma}{V_{\mathrm{f0}}} + \frac{2\tau_{\mathrm{i}}}{r_{\mathrm{f}}}x, & x \in (0, y) \\ \sigma_{\mathrm{f}}(x) = \frac{\sigma}{V_{\mathrm{f0}}} + \frac{2\tau_{\mathrm{i}}}{r_{\mathrm{f}}}(2y - x), & x \in (y, l_{\mathrm{c}}/2) \end{cases} \tag{2.53}$$

Upon reloading, the fiber axial stress distribution can be described using the following equation.

$$\begin{cases} \sigma_{\mathrm{f}}(x) = \frac{\sigma}{V_{\mathrm{f0}}} - \frac{2\tau_{\mathrm{i}}}{r_f}x, & x \in (0, z) \\ \sigma_{\mathrm{f}}(x) = \frac{\sigma}{V_{\mathrm{f0}}} - \frac{2\tau_{\mathrm{i}}}{r_f}(2z - x), & x \in (z, y) \\ \sigma_{\mathrm{f}}(x) = \frac{\sigma}{V_{\mathrm{f0}}} + \frac{2\tau_{\mathrm{i}}}{r_{\mathrm{f}}}2(y - z) - \frac{2\tau_{\mathrm{i}}}{r_{\mathrm{f}}}x, & x \in (y, l_{\mathrm{c}}/2) \end{cases} \tag{2.54}$$

#### 2.4.1.3 Case 3

When $\sigma > \sigma_{\mathrm{tr\_fu}}$, the fiber/matrix interface sliding ratio is $2y(\sigma > \sigma_{\mathrm{tr\_fu}})/l_{\mathrm{c}} < 1$, and the fiber axial stress distribution can be described using Eq. (2.53). When $\sigma <$ $_{\mathrm{tr\_fu}}$, the fiber/matrix interface sliding ratio is $2y(\sigma < \sigma_{\mathrm{tr\_fu}})/l_{\mathrm{c}} = 1$, and the fiber axial stress distribution can be described using Eq. (2.53) by setting $y = l_{\mathrm{c}}/2$.

When $\sigma < \sigma_{\mathrm{tr\_fr}}$, the fiber/matrix interface sliding ratio is $2z(\sigma < \sigma_{\mathrm{tr\_fr}})/l_{\mathrm{c}} < 1$, and the fiber axial stress distribution can be described using Eq. (2.54). When $\sigma_{\mathrm{tr\_fr}} <$ $<\sigma_{\max}$, the fiber/matrix interface sliding ratio is $2z(\sigma > \sigma_{\mathrm{tr\_fr}})/l_{\mathrm{c}} = 1$, and the fiber axial stress distribution can be described using Eq. (2.54) by setting $z = l_{\mathrm{c}}/2$.

#### 2.4.1.4 Stress–Strain Hysteresis Loops

When the fiber/matrix interface partially debonds, the unloading and reloading stress–strain relationships can be described using the following equations.

$$\begin{aligned} \varepsilon = & \frac{\sigma}{E_{\mathrm{c}}} + 3\frac{1 - V_{\mathrm{f0}}}{V_{\mathrm{f0}}}\frac{E_{\mathrm{m-mini}}}{E_{\mathrm{f}}E_{\mathrm{inner0}}}\frac{l_{\mathrm{d}}}{l_{\mathrm{c}}}(\sigma + \sigma_{\mathrm{th}}) \\ & - 2\frac{1 - V_{\mathrm{f0}}}{V_{\mathrm{f0}}}\frac{E_{\mathrm{m-mini}}}{E_{\mathrm{f}}E_{\mathrm{inner0}}}\frac{l_{\mathrm{d}}}{l_{\mathrm{c}}}(\sigma + \sigma_{\mathrm{th}})\left(1 - \frac{y}{l_{\mathrm{d}}}\right)^2 \end{aligned} \tag{2.55}$$

$$\begin{aligned} \varepsilon = & \frac{\sigma}{E_{\mathrm{c}}} + \frac{1 - V_{\mathrm{f0}}}{V_{\mathrm{f0}}}\frac{E_{\mathrm{m-mini}}}{E_{\mathrm{f}}E_{\mathrm{inner0}}}\frac{l_{\mathrm{d}}}{l_{\mathrm{c}}}(\sigma + \sigma_{\mathrm{th}}) \\ & - 2\frac{1 - V_{\mathrm{f0}}}{V_{\mathrm{f0}}}\frac{E_{\mathrm{m-mini}}}{E_{\mathrm{f}}E_{\mathrm{inner0}}}\frac{l_{\mathrm{d}}}{l_{\mathrm{c}}}(\sigma + \sigma_{\mathrm{th}})\left[\left(\frac{z}{l_{\mathrm{d}}} - 1\right)^2 - \left(\frac{y}{l_{\mathrm{d}}} - 1\right)^2\right] \end{aligned} \tag{2.56}$$

When the fiber/matrix interface completely debonds, the unloading and reloading stress–strain relationships can be described using the following equations.

$$\varepsilon = \frac{\sigma}{E_c} + \frac{3}{2}\frac{1 - V_{f0}}{V_{f0}}\frac{E_{m-mini}}{E_f E_{inner0}}(\sigma + \sigma_{th}) - \frac{1 - V_{f0}}{V_{f0}}\frac{E_{m-mini}}{E_f E_{inner0}}(\sigma + \sigma_{th})\left(1 - \frac{2y}{l_c}\right)^2 \tag{2.57}$$

$$\begin{aligned}\varepsilon = &\frac{\sigma}{E_c} + \frac{1}{2}\frac{1 - V_{f0}}{V_{f0}}\frac{E_{m-mini}}{E_f E_{inner0}}(\sigma + \sigma_{th}) \\ &+ \frac{1 - V_{f0}}{V_{f0}}\frac{E_{m-mini}}{E_f E_{inner0}}(\sigma + \sigma_{th})\left[\left(\frac{2z}{l_c} - 1\right)^2 - \left(\frac{2y}{l_c} - 1\right)^2\right]\end{aligned} \tag{2.58}$$

The evolution of the fatigue stress–strain hysteresis loops, area of the hysteresis loops, fiber/matrix interface debonding ratio and sliding ratio versus the fiber/matrix interface shear stress curves of 2.5D C/SiC composite are shown in Fig. 2.54. The stress–strain hysteresis loops corresponding to the interface slip Case I, II, and III

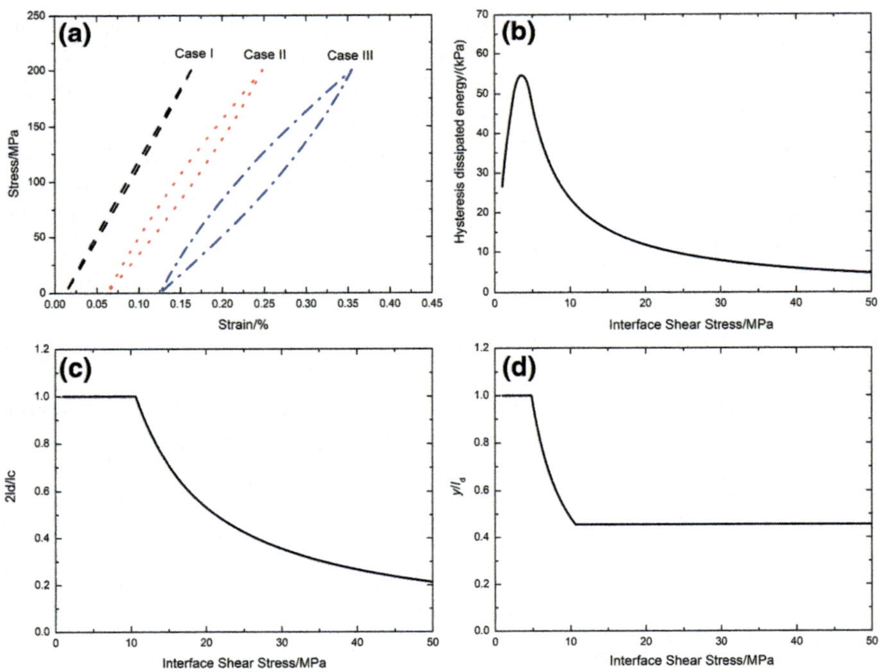


**Fig. 2.54** **a** Fatigue stress–strain hysteresis loops of different fiber/matrix interface slip cases; **b** the fatigue hysteresis dissipated energy versus the fiber/matrix interface shear stress curve; **c** the fiber/matrix interface debonding ratio versus the interface shear stress curve; and **d** the fiber/matrix interface slip ratio versus the interface shear stress curve of 2.5D C/SiC composite. Reprinted with permission from Ref. [57]. Copyright 2014 Elsevier B.V

are shown in Fig. 2.54a. The area, residual strain, and peak strain of the stress–strain hysteresis loops depend on the fiber/matrix interface debonding and slip condition between the fiber and the matrix in the matrix cracking space of CMCs. The fatigue hysteresis dissipated energy versus the fiber/matrix interface shear stress curve is shown in Fig. 2.54b. When the fiber/matrix interface shear stress increases, the fatigue hysteresis dissipated energy increases first and then decreases. When the fiber/matrix interface shear stress increases from $\tau_i = 1$ MPa to $\tau_i = 4.8$ MPa, the fatigue hysteresis dissipated energy increases to the peak value and then decreases with increasing fiber/matrix interface shear stress, corresponding to the fiber/matrix interface debonding ratio of $2l_d/l_c = 1$ and the fiber/matrix interface sliding ratio of $y/l_d = 1$. When the fiber/matrix interface shear stress increases from $\tau_i = 4.8$ MPa to $\tau_i = 10.5$ MPa, the fatigue hysteresis dissipated energy decreases with increasing fiber/matrix interface shear stress, corresponding to the fiber/matrix interface debonding ratio of $2l_d/l_c = 1$ and the fiber/matrix interface sliding ratio of $y/l_d < 1$. When the fiber/matrix interface shear stress increases from $\tau_i = 10.5$ MPa to $\tau_i = 50$ MPa, the fatigue hysteresis dissipated energy decreases with increasing fiber/matrix interface shear stress, corresponding to the fiber/matrix interface debonding ratio of $2l_d/l_c < 1$ and the fiber/matrix interface sliding ratio of $y/l_d < 1$.

### 2.4.2 Experimental Comparisons

Wang et al. [58] investigated the cyclic loading/unloading tensile hysteresis loops of 2.5D C/SiC composite at room temperature. The experimental and theoretical fatigue stress–strain hysteresis loops of 2.5D C/SiC composite at the peak stress of 155 MPa, 195 MPa, and 265 MPa are predicted, as shown in Fig. 2.55.

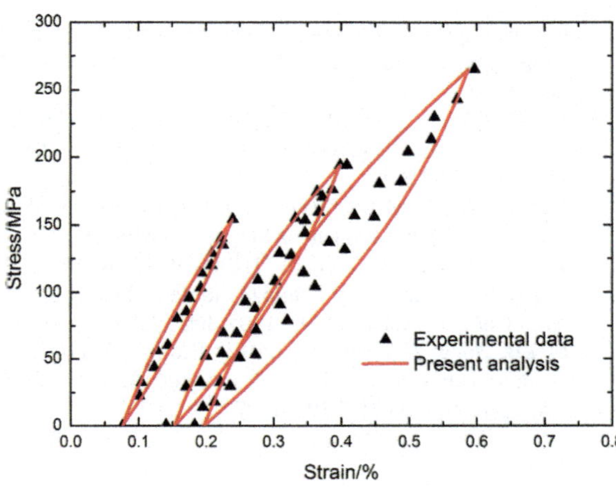


**Fig. 2.55** Experimental and theoretical fatigue stress–strain hysteresis loops of 2.5D C/SiC composite at room temperature. Reprinted with permission from Ref. [8]. Copyright 2013 Elsevier B.V

## 2.5 Conclusions

The fatigue hysteresis behavior of unidirectional, 2D cross-ply and woven, and 2.5D woven fiber-reinforced CMCs have been analyzed. Based on the fiber/matrix interface debonding and sliding behavior, the fiber/matrix interface debonding and sliding lengths have been determined using the fracture mechanics approach. The fiber/matrix interface debonding ratio and interface sliding ratio are determined for different interface slip cases. The effects of fiber volume fraction, peak stress, matrix crack spacing, interface shear stress, interface debonded energy, fibers failure, fiber Poisson contraction, fiber strength, fiber Weibull modulus, matrix cracking mode, applied cycle number and fiber/matrix interface wear on the fatigue stress–strain hysteresis loops and the fiber/matrix interface debonding and sliding have been discussed. The experimental cyclic fatigue stress–strain hysteresis loops of unidirectional SiC/CAS, SiC/1723 and C/SiC, 2D cross-ply SiC/CAS and woven SiC/SiC, and 2.5D woven C/SiC composites under cyclic loading/unloading tensile and tension–tension fatigue loading have been predicted.

## References

1. Rouby D, Reynaud P (1993) Fatigue behavior related to interface modification during load cycling in ceramic-matrix fibre composites. Compos Sci Technol 48(1–4):109–118. https://doi.org/10.1016/0266-3538(93)90126-2
2. McNulty JC, Zok FW (1999) Low cycle fatigue of Nicalon-fiber-reinforced ceramic composites. Compos Sci Technol 59(10):1597–1607. https://doi.org/10.1016/S0266-3538(99)00019-6
3. Reynaud P (1996) Cyclic fatigue of ceramic-matrix composites at ambient and elevated temperatures. Compos Sci Technol 56(7):809–814. https://doi.org/10.1016/0266-3538(96)00025-5
4. Evans AG (1997) Design and life prediction issues for high-temperature engineering ceramics and their composites. Acta Mater 45(1):23–40. https://doi.org/10.1016/S1359-6454(96)00143-7
5. Mall S, Engesser JM (2006) Effects of frequency on fatigue behavior of CVI C/SiC at elevated temperature. Compos Sci Technol 66(7–8):863–874. https://doi.org/10.1016/j.compscitech.2005.06.020
6. Li LB, Song YD (2010) An approach to estimate interface shear stress of ceramic matrix composites from hysteresis loops. Appl Compos Mater 17:309–328. https://doi.org/10.1007/s10443-013-9314-y
7. Ruggles-Wrenn MB, Christensen DT, Chamberlain AL, Lane JE, Cook TS (2011) Effect of frequency and environment on fatigue behaviour of a CVI SiC/SiC ceramic matrix composite at 1200 °C. Compo Sci Technol 71:190–196. https://doi.org/10.1016/j.compscitech.2010.11.008
8. Li LB, Song YD (2013) Estimate interface shear stress of woven ceramic matrix composites. Appl Compos Mater 20:993–1005. https://doi.org/10.1007/s10443-013-9314-y
9. Li LB, Song YD, Sun YC (2013) Estimate interface shear stress of unidirectional C/SiC ceramic matrix composites from hysteresis loops. Appl Compos Mater 20:693–707. https://doi.org/10.1007/s10443-012-9297-0

10. Li LB, Song YD, Sun YC (2013) Modeling loading/unloading hysteresis behaviour of unidirectional C/SiC ceramic matrix composites. Appl Compos Mater 20:655–672. https://doi.org/10.1007/s10443-012-9294-3
11. Ruggles-Wrenn MB, Jones TP (2013) Tension-compression fatigue of a SiC/SiC ceramic matrix composite at 1200 °C in air and in steam. Int J Fatigue 47:154–160. https://doi.org/10.1016/j.ijfatigue.2012.08.006
12. Li LB (2016) Modeling cyclic fatigue hysteresis loops of 2D woven ceramic matrix composites at elevated temperatures in steam. Materials 9:421. https://doi.org/10.3390/ma9060421
13. Li LB (2016) Modeling the effect of multiple matrix cracking modes on cyclic hysteresis loops of 2D woven ceramic-matrix composites. Appl Compos Mater 23:555–581. https://doi.org/10.1007/s10443-016-9474-7
14. Li LB (2016) Modeling cyclic fatigue hysteresis loops of 2D woven ceramic-matrix composite at elevated temperatures in air considering multiple matrix cracking modes. Theoret Appl Fract Mech 85:246–261. https://doi.org/10.1016/j.tafmec.2016.03.010
15. Marshall DB, Evans AG (1985) Failure mechanisms in ceramic-fiber/ceramic-matrix composites. J Am Ceram Soc 68(5):225–231. https://doi.org/10.1111/j.1151-2916.1985.tb15313.x
16. Holmes JW, Cho CD (1992) Experimental observation of frictional heating in fiber-reinforced ceramics. J Am Ceram Soc 75(4):929–938. https://doi.org/10.1111/j.1151-2916.1992.tb04162.x
17. Fantozzi G, Reynaud P (2009) Mechanical hysteresis in ceramic matrix composites. Mater Sci Eng A 521–522:18–23. https://doi.org/10.1016/j.msea.2008.09.128
18. Kotil T, Holmes JW, Comninou M (1990) Origin of hysteresis observed during fatigue of ceramic-matrix composites. J Am Ceram Soc 73(7):1879–1883. https://doi.org/10.1111/j.1151-2916.1990.tb05239.x
19. Cho CD, Holmes JW, Barber JR (1991) Estimate of interfacial shear in ceramic composites from frictional heating measurements. J Am Ceram Soc 74(11):2802–2808. https://doi.org/10.1111/j.1151-2916.1991.tb06846.x
20. Pryce AW, Smith PA (1993) Matrix cracking in unidirectional ceramic matrix composites under quasi-static and cyclic loading. Acta Metall Mater 41(4):1269–1281. https://doi.org/10.1016/0956-7151(93)90178-U
21. Ahn BK, Curtin WA (1997) Strain and hysteresis by stochastic matrix cracking in ceramic matrix composites. J Mech Phys Solids 45(2):177–209. https://doi.org/10.1016/S0022-5096(96)00081-6
22. Solti JP, Mall S, Robertson DD (1995) Modeling damage in unidirectional ceramic-matrix composites. Compos Sci Technol 54(1):55–66. https://doi.org/10.1016/0266-3538(95)00041-0
23. Solti JP, Mall S, Robertson DD (1997) Modeling of fatigue in cross-ply ceramic matrix composites. J Compos Mater 31(19):1921–1943. https://doi.org/10.1111/j.1151-2916.1990.tb05239.x
24. Vagaggini E, Domergue JM, Evans AG (1995) Relationships between hysteresis measurements and the constituent properties of ceramic matrix composites: I. Theory. J Am Ceram Soc 78(10):2709–2720. https://doi.org/10.1111/j.1151-2916.1995.tb08047.x
25. Hutchison JW, Jensen HM (1990) Models of fiber debonding and pullout in brittle composites with friction. Mech Mater 9(2):139–163. https://doi.org/10.1016/0167-6636(90)90037-G
26. Keith WP, Kedward KT (1995) The stress-strain behavior of a porous unidirectional ceramic matrix composite. Composites 26(3):163–174. https://doi.org/10.1016/0010-4361(95)91379-J
27. Li LB, Song YD, Sun ZG (2009) Influence of interface de-bonding on the fatigue hysteresis loops of ceramic matrix composites. Chin J Solids Mech 30:8–14
28. Li LB (2016) Modeling the effect of oxidation on hysteresis loops of carbon fiber-reinfroced ceramic-matrix composites under static fatigue at elevated temperature. J Eur Ceram Soc 36:465–480. https://doi.org/10.1016/j.jeurceramsoc.2015.11.005
29. Li LB, Song YD, Sun ZG (2009) Effect of fiber Poisson contraction on fatigue hysteresis loops of ceramic matrix composites. J Nanjing Univ Aeronaut Astronaut 41:181–186

30. Li LB (2013) Modeling hysteresis behavior of cross-ply C/SiC ceramic matrix composites. Compos B 53:36–45. https://doi.org/10.1016/j.compositesb.2013.04.029
31. Li LB (2013) Fatigue hysteresis behavior of cross-ply C/SiC ceramic matrix composites at room and elevated temperatures. Mater Sci Eng A 586:160–170. https://doi.org/10.1016/j.msea.2013.08.017
32. Li LB, Song YD, Sun YC (2014) Effect of matrix cracking on hysteresis behavior of cross-ply ceramic matrix composites. J Compos Mater 48:1505–1530. https://doi.org/10.1177/0021998313488149
33. Li LB (2015) Micromechanics modeling of fatigue hysteresis loops in carbon fiber-reinforced ceramic-matrix composites. J Compos Mater 49:3471–3495. https://doi.org/10.1177/0021998314566055
34. Li LB, Song YD (2011) Influnece of fiber failure on fatigue hysteresis loops of ceramic matrix composites. J Reinf Plast Compos 30:12–25. https://doi.org/10.1177/0731684410386273
35. Li LB (2014) Modeling fatigue hysteresis behavior of unidirectional C/SiC ceramic-matrix composites. Compos B 66:466–474. https://doi.org/10.1016/j.compositesb.2014.06.014
36. Li LB (2015) Fatigue hysteresis behavior of unidirectional C/SiC ceramic matrix composites at room and elevated temperatures. Mater Sci Eng A 625:1–18. https://doi.org/10.1016/j.msea.2014.11.086
37. Li LB (2015) Synergistic effect of arbitrary loading sequence and interface wear on the fatigue hysteresis loops of carbon fiber-reinforced ceramic-matrix composites. Eng Fract Mech 146:67–88. https://doi.org/10.1016/j.engfracmech.2015.07.060
38. Li LB (2015) Modeling the effect of interface wear on fatigue hysteresis behavior of carbon fiber-reinforced ceramic-matrix composites. Appl Compos Mater 22:887–920. https://doi.org/10.1007/s10443-015-9442-7
39. Li LB (2017) Effects of loading type, temperature and oxidation on mechanical hysteresis behavior of carbon fiber-reinforced ceramic-matrix composites. Eng Fract Mech 169:336–353. https://doi.org/10.1016/j.engfracmech.2016.10.010
40. Li LB (2015) Modeling for fatigue hysteresis loops of carbon fiber-reinforced ceramic-matrix composites under multiple loading. Appl Compos Mater 22:945–959. https://doi.org/10.1007/s10443-015-9444-5
41. Li LB (2016) Hysteresis loops of carbon fiber-reinforced ceramic-matrix composites with different fiber preforms. Ceram Int 42:16535–16551. https://doi.org/10.1016/j.ceramint.2016.07.073
42. Li LB (2016) Comparison of cyclic hysteresis behavior between cross-ply C/SiC and SiC/SiC ceramic-matrix composites. Materials 9:62. https://doi.org/10.3390/ma9010062
43. Budiansky B, Hutchinson JW, Evans AG (1986) Matrix fracture in fiber-reinforced ceramics. J Mech Phys Solids 34(2):167–189. https://doi.org/10.1016/0022-5096(86)90035-9
44. Evans AG, Zok FW, McMeeking RM (1995) Fatigue of ceramic matrix composites. Acta Metall Mater 43(3):859–875. https://doi.org/10.1016/0956-7151(94)00304-Z
45. Zawada LP, Butkus LM, Hartman GA (1991) Tensile and fatigue behavior of silicon carbide fiber-reinforced aluminosilicate glass. J Am Ceram Soc 74(11):2851–2858. https://doi.org/10.1111/j.1151-2916.1991.tb06854.x
46. Kuo WS, Chou TW (1995) Multiple cracking of unidirectional and cross-ply ceramic matrix composites. J Am Ceram Soc 78(3):745–755. https://doi.org/10.1111/j.1151-2916.1995.tb08242.x
47. Takeda N, Kiriyama M (1999) Matrix crack evolution in SiC fiber/glass matrix cross-ply laminates. Compos A 30(4):593–597. https://doi.org/10.1016/S1359-835X(98)00155-9
48. Li LB, Song YD (2010) Fatigue behavior of cross-ply C/SiC ceramic matrix composites at ambient and elevated temperatures. In: The 7th International Conference on High Temperature Ceramic Matrix Composites, 20–22 Sep 2010, Bayreuth, Germany, pp 314–319
49. Fantozzi G, Reynaud P, Rouby D (2001) Thermomechanical behavior of long fibers ceramic-ceramic composites. Silic Indus 66(9–10):109–119
50. Lee JW, Daniel IM (1990) Progressive transverse cracking of crossply composite laminates. J Compos Mater 24(11):1225–1243. https://doi.org/10.1177/002199839002401108

51. Wang SW, Parvizi-Majidi A (1992) Experimental characterization of the tensile behavior of Nicalon fiber-reinforced calcium aluminosilicate composites. J Mater Sci 27(20):5483–5496. https://doi.org/10.1007/BF00541610
52. Opalski FA, Mall S (1994) Tension-compression fatigue behavior of a silicon carbide calcium-aluminosilicate ceramic matrix composites. J Reinf Plast Compos 13(5):420–438. https://doi.org/10.1177/073168449401300503
53. Li P, Wang B, Zhen WQ, Jiao GQ (2014) Tensile loading/unloading stress-strain behavior of 2D-SiC/SiC composites. Acta Mater Compos Sinica 31:676–682
54. Michael K (2010) Fatigue behavior of a SiC/SiC composite at 1000 °C in air and steam. Master Thesis, Air Force Institute of Technology
55. Jacob D (2010) Fatigue behavior of an advanced SiC/SiC composite with an oxidation inhibited matrix at 1200 °C in air and in steam. Master Thesis, Air Force Institute of Technology
56. Zhu SJ, Mizuno M, Nagano Y, Cao JW, Kagawa Y et al (1998) Creep and fatigue behavior in an enhanced SiC/SiC composite at high temperature. J Am Ceram Soc 81:2269–2277. https://doi.org/10.1111/j.1151-2916.1998.tb02621.x
57. Li LB (2014) Assessment of the interfacial properties from fatigue hysteresis loss energy in ceramic-matrix composites with different fiber preforms. Mater Sci Eng A 613:17–36. https://doi.org/10.1016/j.msea.2014.06.092
58. Wang YQ, Zhang LT, Cheng LF, Ma JQ, Zhang WH (2008) Tensile performance and damage evolution of a 2.5-D C/SiC composite characterized by acoustic emission. Appl Compos Mater 15:183–188. https://doi.org/10.1007/s10443-008-9066-2

# Chapter 3
# Interface Damage of Ceramic-Matrix Composites

## 3.1 Introduction

Under cyclic fatigue loading, the fiber/matrix interface shear stress affects the fatigue damage and fracture, and the fatigue hysteresis energy dissipation. The repeated sliding between the fiber and the matrix degrades the fiber/matrix interface shear stress, which has already been proved by the fiber push out and push back tests [1–3]. During the proceeding of frictional sliding, the temperature rising occurs at the interface between the fiber and the matrix, and also the surface of composite [4–6]. At elevated temperature, the oxidation of carbon or PyC interphase leads to the rapidly degradation for the fiber/matrix interface shear stress [7]. Under a non-stress oxidation test of 2D C/SiC at 700 °C in air condition, the fiber diameter reduced uniformly and longer fiber pullout length occurred at the fracture surface [8]. The fatigue stress–strain hysteresis loops can be used to obtain the fiber/matrix interface damage condition or the fiber/matrix interface shear stress for fiber-reinforced CMCs under cyclic fatigue loading at elevated temperatures [2, 9–13]. Cho et al. [14] developed an approach to estimate the fiber/matrix interface shear stress from frictional heating measurement. It was found that the fiber/matrix interfacial shear stress of unidirectional SiC/CAS-II composite undergoes an initially rapid decrease at the cyclic initial fatigue loading [4]. Evans et al. [2] developed an approach to evaluate the fiber/matrix interface shear stress through analyzing the parabolic regions of the fatigue stress–strain hysteresis loops using the Vagaggini's hysteresis loops models [15]. Solti et al. [16] proposed a means of inferring the state of the fiber/matrix interface through comparison of experimental and theoretical fatigue hysteresis energy dissipation. Li et al. developed the theoretical relationship between the fatigue hysteresis energy dissipation and the fiber/matrix interface shear stress of unidirectional [17, 18], cross-ply [19–23] and 2.5D woven [24, 25] fiber-reinforced CMCs.

In this chapter, the fiber/matrix interface shear stress of fiber-reinforced CMCs with different fiber preforms, i.e., unidirectional, 2D cross-ply and woven, 2.5D



L. Li, *Damage, Fracture, and Fatigue of Ceramic-Matrix Composites*,
https://doi.org/10.1007/978-981-13-1783-5_3

woven and 3D braided, is estimated from the fatigue hysteresis dissipated energy at room and elevated temperatures. The experimental fatigue hysteresis dissipated energy versus the applied cycles and the theoretical fatigue hysteresis dissipated energy versus the fiber/matrix interface shear stress relationship are analyzed. With decreasing fiber/matrix interface shear stress, the fatigue hysteresis dissipated energy increases to the peak value, and then decreases to zero, corresponding to the fiber/matrix interface slip Case I, II, III and IV. Using the experimental fatigue hysteresis dissipated energy, the fiber/matrix interface shear stress of unidirectional SiC/CAS, SiC/$Si_3N_4$ with the strong and weak fiber/matrix interface bonding, C/SiC at room temperature and 800 °C in air condition, cross-ply SiC/CAS and C/SiC at room temperature, 700, 800, and 850 °C in air condition, 2D C/SiC at room temperature, 550 °C in air and 1200 °C in vacuum conditions, 2D SiC/SiC at room temperature, 800 °C in air, 600, 800, and 1000 °C in inert, 1000, 1100, and 1200 °C in air and steam, 1300 °C in air conditions, 2.5D C/SiC at room temperature, 800 °C in air and 600 °C in inert conditions, and 3D braided SiC/SiC at 1300 °C in air conditions are obtained.

## 3.2 Interface Shear Stress Estimation Approach

Based on the interface debonding and frictional sliding case between the fiber and the matrix upon unloading/reloading, the fatigue stress–strain hysteresis loops of fiber-reinforced CMCs can be divided into four different cases [26, 27] as follows:

(1) Case I, the fiber/matrix interface partially debonding ($l_d < l_c/2$), and the fiber sliding completely relative to the matrix in the interface debonded region ($y/l_d = z/l_d = 1$);
(2) Case II, the fiber/matrix interface partially debonding ($l_d < l_c/2$), and the fiber sliding partially relative to the matrix in the interface debonded region ($y/l_d = z/l_d < 1$);
(3) Case III, the fiber/matrix interface completely debonding ($l_d = l_c/2$), and the fiber sliding partially relative to the matrix in the interface debonded region ($y/l_d = z/l_d < 1$);
(4) Case IV, the fiber/matrix interface completely debonding ($l_d = l_c/2$), and the fiber sliding completely relative to the matrix in the interface debonded region ($y/l_d = z/l_d = 1$).

The relationships among the fatigue stress–strain hysteresis loops, fatigue hysteresis dissipated energy, fiber/matrix interface slip and the fiber/matrix interface shear stress of unidirectional, 2D cross-ply and woven, 2.5D woven, and 3D braided CMCs are analyzed.

### 3.2.1 Unidirectional Ceramic-Matrix Composites

The fatigue stress–strain hysteresis loops for the fiber/matrix interface slip Case II, III, and IV, the fatigue hysteresis dissipated energy, fiber/matrix interface debonding ratio of $2l_d/l_c$ and fiber/matrix interface sliding ratio of $y/l_d$ versus the fiber/matrix interface shear stress curves of unidirectional C/SiC composite at the fatigue peak stress of $\sigma_{max} = 240$ MPa are shown in Fig. 3.1.

When the fiber/matrix interface shear stress increases from $\tau_i = 0\,\text{MPa}$ to $\tau_i = 6.3\,\text{MPa}$, the fatigue hysteresis dissipated energy increases from zero to the peak value, and then decreases with increasing fiber/matrix interface shear stress, and the fiber/matrix interface debonding ratio is $2l_d/l_c = 1$, and the fiber/matrix interface sliding ratio is $y/l_d = 1$. When the fiber/matrix interface shear stress increases from $\tau_i = 6.3\,\text{MPa}$ to $\tau_i = 11.2\,\text{MPa}$, the fatigue hysteresis dissipated energy decreases with increasing of fiber/matrix interface shear stress, and the fiber/matrix interface debonding ratio is $2l_d/l_c = 1$, and the fiber/matrix interface sliding ratio is $y/l_d < 1$. When the fiber/matrix interface shear stress increases from $\tau_i = 11.2\,\text{MPa}$ to $\tau_i = 50\,\text{MPa}$, the fatigue hysteresis dissipated energy decreases

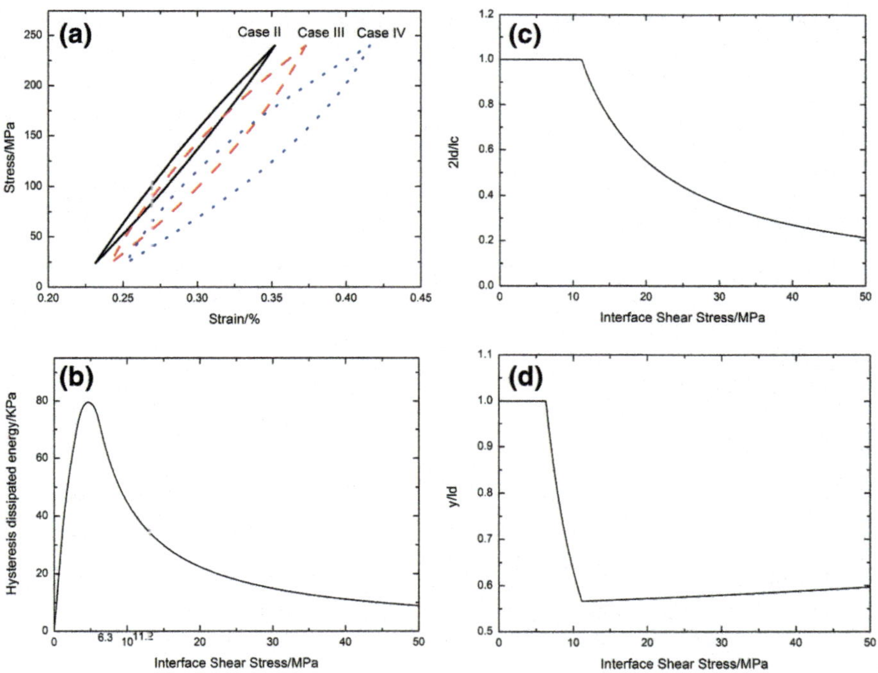


**Fig. 3.1** **a** Fatigue stress–strain hysteresis loops of interface slip Case II, III, and IV; **b** the fatigue hysteresis dissipated energy versus the fiber/matrix interface shear stress; **c** the fiber/matrix interface debonding ratio versus the interface shear stress; and **d** the fiber/matrix interface sliding ratio versus the fiber/matrix interface shear stress of unidirectional C/SiC composite at fatigue peak stress of $\sigma_{max} = 240$ MPa. Reprinted with permission from Ref. [25]. Copyright 2014 Elsevier B.V

with increasing of the fiber/matrix interface shear stress, and the fiber/matrix interface debonding ratio is $2l_d/l_c < 1$, and the fiber/matrix interface sliding ratio is $y/l_d < 1$.

### 3.2.2 Cross-Ply and 2D Woven Ceramic-Matrix Composites

The fatigue stress–strain hysteresis loops for the fiber/matrix interface slip Case II, III, and IV, the fatigue hysteresis dissipated energy, fiber/matrix interface debonding ratio of $2l_d/l_c$ and fiber/matrix interface sliding ratio of $y/l_d$ versus the fiber/matrix interface shear stress curves of cross-ply C/SiC composite for the matrix cracking mode III at the fatigue peak stress of $\sigma_{\max} = 105$ MPa are shown in Fig. 3.2.

When the fiber/matrix interface shear stress increases from $\tau_i = 0$ MPa to $\tau_i = 8.3$ MPa, the fatigue hysteresis dissipated energy increases from zero to the

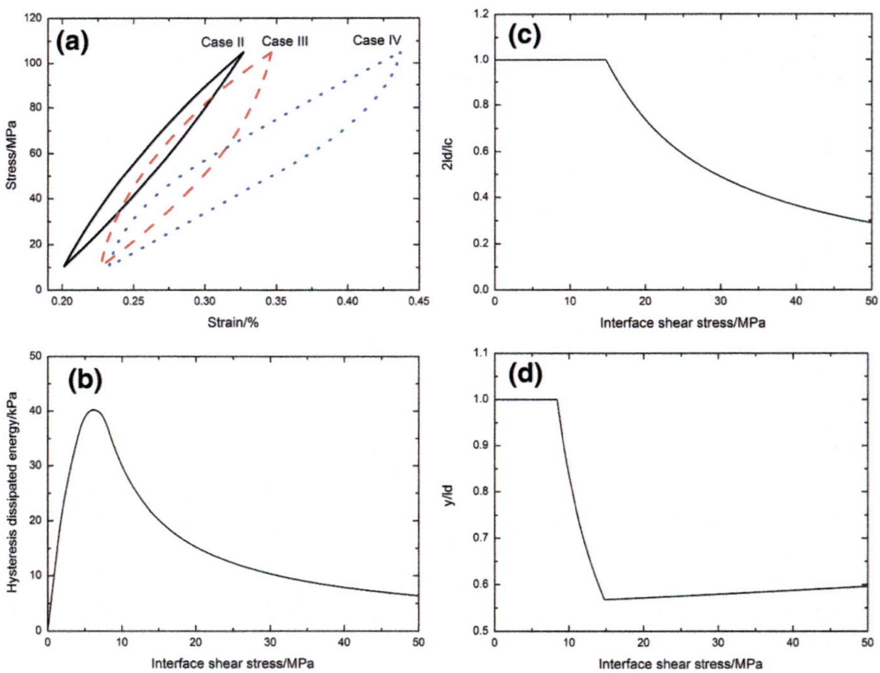


**Fig. 3.2** **a** Fatigue stress–strain hysteresis loops of interface slip Case II, III, and IV; **b** the fatigue hysteresis dissipated energy versus the fiber/matrix interface shear stress; **c** the fiber/matrix interface debonding ratio versus the fiber/matrix interface shear stress; and **d** the fiber/matrix interface sliding ratio versus the fiber/matrix interface shear stress of cross-ply C/SiC composite for the matrix cracking mode III at fatigue peak stress of $\sigma_{\max}$ = 105 MPa. Reprinted with permission from Ref. [25]. Copyright 2014 Elsevier B.V

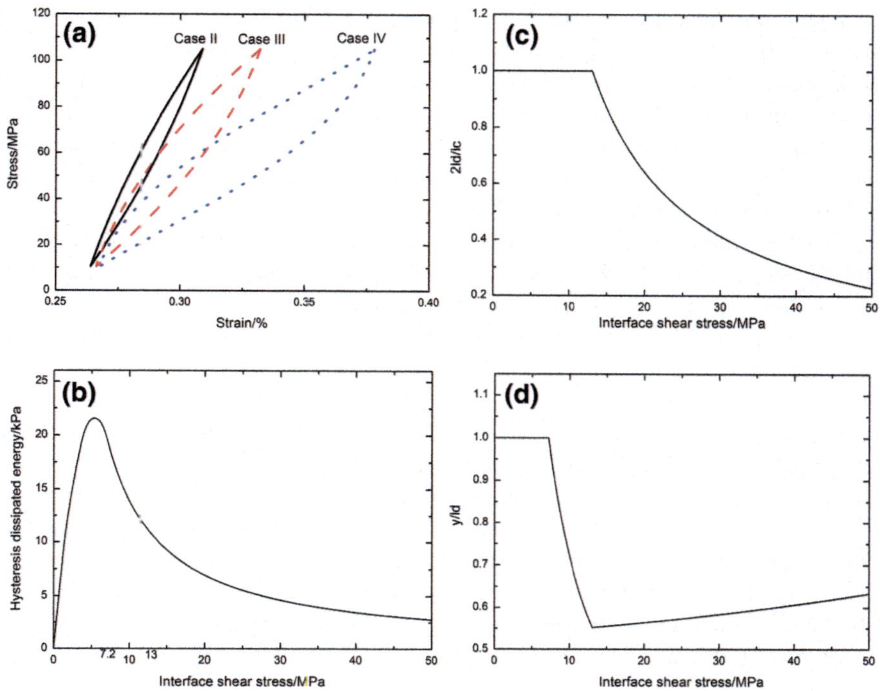


**Fig. 3.3** **a** Fatigue stress–strain hysteresis loops of interface slip Case II, III, and IV; **b** the fatigue hysteresis dissipated energy versus the fiber/matrix interface shear stress; **c** the fiber/matrix interface debonding ratio of $2l_d/l_c$ versus the fiber/matrix interface shear stress; and **d** the fiber/matrix interface sliding ratio of $y/l_d$ versus the fiber/matrix interface shear stress of cross-ply C/SiC composite for matrix cracking mode V at the fatigue peak stress of $\sigma_{max}$ = 105 MPa. Reprinted with permission from Ref. [25]. Copyright 2014 Elsevier B.V

peak value, and then decreases with increasing of fiber/matrix interface shear stress, and the fiber/matrix interface debonding ratio is $2l_d/l_c = 1$, and the fiber/matrix interface sliding ratio is $y/l_d = 1$. When the fiber/matrix interface shear stress increases from $\tau_i = 8.3\,\text{MPa}$ to $\tau_i = 14.7\,\text{MPa}$, the fatigue hysteresis dissipated energy decreases with increasing of fiber/matrix interface shear stress, and the fiber/matrix interface debonding ratio is $2l_d/l_c = 1$, and the fiber/matrix interface sliding ratio is $y/l_d < 1$. When the fiber/matrix interface shear stress increases from $\tau_i = 14.7\,\text{MPa}$ to $\tau_i = 50\,\text{MPa}$, the fatigue hysteresis dissipated energy decreases with increasing of fiber/matrix interface shear stress, and the fiber/matrix interface debonding ratio is $2l_d/l_c < 1$, and the fiber/matrix interface sliding ratio is $y/l_d < 1$.

The fatigue stress–strain hysteresis loops for the fiber/matrix interface slip Case II, III, and IV, the fatigue hysteresis dissipated energy, fiber/matrix interface debonding ratio of $2l_d/l_c$ and fiber/matrix interface sliding ratio of $y/l_d$ versus the fiber/matrix interface shear stress curves of cross-ply C/SiC composite for cracking mode V at the fatigue peak stress of $\sigma_{max} = 105\,\text{MPa}$ are shown in Fig. 3.3.

When the fiber/matrix interface shear stress increases from $\tau_i = 0\,\text{MPa}$ to $\tau_i = 7.2\,\text{MPa}$, the fatigue hysteresis dissipated energy increases from zero to the peak value, and then decreases with increasing of the fiber/matrix interface shear stress, and the fiber/matrix interface debonding ratio is $2l_d/l_c = 1$, and the fiber/matrix interface sliding ratio is $y/l_d = 1$. When the fiber/matrix interface shear stress increases from $\tau_i = 7.2\,\text{MPa}$ to $\tau_i = 13\,\text{MPa}$, the fatigue hysteresis dissipated energy decreases with increasing of the fiber/matrix interface shear stress, and the fiber/matrix interface debonding ratio is $2l_d/l_c = 1$, and the fiber/matrix interface sliding ratio is $y/l_d < 1$. When the fiber/matrix interface shear stress increases from $\tau_i = 13\,\text{MPa}$ to $\tau_i = 50\,\text{MPa}$, the fatigue hysteresis dissipated energy decreases with increasing of the fiber/matrix interface shear stress, and the fiber/matrix interface debonding ratio is $2l_d/l_c < 1$, and the fiber/matrix interface sliding ratio is $y/l_d < 1$.

The fatigue stress–strain hysteresis loops and the fatigue hysteresis dissipated energy versus the fiber/matrix interface shear stress curves of matrix cracking mode III, mode V and composite are shown in Fig. 3.4. The composite fatigue stress–strain hysteresis loops lie between that of the matrix cracking mode III and mode V. For the composite, the fatigue hysteresis dissipated energy increases with fiber/matrix interface shear stress to the peak value, and then decreases with increasing fiber/matrix interface shear stress, and the composite fatigue hysteresis dissipated energy curve also lies between that of matrix cracking mode III and mode V.

### 3.2.3 2.5D Woven Ceramic-Matrix Composites

The evolution of the fatigue stress–strain hysteresis loops, fatigue hysteresis dissipated energy, fiber/matrix interface debonding ratio of $2l_d/l_c$ and the fiber/matrix interface sliding ratio of $y/l_d$ versus the fiber/matrix interface shear stress curves of 2.5D C/SiC composite are shown in Fig. 3.5. The stress–strain hysteresis loops corresponding to the interface slip Case I, II and III are shown in Fig. 3.5a. The area, residual strain, and peak strain of the fatigue stress–strain hysteresis loops depend on the fiber/matrix interface debonding and slip condition between the fiber and the matrix in the matrix cracking space of CMCs. The fatigue hysteresis dissipated energy versus the fiber/matrix interface shear stress curve is shown in Fig. 3.5b. When the fiber/matrix interface shear stress increases, the fatigue hysteresis dissipated energy increases first and then decreases. When the fiber/matrix interface shear stress increases from $\tau_i = 1\,\text{MPa}$ to $\tau_i = 4.8\,\text{MPa}$, the fatigue hysteresis dissipated energy increases to the peak value and then decreases with increasing of the fiber/matrix interface shear stress, corresponding to the fiber/matrix interface debonding ratio of $2l_d/l_c = 1$ and the fiber/matrix interface sliding ratio of $y/l_d = 1$. When the fiber/matrix interface shear stress increases from $\tau_i = 4.8\,\text{MPa}$ to $\tau_i = 10.5\,\text{MPa}$, the fatigue hysteresis dissipated energy decreases with increasing of the fiber/matrix interface shear stress, corresponding to the fiber/matrix interface debonding ratio of $2l_d/l_c = 1$ and the fiber/matrix interface sliding

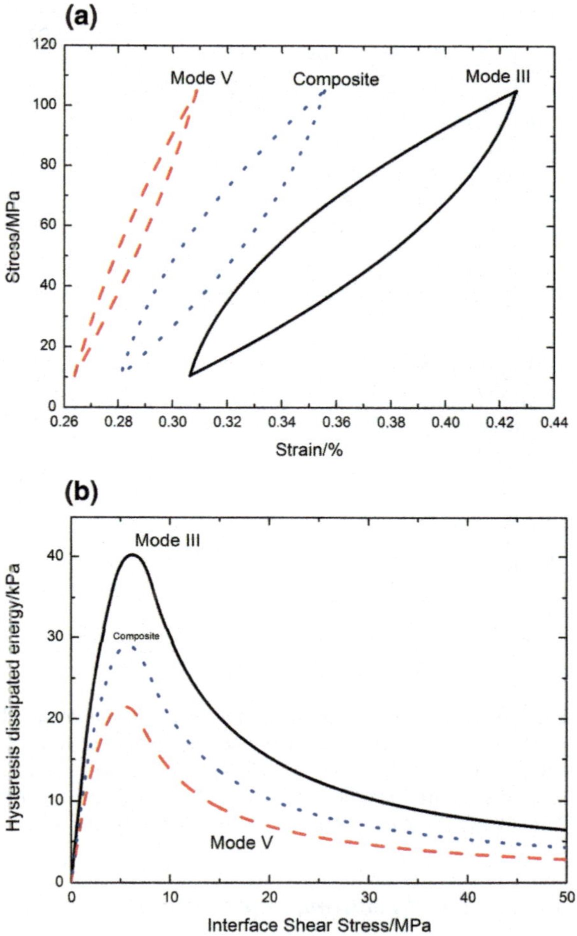


**Fig. 3.4** **a** Fatigue stress–strain hysteresis loops of matrix cracking mode III, mode V and composite; and **b** the fatigue hysteresis dissipated energy versus the fiber/matrix interface shear stress curves for matrix cracking mode III, mode V, and composite. Reprinted with permission from Ref. [25]. Copyright 2014 Elsevier B.V

ratio of $y/l_d < 1$. When the fiber/matrix interface shear stress increases from $\tau_i = 10.5$ MPa to $\tau_i = 50$ MPa, the fatigue hysteresis dissipated energy decreases with increasing of the fiber/matrix interface shear stress, corresponding to the fiber/matrix interface debonding ratio of $2l_d/l_c < 1$ and the fiber/matrix interface sliding ratio of $y/l_d < 1$.

## 3.3 Experimental Comparisons

Comparing the experimental fatigue hysteresis dissipated energy with theoretical values of unidirectional, 2D cross-ply and woven, 2.5D woven and 3D braided fiber-reinforced CMCs, the fiber/matrix interface shear stress can be estimated for different fatigue peak stresses and applied cycles at room and elevated temperatures.

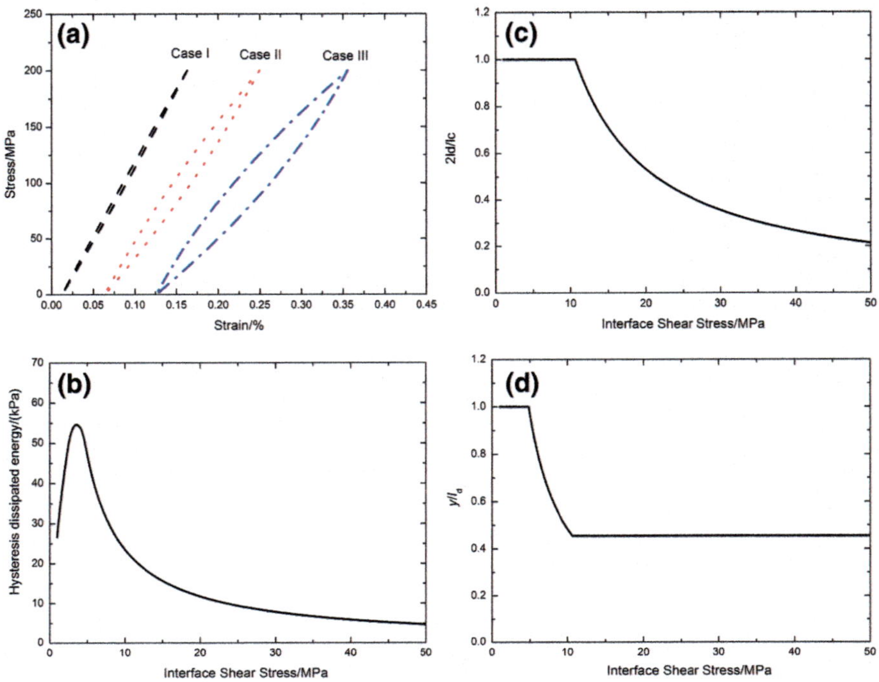


**Fig. 3.5** **a** Fatigue stress–strain hysteresis loops of different fiber/matrix interface slip cases; **b** the fatigue hysteresis dissipated energy versus the fiber/matrix interface shear stress curve; **c** the fiber/matrix interface debonding ratio of $2l_d/l_c$ versus the fiber/matrix interface shear stress curve; and **d** the fiber/matrix interface slip ratio of $y/l_d$ versus the fiber/matrix interface shear stress curve of 2.5D C/SiC composite. Reprinted with permission from Ref. [25]. Copyright 2014 Elsevier B.V

## *3.3.1 Unidirectional Ceramic-Matrix Composites*

### 3.3.1.1 SiC/CAS Composite

Evans et al. [2] investigated the cyclic tension–tension fatigue behavior of unidirectional SiC/CAS composite at room temperature. The experimental fatigue stress–strain hysteresis loops corresponding to the applied cycle number of $N = 1$, 5 and 109 under the fatigue peak stress of $\sigma_{max} = 280$ MPa are shown in Fig. 3.6a. The fatigue hysteresis dissipated energy as a function of the fiber/matrix interface shear stress curve is shown in Fig. 3.6b. With decreasing of the fiber/matrix interface shear stress, the fiber/matrix interface debonding ratio of $2l_d/l_c$ changes from $2l_d/l_c < 1$ to $2l_d/l_c = 1$, and the fiber/matrix interface sliding ratio of $y/l_d$ changes from $y/l_d = 1$ to $y/l_d < 1$, and then $y/l_d = 1$, corresponding to the fiber/matrix interface slip Case I, II, III, and IV. The experimental fatigue hysteresis dissipated energy at $N = 1$, 5, 9 and 109 are 25, 55, 80, and 22 kPa, and the corresponding fiber/matrix interface shear stress are 27, 12, 8 and 1 MPa.

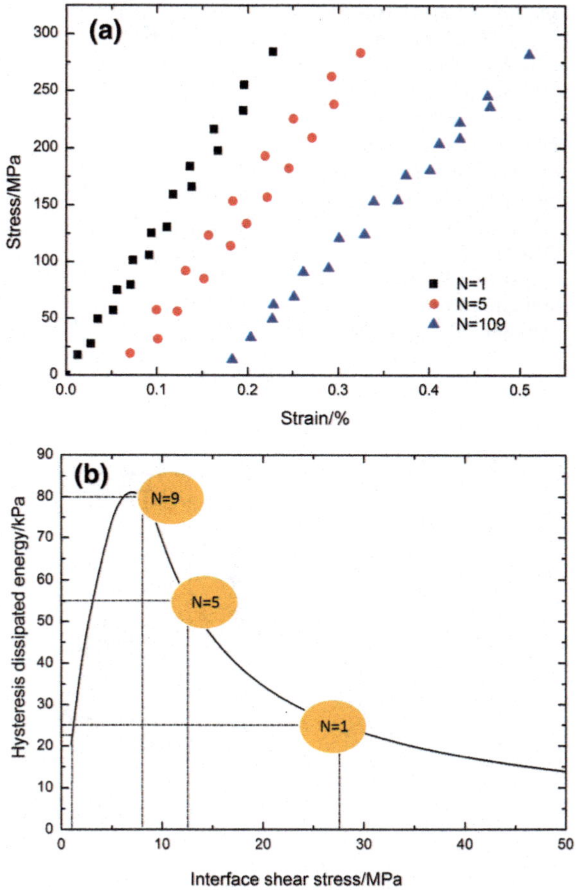


**Fig. 3.6** **a** Experimental fatigue stress–strain hysteresis loops; **b** the theoretical fatigue hysteresis dissipated energy versus the fiber/matrix interface shear stress of unidirectional SiC/CAS composite under the fatigue peak stress of $\sigma_{max}$ = 280 MPa at room temperature. Reprinted with permission from Ref. [25]. Copyright 2014 Elsevier B.V

Holmes and Cho [4] investigated the cyclic tension–tension fatigue behavior of unidirectional SiC/CAS-II composite at room temperature. The experimental fatigue stress–strain hysteresis loops corresponding to the applied cycle number of $N$ = 10, 20 and 3200 under the fatigue peak stress of $\sigma_{max}$ = 180 MPa are shown in Fig. 3.7a. The fatigue hysteresis dissipated energy versus the fiber/matrix interface shear stress curve is shown in Fig. 3.7b. With decreasing of the fiber/matrix interface shear stress, the fiber/matrix interface debonding ratio of $2l_d/l_c$ changes from $2l_d/l_c < 1$ to $2l_d/l_c = 1$, and the fiber/matrix interface sliding ratio of $y/l_d$ changes from $y/l_d = 1$ to $y/l_d < 1$, and then $y/l_d = 1$, corresponding to the fiber/matrix interface slip Case I, II, III, and IV. The experimental fatigue hysteresis dissipated energy at the applied cycle number of $N$ = 10, 20, and 3200 are 9, 16, and 25 kPa, and the corresponding fiber/matrix interface shear stress are 19, 11, and 7 MPa.

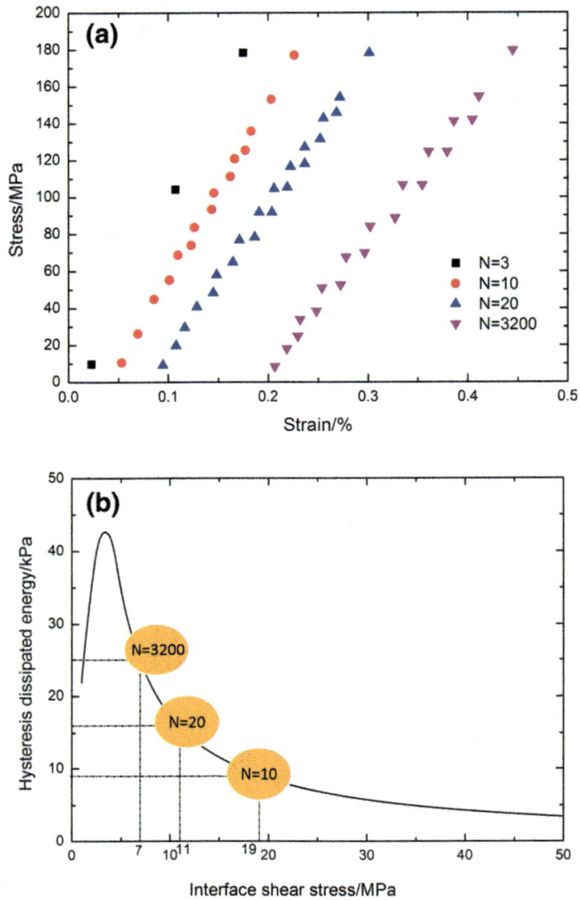


**Fig. 3.7** **a** Experimental fatigue stress–strain hysteresis loops; **b** the theoretical fatigue hysteresis dissipated energy versus the fiber/matrix interface shear stress of unidirectional SiC/CAS-II composite under the fatigue peak stress of $\sigma_{max}$ = 180 MPa MPa at room temperature. Reprinted with permission from Ref. [25]. Copyright 2014 Elsevier B.V

#### 3.3.1.2 SiC/$Si_3N_4$ Composite

The experimental fatigue hysteresis dissipated energy versus the applied cycle curves, the theoretical fatigue hysteresis dissipated energy versus the fiber/matrix interface shear stress, and the estimated fiber/matrix interface shear stress versus applied cycle curves of unidirectional SiC/$Si_3N_4$ composite with strong and weak fiber/matrix interface bonding under the fatigue peak stress of $\sigma_{max}$ = 460 MPa are analyzed.

Strong Interface Bonding

For the strong interface bonding unidirectional SiC/$Si_3N_4$ composite under the fatigue peak stress of $\sigma_{max}$ = 460 MPa, the experimental fatigue hysteresis dissipated energy versus the applied cycles number curve is shown in Fig. 3.8a.

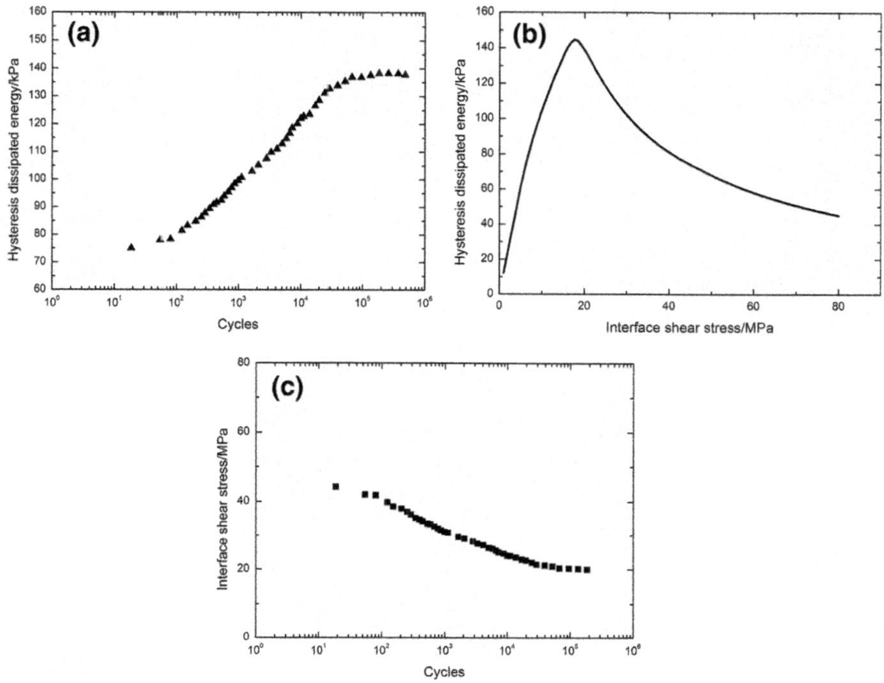


**Fig. 3.8** **a** Experimental fatigue hysteresis dissipated energy versus the applied cycle number; **b** the theoretical fatigue hysteresis dissipated energy versus the fiber/matrix interface shear stress; and **c** the estimated fiber/matrix interface shear stress versus the applied cycle number curve of strong interface bonding $SiC/Si_3N_4$ composite under the fatigue peak stress of $\sigma_{max} = 460$ MPa at room temperature

The theoretical fatigue hysteresis dissipated energy versus the fiber/matrix interface shear stress curve is shown in Fig. 3.8b. With decreasing of the fiber/matrix interface shear stress, the fiber/matrix interface debonding ratio of $2l_d/l_c$ changes from $2l_d/l_c < 1$ to $2l_d/l_c = 1$, and the fiber/matrix interface sliding ratio of $y/l_d$ changes from $y/l_d = 1$ to $y/l_d < 1$, and then $y/l_d = 1$, corresponding to the fiber/matrix interface slip Case I, II, III and IV. The experimental fatigue hysteresis dissipated energy increases from 75 kPa at the applied cycle number of $N = 18$–138 kPa at the applied cycle number of $N = 183{,}382$. The fiber/matrix interface shear stress decreases from $\tau_i = 44$ MPa at the applied cycle number of $N = 18$ to $\tau_i = 20$ MPa at the applied cycle number of $N = 183{,}382$. The estimated fiber/matrix interface shear stress versus cycle number is shown in Fig. 3.8c.

## Weak Interface Bonding

For the weak interface bonding $SiC/Si_3N_4$ composite under the fatigue peak stress of $\sigma_{max} = 460$ MPa, the experimental fatigue hysteresis dissipated energy versus

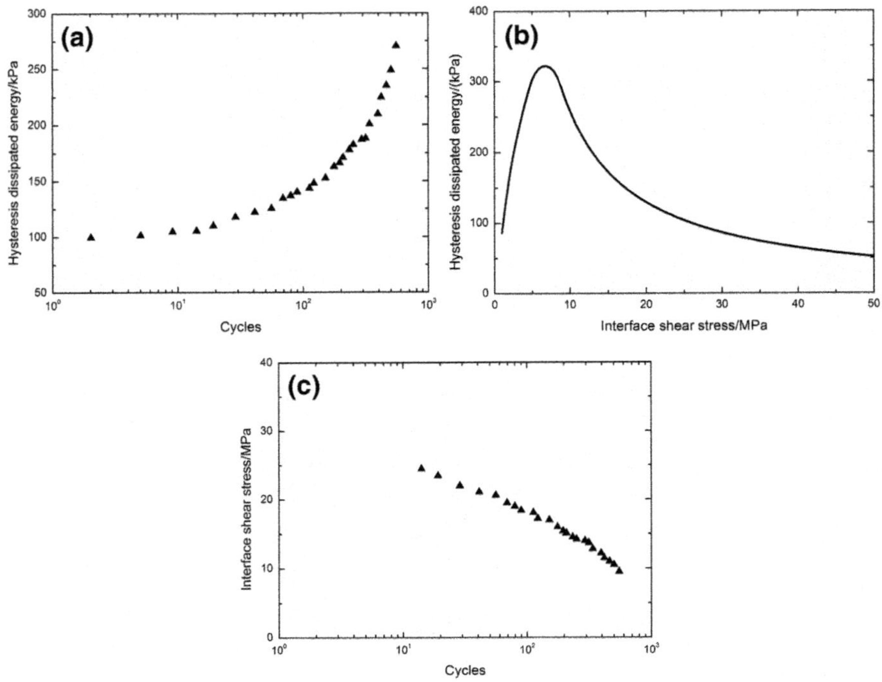


**Fig. 3.9** **a** Experimental fatigue hysteresis dissipated energy versus the applied cycle number; **b** the theoretical fatigue hysteresis dissipated energy versus the fiber/matrix interface shear stress; and **c** the estimated fiber/matrix interface shear stress versus the applied cycle number of weak interface bonding SiC/$Si_3N_4$ composite under the fatigue peak stress of $\sigma_{max}$ = 460 MPa at room temperature

the applied cycles is shown in Fig. 3.9a. The theoretical fatigue hysteresis dissipated energy versus the fiber/matrix interface shear stress curve is shown in Fig. 3.9b. With decreasing fiber/matrix interface shear stress, the fiber/matrix interface debonding ratio of $2l_d/l_c$ changes from $2l_d/l_c < 1$ to $2l_d/l_c = 1$, and the fiber/matrix interface sliding ratio of $y/l_d$ changes from $y/l_d = 1$ to $y/l_d < 1$, and then $y/l_d = 1$, corresponding to the fiber/matrix interface slip Case I, II, III and IV. The experimental fatigue hysteresis dissipated energy increases from 105 kPa at the applied cycle number of $N$ = 14–270 kPa at the applied cycle number of $N$ = 552. The fiber/matrix interface shear stress decreases from $\tau_i$ = 24 MPa at the applied cycle number of $N$ = 14 to $\tau_i$ = 9.5 MPa at the applied cycle number of $N$ = 552. The estimated fiber/matrix interface shear stress versus cycle number is shown in Fig. 3.9c.

#### 3.3.1.3 C/SiC Composite

Room Temperature

When the fatigue peak stress is $\sigma_{max} = 240$ MPa, the experimental fatigue stress–strain hysteresis loops for the cycle number of $N = 1$, 10, 10,000, 100,000, and 1,000,000 are shown in Fig. 3.10a. The experimental fatigue hysteresis dissipated energy for the cycle number of $N = 1$, 10, 10,000, 100,000, and 1,000,000 are 56, 45, 30, 12.7, 12, 10.4, 8.8, and 8 kPa, respectively. The theoretical fatigue hysteresis dissipated energy as a function of the fiber/matrix interface shear stress is shown in Fig. 3.10b. With decreasing fiber/matrix interface shear stress, the fiber/matrix interface debonding ratio of $2l_d/l_c$ changes from $2l_d/l_c < 1$ to $2l_d/l_c = 1$, and the fiber/matrix interface sliding ratio of $y/l_d$ changes from $y/l_d = 1$ to $y/l_d < 1$, and then $y/l_d = 1$, corresponding to the fiber/matrix interface slip Case I, II, III and IV.

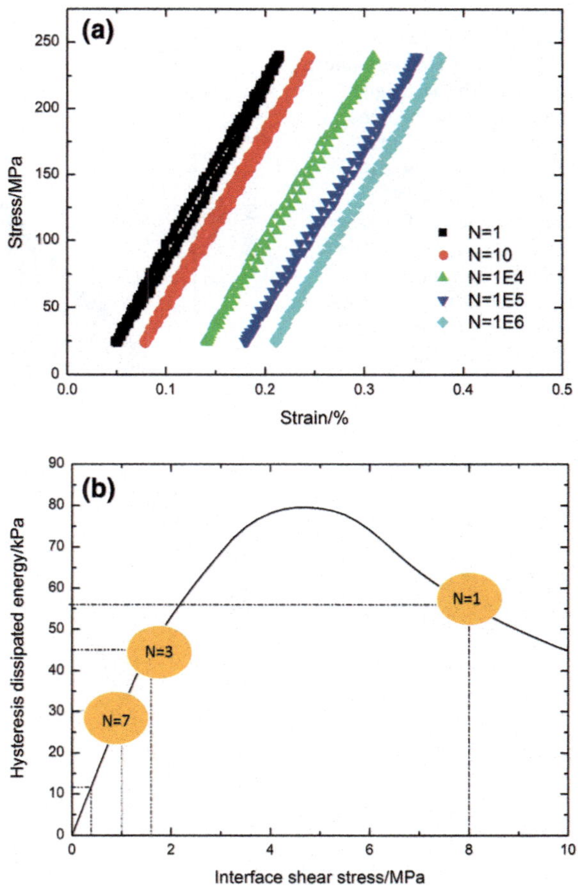


**Fig. 3.10** **a** Experimental fatigue stress–strain hysteresis loops; and **b** the fatigue hysteresis dissipated energy as a function of the fiber/matrix interface shear stress of unidirectional C/SiC composite under fatigue peak stress of $\sigma_{max} = 240$ MPa at room temperature. Reprinted with permission from Ref. [25]. Copyright 2014 Elsevier B.V

The estimated fiber/matrix interface shear stress for the applied cycle number of $N = 1$, 10, 10,000, 100,000, and 1,000,000 are 8, 1.5, 1.0, 0.4, 0.38, 0.35, 0.32, and 0.3 MPa, respectively.

Elevated Temperature

When the fatigue peak stress is $\sigma_{max} = 250\,\text{MPa}$, the experimental fatigue stress–strain hysteresis loops corresponding to the applied cycle number of $N = 1$, 1000, 5000, 10,000, 15,000, 20,000, and 24,000 are shown in Fig. 3.11a. The experimental fatigue hysteresis dissipated energy for the applied cycle number of $N = 1$, 1000, 5000, 10,000, 15,000, 20,000, and 24,000 are 62, 50, 24, 16, 12, 8.0, 7.8, and 7.2 kPa, respectively. The theoretical fatigue hysteresis dissipated energy as a function of the fiber/matrix interface shear stress is shown in Fig. 3.11b. With decreasing of the fiber/matrix interface shear stress, the fiber/matrix interface

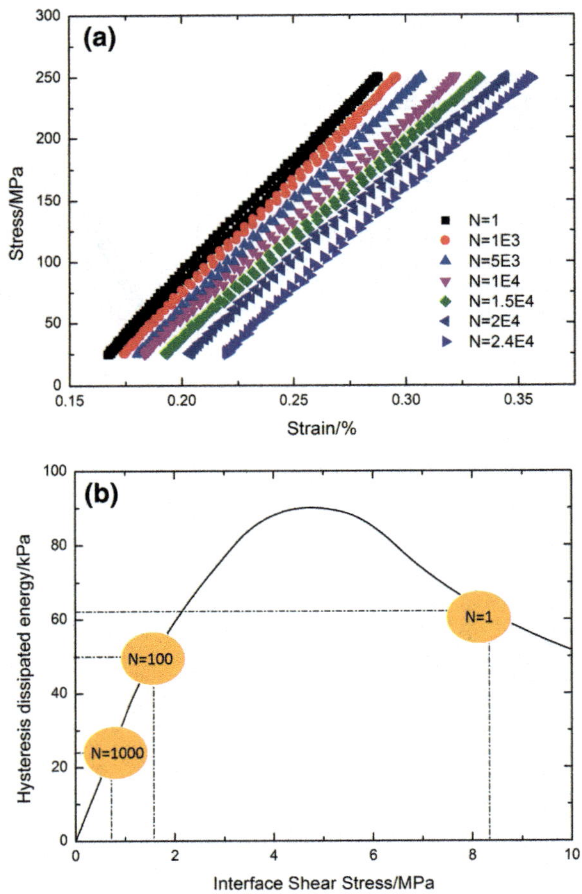


**Fig. 3.11** **a** Experimental fatigue stress–strain hysteresis loops; and **b** the fatigue hysteresis dissipated energy versus the fiber/matrix interface shear stress of unidirectional C/SiC composite under fatigue peak stress of $\sigma_{max} = 250\,\text{MPa}$ at 800 °C in air atmosphere. Reprinted with permission from Ref. [25]. Copyright 2014 Elsevier B.V

debonding ratio of $2l_d/l_c$ changes from $2l_d/l_c < 1$ to $2l_d/l_c = 1$, and the fiber/matrix interface sliding ratio of $y/l_d$ changes from $y/l_d = 1$ to $y/l_d < 1$, and then $y/l_d = 1$, corresponding to the fiber/matrix interface slip Case I, II, III, and IV. The estimated fiber/matrix interface shear stress of $N = 1$, 1000, 5000, 10,000, 15,000, 20,000, and 24,000 are 8.3, 1.5, 0.7, 0.47, 0.35, 0.24, 0.23, and 0.21 MPa, respectively.

## 3.3.2 *Cross-Ply and 2D Woven Ceramic-Matrix Composites*

The experimental fatigue stress–strain hysteresis loops for different applied cycles, the theoretical fatigue hysteresis dissipated energy versus the fiber/matrix interface shear stress, and the fiber/matrix interface shear stress versus applied cycles of cross-ply SiC/CAS and C/SiC, 2D woven C/SiC and SiC/SiC composites under different fatigue peak stresses at room and elevated temperatures are analyzed.

### 3.3.2.1 Cross-Ply SiC/CAS Composite

Room Temperature

The experimental fatigue hysteresis dissipated energy versus applied cycle number curves at the fatigue peak stresses of $\sigma_{max} = 140$, 160, and 180 MPa of cross-ply SiC/CAS composite are shown in Fig. 3.12. When the fatigue peak stress is $\sigma_{max} = 140$ MPa, the experimental fatigue hysteresis dissipated energy decreases from 47 kPa at the applied cycle number of $N = 1$–12 kPa at the applied cycle number of $N = 1{,}023{,}410$, and the fiber/matrix interface shear stress decreases from $\tau_i = 12.5$ MPa at the applied cycle number of $N = 1$ to $\tau_i = 2$ MPa at the applied cycle number of $N = 1{,}023{,}410$; when the fatigue peak stress is $\sigma_{max} = 160$ MPa, the experimental fatigue hysteresis dissipated energy decreases from 67 kPa at the applied cycle number of $N = 1$–18 kPa at the applied cycle number of $N = 19{,}426$, and the fiber/matrix interface shear stress decreases from $\tau_i = 12.8$ MPa at the applied cycle number of $N = 1$ to $\tau_i = 2.3$ MPa at the applied cycle number of $N = 19{,}426$; and when the fatigue peak stress is $\sigma_{max} = 180$ MPa, the experimental fatigue hysteresis dissipated energy decreases from 127 kPa at the applied cycle number of $N = 1$–40 kPa at the applied cycle number of $N = 6$, and the fiber/matrix interface shear stress decreases from $\tau_i = 14.7$ MPa at the applied cycle number of $N = 1$ to $\tau_i = 3.2$ MPa at the applied cycle number of $N = 6$.

700 °C in Air Condition

The experimental fatigue hysteresis dissipated energy versus the applied cycle number curves at the fatigue peak stresses of $\sigma_{max} = 69$, 83, and 110 MPa of cross-ply SiC/CAS composite are shown in Fig. 3.13. When the fatigue peak stress

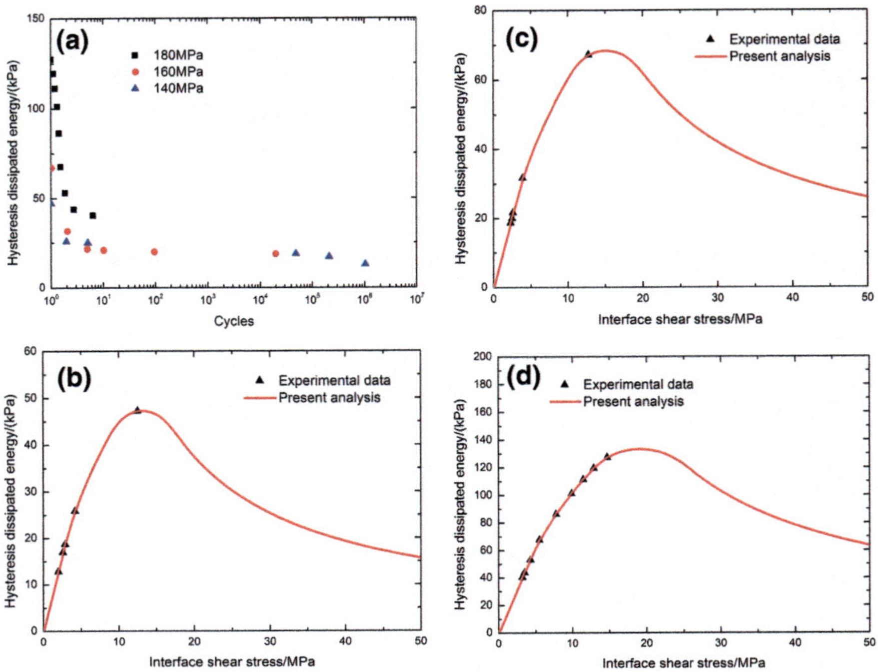


**Fig. 3.12** **a** Experimental fatigue hysteresis dissipated energy versus the applied cycle number curves when the fatigue peak stresses are $\sigma_{max}$ = 140, 160 and 180 MPa; **b** the experimental and theoretical fatigue hysteresis dissipated energy versus the fiber/matrix interface shear stress curves when the fatigue peak stress is $\sigma_{max}$ = 140 MPa; **c** the experimental and theoretical fatigue hysteresis dissipated energy versus the fiber/matrix interface shear stress curves when the fatigue peak stress is $\sigma_{max}$ = 160 MPa; and **d** the experimental and theoretical fatigue hysteresis dissipated energy versus the fiber/matrix interface shear stress curves when the fatigue peak stress is $\sigma_{max}$ = 180 MPa of cross-ply SiC/CAS composite at room temperature. Reprinted with permission from Australian Ceramic Society [23]

is $\sigma_{max}$ = 69 MPa, the experimental fatigue hysteresis dissipated energy decreases from 13 kPa at the applied cycle number of $N$ = 1–4.2 kPa at the applied cycle number of $N$ = 989,602, and the fiber/matrix interface shear stress decreases from $\tau_i$ = 9.8 MPa at the applied cycle number of $N$ = 1 to $\tau_i$ = 2 MPa at the applied cycle number of $N$ = 989,602; when the fatigue peak stress is $\sigma_{max}$ = 83 MPa, the experimental fatigue hysteresis dissipated energy decreases from 14.5 kPa at the applied cycle number of $N$ = 1–7.7 kPa at the applied cycle number of $N$ = 1953, and the fiber/matrix interface shear stress decreases from $\tau_i$ = 8.6 MPa at the applied cycle number of $N$ = 1 to $\tau_i$ = 3.2 MPa at the applied cycle number of $N$ = 1953; and when the fatigue peak stress is $\sigma_{max}$ = 110 MPa, the experimental fatigue hysteresis dissipated energy decreases from 28.5 kPa at the applied cycle number of $N$ = 1–16.1 kPa at the applied cycle number of $N$ = 10, and the fiber/matrix interface shear stress decreases from $\tau_i$ = 9.9 MPa at the applied cycle number of $N$ = 1 to $\tau_i$ = 3.6 MPa at the applied cycle number of $N$ = 10.

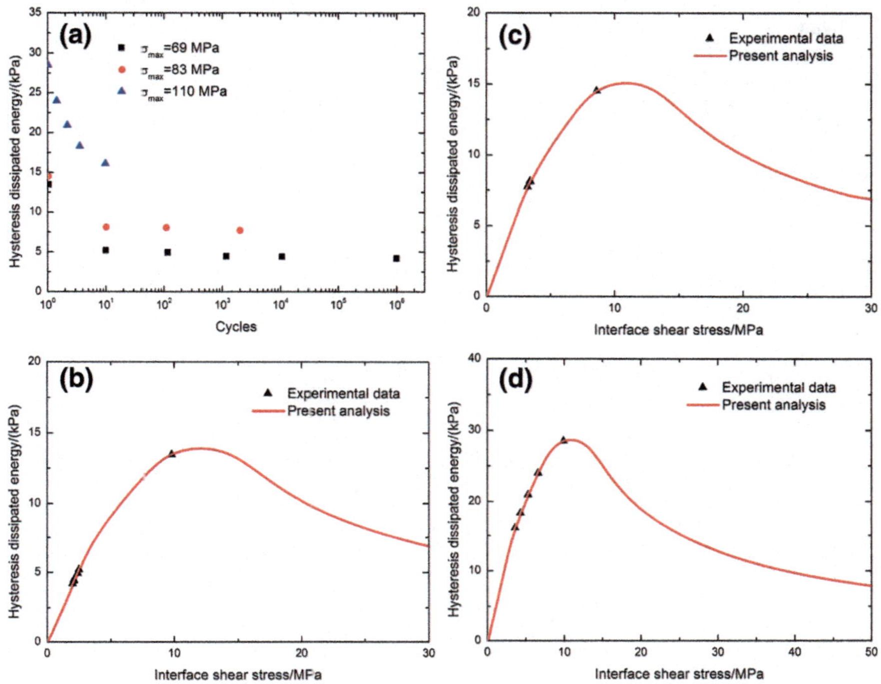


**Fig. 3.13** **a** Experimental fatigue hysteresis dissipated energy versus the applied cycle number curves when the fatigue peak stresses are $\sigma_{max}$ = 69, 83 and 110 MPa; **b** the experimental and theoretical fatigue hysteresis dissipated energy versus the fiber/matrix interface shear stress curves when the fatigue peak stress is $\sigma_{max}$ = 69 MPa; **c** the experimental and theoretical fatigue hysteresis dissipated energy versus the fiber/matrix interface shear stress curves when the fatigue peak stress is $\sigma_{max}$ = 83 MPa; and **d** the experimental and theoretical fatigue hysteresis dissipated energy versus the fiber/matrix interface shear stress curves when the fatigue peak stress is $\sigma_{max}$ = 110 MPa of cross-ply SiC/CAS composite at 700 °C in air condition. Reprinted with permission from Australian Ceramic Society [23]

### 850 °C in air condition

The experimental fatigue hysteresis dissipated energy versus the applied cycle number curves at the fatigue peak stresses of $\sigma_{max}$ = 52 and 55 MPa of cross-ply SiC/CAS composite are shown in Fig. 3.14. When the fatigue peak stress is $\sigma_{max}$ = 52 MPa, the experimental fatigue hysteresis dissipated energy increases from 1.3 kPa at the applied cycle number of $N = 1$ to the peak value of 2.9 kPa at the applied cycle number of $N = 38$, and then decreases to 2.2 kPa at the applied cycle number of $N = 929{,}854$, and the fiber/matrix interface shear stress decreases from $\tau_i = 11.6$ MPa at the applied cycle number of $N = 1$ to $\tau_i = 2.2$ MPa at the applied cycle number of $N = 929{,}854$; when the fatigue peak stress is $\sigma_{max}$ = 55 MPa, the experimental fatigue hysteresis dissipated energy increases from 2.9 kPa at the applied cycle number of $N = 1$–3.3 kPa at the applied cycle

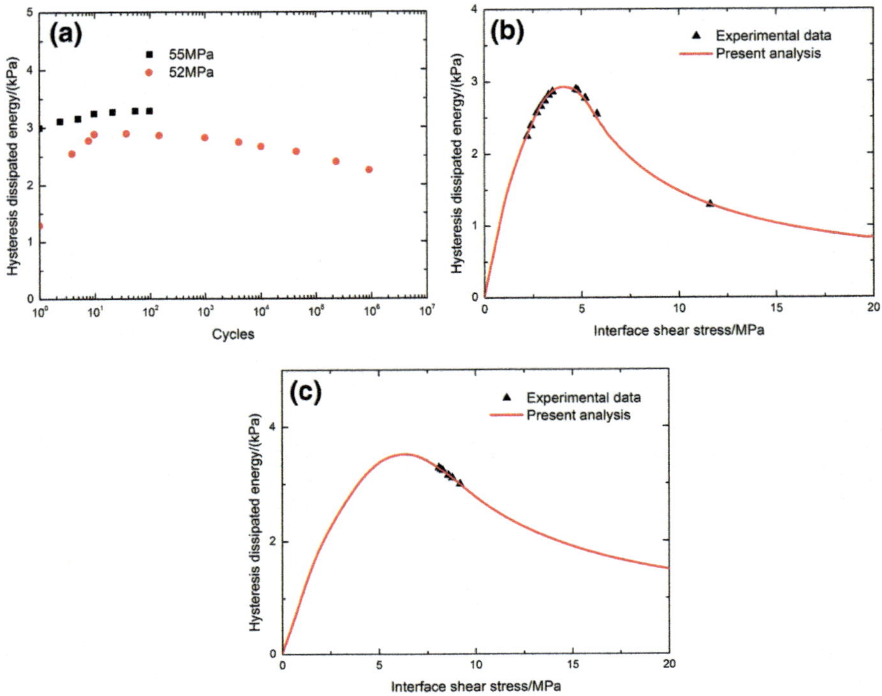


**Fig. 3.14** **a** Experimental fatigue hysteresis dissipated energy versus the applied cycle number curves when the fatigue peak stresses are $\sigma_{max} = 52$ and 55 MPa; **b** the experimental and theoretical fatigue hysteresis dissipated energy versus the fiber/matrix interface shear stress curves when the fatigue peak stress is $\sigma_{max} = 52$ MPa; and **c** the experimental and theoretical fatigue hysteresis dissipated energy versus interface shear stress curves when the fatigue peak stress is $\sigma_{max} = 55$ MPa of cross-ply SiC/CAS composite at 850 °C in air condition. Reprinted with permission from Australian Ceramic Society [23]

number of $N = 98$, and the fiber/matrix interface shear stress decreases from $\tau_i = 9.2$ MPa at the applied cycle number of $N = 1$ to $\tau_i = 8.1$ MPa at the applied cycle number of $N = 98$.

#### 3.3.2.2 Cross-Ply C/SiC Composite

The experimental fatigue stress–strain hysteresis loops for different applied cycles, the theoretical fatigue hysteresis dissipated energy versus the fiber/matrix interface shear stress, and the fiber/matrix interface shear stress versus applied cycles of cross-ply C/SiC composite under the fatigue peak stress of $\sigma_{max} = 105$ MPa at room temperature and 800 °C in air condition are analyzed.

## Room Temperature

At room temperature, when the fatigue peak stress is $\sigma_{max} = 105\,\text{MPa}$, the experimental fatigue stress–strain hysteresis loops at the applied cycle number of $N = 4000$, 10,000, 100,000, and 1,000,000 are shown in Fig. 3.15a. The experimental fatigue hysteresis dissipated energy are 35, 32, 28, 26, 21.5, 19.4, 18.2, 16.9, 12.8, and 10.7 kPa for the applied cycle number of $N = 1, 3, 5, 7, 100, 1000, 4000$, 10,000, 100,000, and 1,000,000, respectively. The theoretical fatigue hysteresis dissipated energy versus the fiber/matrix interface shear stress curve is shown in Fig. 3.15b. With decreasing of the fiber/matrix interface shear stress, the fiber/matrix interface debonding ratio of $2l_d/l_c$ changes from $2l_d/l_c < 1$ to $2l_d/l_c = 1$, and the fiber/matrix interface sliding ratio of $y/l_d$ changes from $y/l_d = 1$ to $y/l_d < 1$, and then $y/l_d = 1$, corresponding to the fiber/matrix interface slip Case I, II, III and IV. The estimated fiber/matrix interface shear stress for the applied cycle number of $N = 1, 3, 5, 7, 100, 1000, 4000$, 10,000, 100,000, and 1,000,000 are 7.3, 4, 3.2, 2.8, 2.1, 1.9, 1.8, 1.6, 1.2, and 1 MPa, respectively, as shown in Fig. 3.15c.

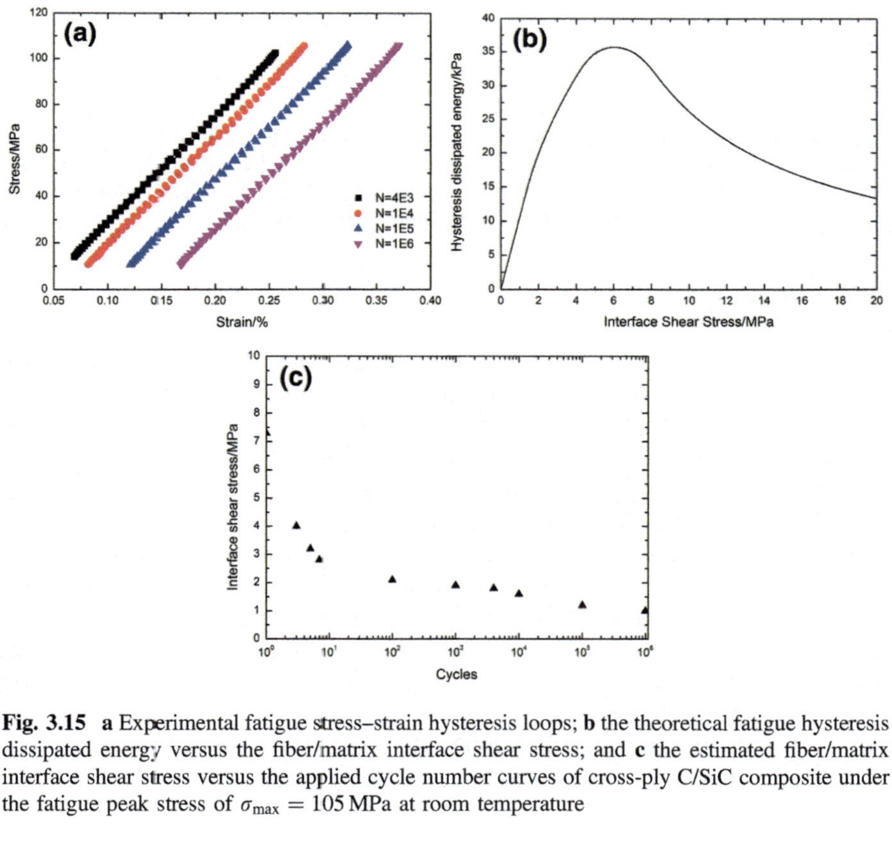


**Fig. 3.15** **a** Experimental fatigue stress–strain hysteresis loops; **b** the theoretical fatigue hysteresis dissipated energy versus the fiber/matrix interface shear stress; and **c** the estimated fiber/matrix interface shear stress versus the applied cycle number curves of cross-ply C/SiC composite under the fatigue peak stress of $\sigma_{max} = 105\,\text{MPa}$ at room temperature

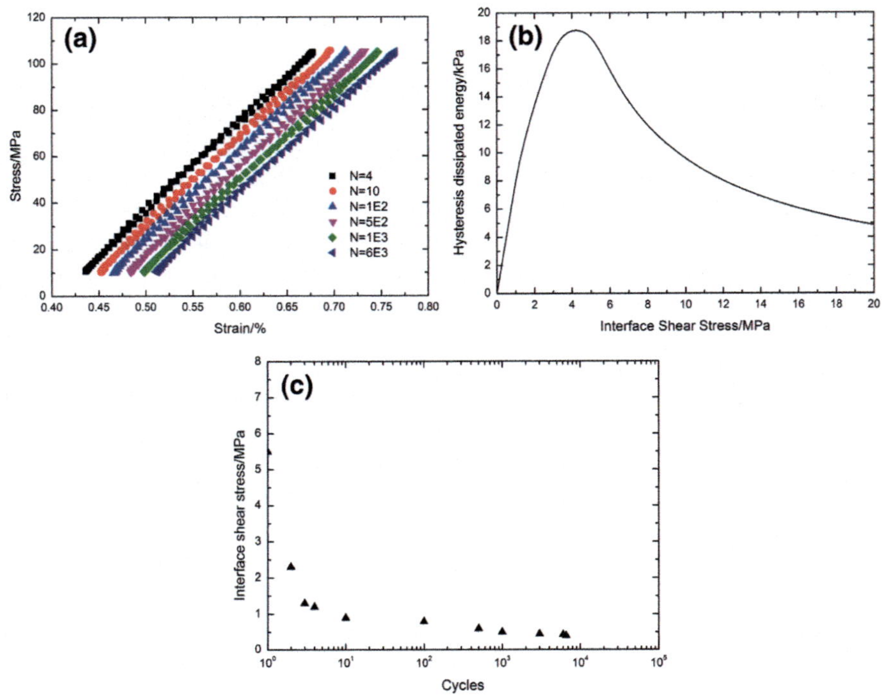


**Fig. 3.16** **a** Experimental fatigue stress–strain hysteresis loops; **b** the theoretical fatigue hysteresis dissipated energy versus the fiber/matrix interface shear stress; and **c** the estimated fiber/matrix interface shear stress versus cycle number curves of cross-ply C/SiC composite under the fatigue peak stress of $\sigma_{\max} = 105$ MPa at 800 °C in air atmosphere

800 °C in Air Condition

At elevated temperature of 800 °C in air condition, when the fatigue peak stress is $\sigma_{\max} = 105$ MPa, the experimental fatigue stress–strain hysteresis loops at the cycle number of $N = 4$, 10, 100, 500, 1000 and 6000 are shown in Fig. 3.16a. The experimental fatigue hysteresis dissipated energy are 24.3, 20, 13, 12, 9.7, 8.6, 7.1, 6.1, 5.4, 5.2, and 5.1 kPa corresponding to the applied cycle number of $N = 1$, 2, 3, 4, 10, 100, 500, 1000, 3000, 6000, and 6600, respectively. The theoretical fatigue hysteresis dissipated energy versus the fiber/matrix interface shear stress curve is shown in Fig. 3.16b. With decreasing fiber/matrix interface shear stress, the fiber/matrix interface debonding ratio of $2l_d/l_c$ changes from $2l_d/l_c < 1$ to $2l_d/l_c = 1$, and the fiber/matrix interface sliding ratio of $y/l_d$ changes from $y/l_d = 1$ to $y/l_d < 1$, and then $y/l_d = 1$, corresponding to the fiber/matrix interface slip Case I, II, III, and IV. The estimated fiber/matrix interface shear stress for the applied cycle number of $N = 1$, 2, 3, 4, 10, 100, 500, 1000, 3000, 6000, and 6600 are 5.5, 2.3, 1.3, 1.2, 0.9, 0.8, 0.6, 0.5, 0.45, 0.43, and 0.4 MPa, respectively, as shown in Fig. 3.16c.

#### 3.3.2.3 2D Woven C/SiC Composite

The experimental fatigue hysteresis dissipated energy versus the applied cycle number curves, the theoretical fatigue hysteresis dissipated energy versus the fiber/matrix interface shear stress curve, and the estimated fiber/matrix interface shear stress versus the applied cycle number curve of 2D woven C/SiC composite under different peak stresses at room temperature, 550 °C in air condition and 1200 °C in vacuum condition are analyzed.

Room Temperature

Shuler et al. [28] investigated the tension–tension cyclic fatigue behavior of 2D C/SiC composite at room temperature. When the fatigue peak stress is $\sigma_{\max} = 335$ MPa, the experimental fatigue hysteresis dissipated energy versus the applied cycle number curves is shown in Fig. 3.17a. The experimental fatigue hysteresis dissipated energy for the applied cycle number of $N = 1$, 1000, and 1,000,000 are 28.2, 20.3, and 17.1 kPa. The theoretical fatigue hysteresis dissipated energy versus the fiber/matrix

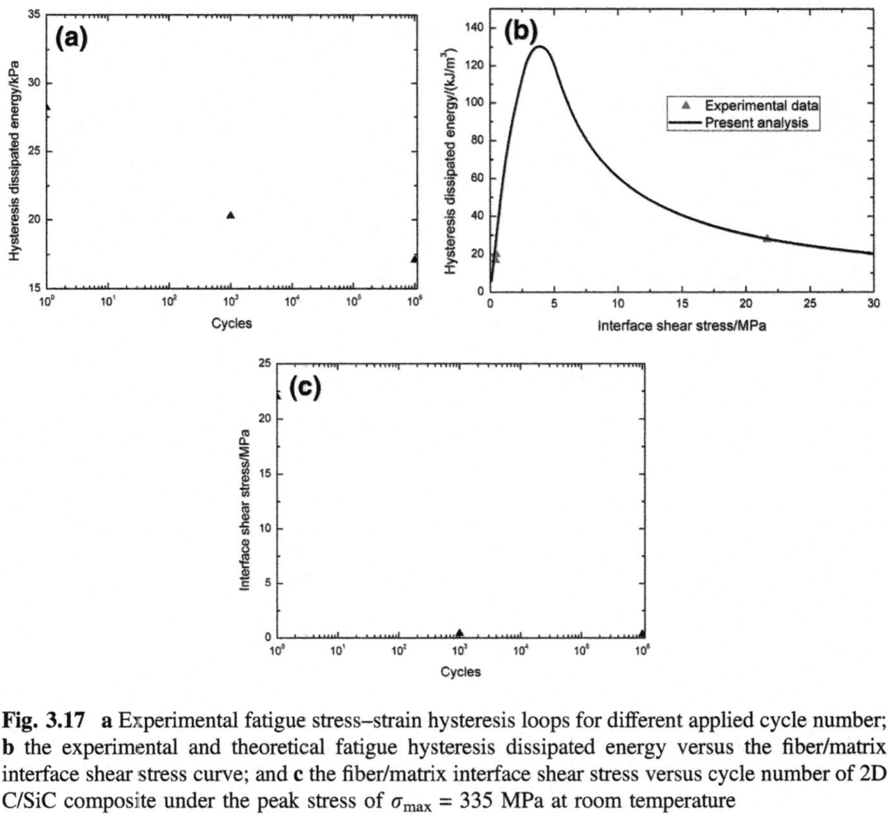


**Fig. 3.17** **a** Experimental fatigue stress–strain hysteresis loops for different applied cycle number; **b** the experimental and theoretical fatigue hysteresis dissipated energy versus the fiber/matrix interface shear stress curve; and **c** the fiber/matrix interface shear stress versus cycle number of 2D C/SiC composite under the peak stress of $\sigma_{\max} = 335$ MPa at room temperature

interface shear stress curve of 2D C/SiC composite at the peak stress of $\sigma_{max}$ = 335 MPa is shown in Fig. 3.17b. With decreasing fiber/matrix interface shear stress, the fiber/matrix interface debonding ratio of $2l_d/l_c$ changes from $2l_d/l_c < 1$ to $2l_d/l_c = 1$, and the fiber/matrix interface sliding ratio of $y/l_d$ changes from $y/l_d = 1$ to $y/l_d < 1$, and then $y/l_d = 1$, corresponding to the fiber/matrix interface slip Case I, II, III, and IV. The estimated fiber/matrix interface shear stress versus applied cycle number curves is shown in Fig. 3.17c. The estimated fiber/matrix interface shear stress at the cycle number of $N$ = 1, 1000, and 1,000,000 are 22, 0.4 and 0.35 MPa, respectively.

Staehler et al. [29] investigated the tension–tension cyclic fatigue behavior of 2D woven C/SiC composite at room temperature. When the fatigue peak stress is $\sigma_{max}$ = 387 MPa, the experimental fatigue hysteresis dissipated energy versus the applied cycle number curve is shown in Fig. 3.18a. The experimental fatigue hysteresis dissipated energy corresponding to the cycle number of $N$ = 100, 22,000, 45,000, 95,000, 300,000, 720,000, and 1,000,000 are 50, 40, 34, 32, 28, 26, and 24 kPa. The theoretical fatigue hysteresis dissipated energy versus the fiber/matrix interface shear stress is shown in Fig. 3.18b. With decreasing of the fiber/matrix interface shear stress, the fiber/matrix interface debonding ratio of $2l_d/l_c$ changes

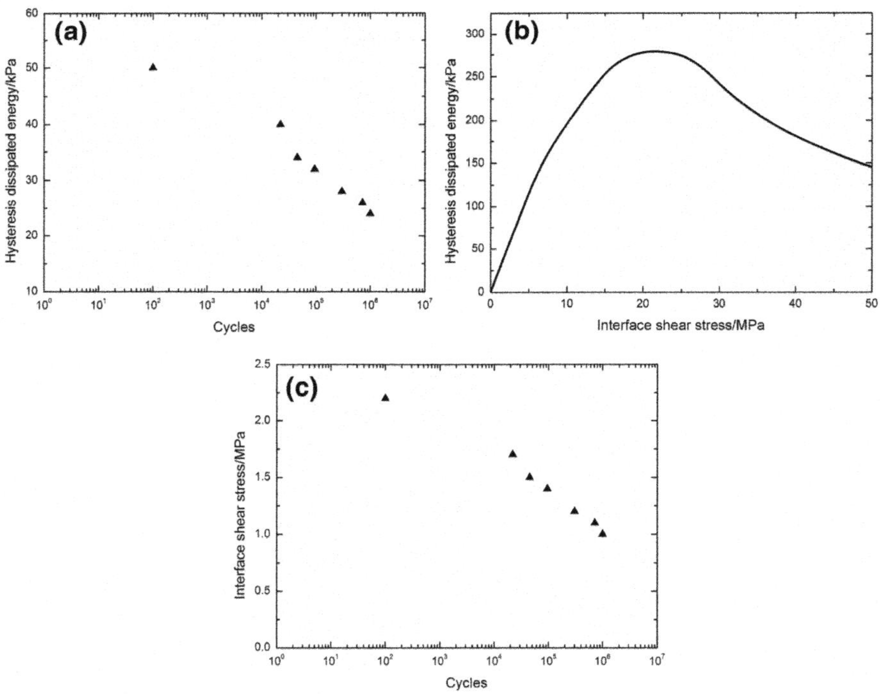


**Fig. 3.18** **a** Experimental fatigue hysteresis dissipated energy versus the applied cycle number; **b** the theoretical fatigue hysteresis dissipated energy versus the fiber/matrix interface shear stress; and **c** the fiber/matrix interface shear stress versus the applied cycle number curve of 2D woven C/SiC composite under the fatigue peak stress of $\sigma_{max}$ = 387 MPa at room temperature

from $2l_d/l_c < 1$ to $2l_d/l_c = 1$, and the fiber/matrix interface sliding ratio of $y/l_d$ changes from $y/l_d = 1$ to $y/l_d < 1$, and then $y/l_d = 1$, corresponding to the fiber/matrix interface slip Case I, II, III, and IV. The estimated fiber/matrix interface shear stress versus applied cycle number curve is shown in Fig. 3.18c. The estimated fiber/matrix interface shear stress corresponding to the applied cycle number of $N$ = 100, 22,000, 45,000, 95,000, 300,000, 720,000, and 1,000,000 are 2.2, 1.7, 1.5, 1.4, 1.2, 1.1, and 1 MPa, respectively.

When the fatigue peak stress is $\sigma_{max} = 425$ MPa, the experimental fatigue hysteresis dissipated energy versus the applied cycle number curve is shown in Fig. 3.19a. The experimental fatigue hysteresis dissipated energy corresponding to the applied cycle number of $N$ = 2, 10, 50, 200, 1000, 5000, 20,000, 60,000, 120,000, and 1,000,000 are 117, 70, 56, 50, 42, 38, 30, 26, 25, and 24 kPa. The theoretical fatigue hysteresis dissipated energy versus the fiber/matrix interface shear stress is shown in Fig. 3.19b. With decreasing of the fiber/matrix interface shear stress, the fiber/matrix interface debonding ratio of $2l_d/l_c$ changes from $2l_d/l_c < 1$ to $2l_d/l_c = 1$, and the fiber/matrix interface sliding ratio of $y/l_d$ changes from $y/l_d = 1$ to $y/l_d < 1$, and then $y/l_d = 1$, corresponding to the fiber/matrix interface

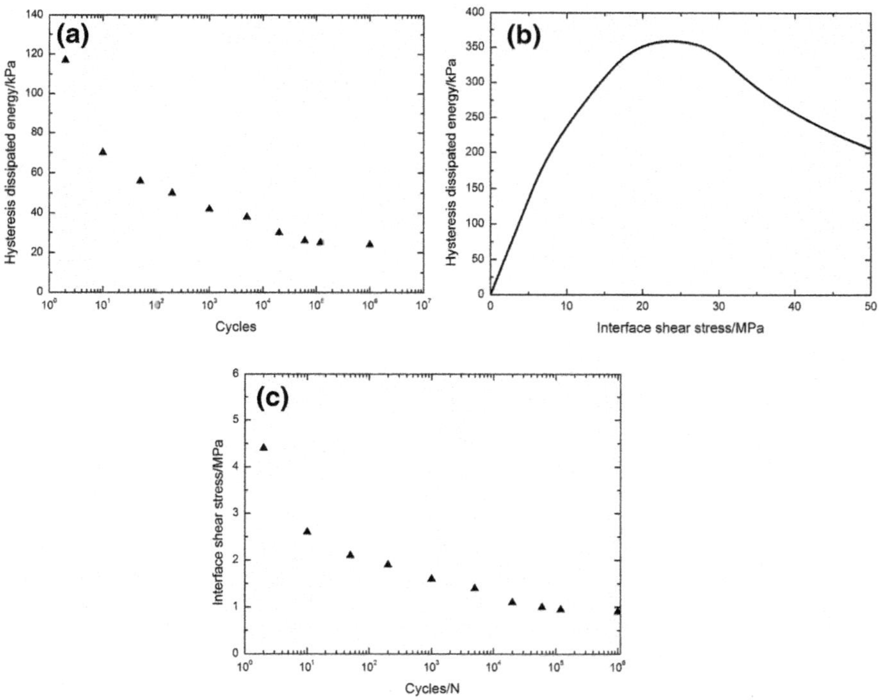


**Fig. 3.19** **a** Experimental fatigue hysteresis dissipated energy versus the applied cycle number; **b** the theoretical fatigue hysteresis dissipated energy versus the fiber/matrix interface shear stress curve; and **c** the fiber/matrix interface shear stress versus the applied cycle number curve of 2D C/SiC composite under the fatigue of $\sigma_{max}$ = 425 MPa at room temperature

slip Case I, II, III, and IV. The estimated fiber/matrix interface shear stress versus applied cycle number curve is shown in Fig. 3.19c. The estimated fiber/matrix interface shear stress corresponding to the applied cycle number of $N$ = 2, 10, 50, 200, 1000, 5000, 20,000, 60,000, 120,000, and 1,000,000 are 4.4, 2.6, 2.1, 1.9, 1.6, 1.4, 1.1, 1, 0.95, and 0.9 MPa, respectively.

Li et al. [30] investigated the tension–tension cyclic fatigue behavior of 2D woven C/SiC composite at room temperature. When the fatigue peak stress is $\sigma_{max}$ = 57 MPa, the experimental fatigue hysteresis dissipated energy versus the applied cycle number curve is shown in Fig. 3.20a. The experimental fatigue hysteresis dissipated energy for the applied cycle number of $N$ = 11,104, 33,262, 55,493, and 100,000 are 1.8, 2.4, 1.7, and 1.6 kPa. The theoretical fatigue hysteresis dissipated energy versus the fiber/matrix interface shear stress is shown in Fig. 3.20b. With decreasing of the fiber/matrix interface shear stress, the fiber/matrix interface debonding ratio of $2l_d/l_c$ changes from $2l_d/l_c < 1$ to $2l_d/l_c = 1$, and the fiber/matrix interface sliding ratio of $y/l_d$ changes from $y/l_d = 1$ to $y/l_d < 1$, and then $y/l_d = 1$, corresponding to the fiber/matrix interface slip Case I, II, III, and IV.

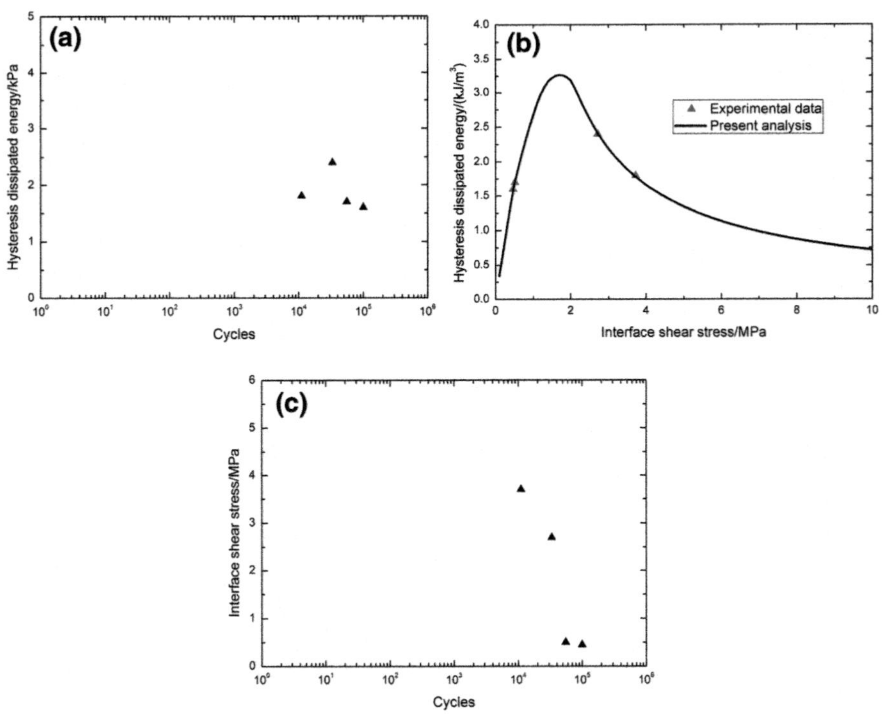


**Fig. 3.20** **a** Experimental fatigue hysteresis dissipated energy for different applied cycle number; **b** the experimental and theoretical fatigue hysteresis dissipated energy versus the fiber/matrix interface shear stress curve; and **c** the estimated fiber/matrix interface shear stress versus applied cycle number curve of 2D C/SiC composite under the fatigue peak stress of $\sigma_{max}$ = 57 MPa at room temperature

The estimated fiber/matrix interface shear stress versus applied cycle number curve is shown in Fig. 3.20c. The estimated fiber/matrix interface shear stress for the applied cycle number of $N$ = 11,104, 33,262, 55,493, and 100,000 are 3.7, 2.7, 0.5, and 0.45 MPa, respectively.

### 550 °C in Air Condition

Mall and Engesser [12] investigated the tension–tension cyclic fatigue behavior of 2D woven C/SiC composite at 550 °C in air condition. When the fatigue peak stress is $\sigma_{max}$ = 350 MPa, the experimental fatigue hysteresis dissipated energy versus the applied cycle number curve is shown in Fig. 3.21a. The experimental fatigue hysteresis dissipated energy for the applied cycle number of $N$ = 100, 200, and 212 are 50.5, 53.4, and 93.8 kPa. The theoretical fatigue hysteresis dissipated energy versus the fiber/matrix interface shear stress is shown in Fig. 3.21b. With decreasing of the fiber/matrix interface shear stress, the fiber/matrix interface debonding ratio of $2l_d/l_c$ changes from $2l_d/l_c < 1$ to $2l_d/l_c = 1$, and the fiber/matrix

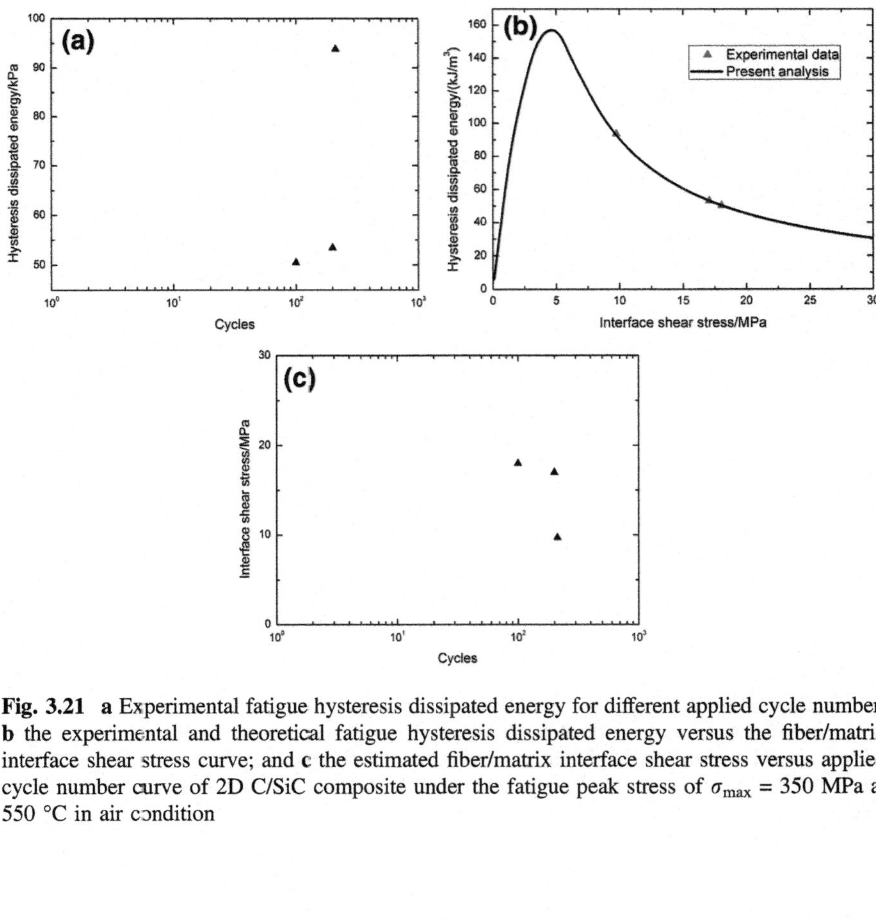


**Fig. 3.21** **a** Experimental fatigue hysteresis dissipated energy for different applied cycle number; **b** the experimental and theoretical fatigue hysteresis dissipated energy versus the fiber/matrix interface shear stress curve; and **c** the estimated fiber/matrix interface shear stress versus applied cycle number curve of 2D C/SiC composite under the fatigue peak stress of $\sigma_{max}$ = 350 MPa at 550 °C in air condition

interface sliding ratio of $y/l_d$ changes from $y/l_d = 1$ to $y/l_d < 1$, and then $y/l_d = 1$, corresponding to the fiber/matrix interface slip Case I, II, III, and IV. The estimated fiber/matrix interface shear stress versus applied cycle number curve is shown in Fig. 3.21c. The estimated fiber/matrix interface shear stress corresponding to $N = 100$, 200, and 212 are 18, 17, and 9.7 MPa, respectively.

### 1200 °C in Vacuum Condition

Rodrigues et al. [31] investigated the tension–tension cyclic fatigue behavior of 2D C/SiC composite at 1200 °C in vacuum condition. When the fatigue peak stress is $\sigma_{\max} = 300\,\text{MPa}$, the experimental fatigue hysteresis dissipated energy versus the applied cycle number curve is shown in Fig. 3.22a. The experimental fatigue hysteresis dissipated energy for the applied cycle number of $N = 1000$, 10,000, 100,000, 500,000, 1,000,000, 2,100,000, and 2,600,000 are 30.7, 32.1, 38.6, 30.8, 29.1, 28.6, and 25 kPa. The theoretical fatigue hysteresis dissipated energy versus the fiber/matrix interface shear stress is shown in Fig. 3.22b. With decreasing of the

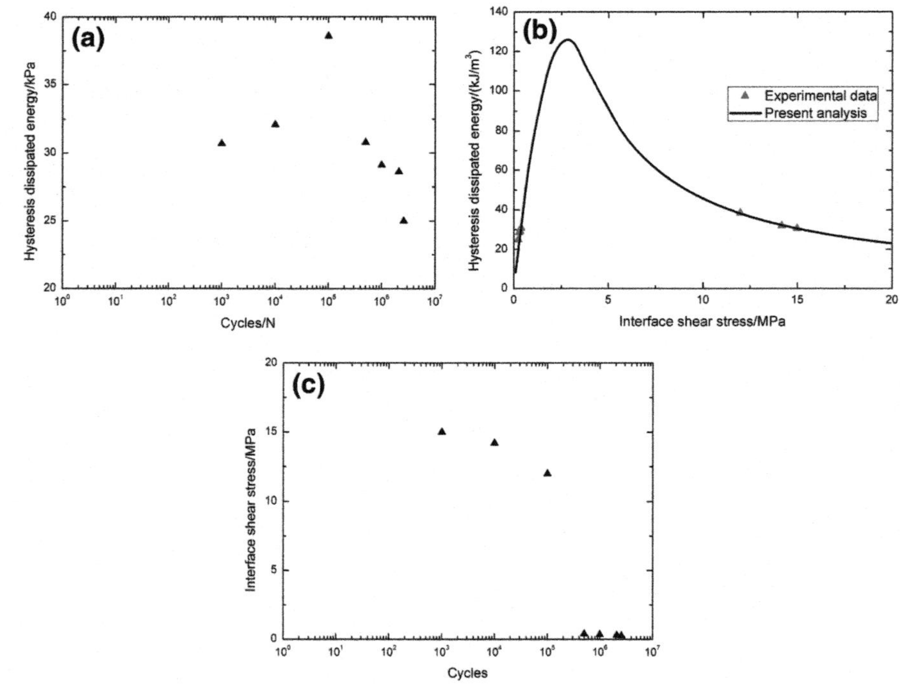


**Fig. 3.22** **a** Experimental fatigue hysteresis dissipated energy for different applied cycle number; **b** the experimental and theoretical fatigue hysteresis dissipated energy versus the fiber/matrix interface shear stress curve; and **c** the estimated fiber/matrix interface shear stress versus applied cycle number curve of 2D C/SiC composite under the fatigue peak stress of $\sigma_{\max} = 300$ MPa at 1200 °C in vacuum condition

fiber/matrix interface shear stress, the fiber/matrix interface debonding ratio of $2l_d/l_c$ changes from $2l_d/l_c < 1$ to $2l_d/l_c = 1$, and the fiber/matrix interface sliding ratio of $y/l_d$ changes from $y/l_d = 1$ to $y/l_d < 1$, and then $y/l_d = 1$, corresponding to the fiber/matrix interface slip Case I, II, III, and IV. The estimated fiber/matrix interface shear stress versus applied cycle number curve is shown in Fig. 3.22c. The estimated fiber/matrix interface shear stress for the applied cycle number of $N = 1000$, 10,000, 100,000, 500,000, 1,000,000, 2,100,000, and 2,600,000 are 15, 14.2, 12, 0.4, 0.35, 0.3, and 0.25 MPa, respectively.

#### 3.3.2.4 2D Woven SiC/SiC Composite

The experimental fatigue hysteresis dissipated energy versus the applied cycle number curves, the theoretical fatigue hysteresis dissipated energy versus the fiber/matrix interface shear stress curve, and the estimated fiber/matrix interface shear stress versus the applied cycle number curve of 2D woven SiC/SiC composite under different peak stresses at room temperature, 600, 800, and 1000 °C in inert condition, 1000, 1100, and 1200 °C in air and steam conditions, and 1300 °C in air condition are analyzed.

Room Temperature

Shi [32] investigated the tension–tension cyclic fatigue behavior of 2D woven SiC/SiC composite at room temperature. When the fatigue peak stress is $\sigma_{max} = 150\,\text{MPa}$, the experimental fatigue hysteresis dissipated energy versus the applied cycle number curve is shown in Fig. 3.23a. The experimental fatigue hysteresis dissipated energy for the applied cycle number of $N = 121$, 331, 661, 1241, 1601, 253,561, 405,451, 757,531, and 1,200,331 are 10.4, 10.6, 10.9, 12.1, 12.7, 13.3, 14.6, 15.5, and 18 kPa. The theoretical fatigue hysteresis dissipated energy versus the fiber/matrix interface shear stress is shown in Fig. 3.23b. With decreasing fiber/matrix interface shear stress, the fiber/matrix interface debonding ratio of $2l_d/l_c$ changes from $2l_d/l_c < 1$ to $2l_d/l_c = 1$, and the fiber/matrix interface sliding ratio of $y/l_d$ changes from $y/l_d = 1$ to $y/l_d < 1$, and then $y/l_d = 1$, corresponding to the fiber/matrix interface slip Case I, II, III, and IV. The estimated fiber/matrix interface shear stress versus applied cycle number curve is shown in Fig. 3.23c. The estimated fiber/matrix interface shear stress for the applied cycle number of $N = 121$, 331, 661, 1241, 1601, 253,561, 405,451, 757,531, and 1,200,331 are 17, 16.8, 16.3, 14.7, 14, 13.3, 12.1, 11.4, and 10 MPa, respectively.

When the fatigue peak stress is $\sigma_{max} = 250\,\text{MPa}$, the experimental fatigue hysteresis dissipated energy versus the applied cycle number curve is shown in Fig. 3.24a. The experimental fatigue hysteresis dissipated energy for the applied cycle number of $N = 81$, 401, 641, 761, 1281, 1441, 23,361, 83,161, 174,561, 217,761, 239,361, 296,961, and 332,961 are 106, 107, 113, 107, 105, 102, 93, 87, 86, 85, 83, 82, and 80 kPa. The theoretical fatigue hysteresis dissipated energy

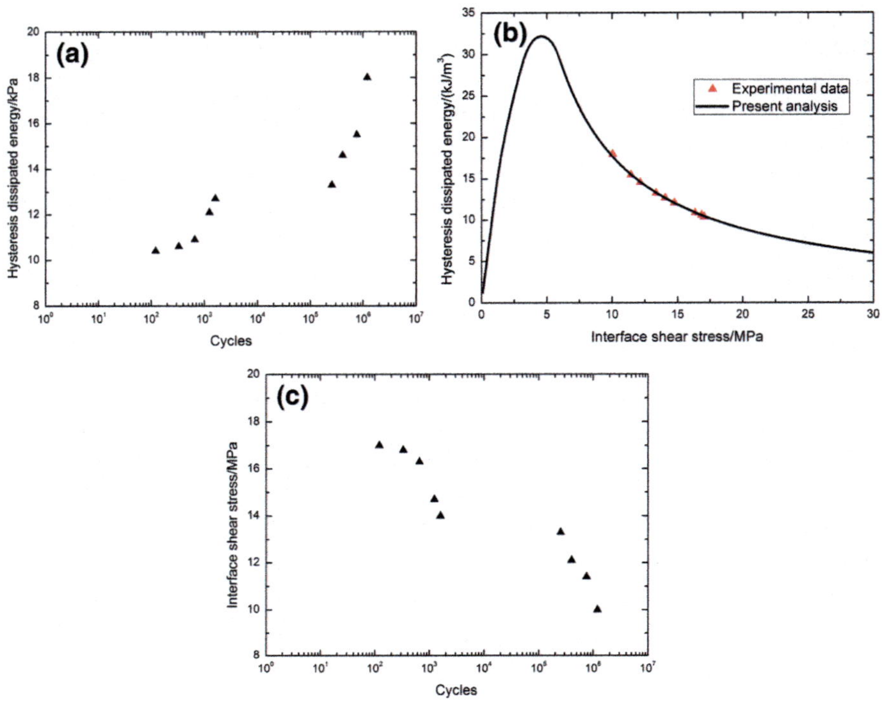


**Fig. 3.23** **a** Experimental fatigue stress–strain hysteresis loops for different applied cycle number; **b** the experimental and theoretical fatigue hysteresis dissipated energy versus the fiber/matrix interface shear stress curve; and **c** the estimated fiber/matrix interface shear stress versus the applied cycle number curve of 2D woven SiC/SiC composite under the fatigue peak stress of $\sigma_{max}$ = 150 MPa at room temperature

versus the fiber/matrix interface shear stress is shown in Fig. 3.24b. With decreasing fiber/matrix interface shear stress, the fiber/matrix interface debonding ratio of $2l_d/l_c$ changes from $2l_d/l_c < 1$ to $2l_d/l_c = 1$, and the fiber/matrix interface sliding ratio of $y/l_d$ changes from $y/l_d = 1$ to $y/l_d < 1$, and then $y/l_d = 1$, corresponding to the fiber/matrix interface slip Case I, II, III and IV. The estimated fiber/matrix interface shear stress versus applied cycle number curve is shown in Fig. 3.24c. The estimated fiber/matrix interface shear stress for the applied cycle number of $N$ = 81, 401, 641, 761, 1281, 1441, 23,361, 83,161, 174,561, 217,761, 239,361, 296,961, and 332,961 are 11.1, 11, 10,1, 5.8, 5.6, 5.4, 4.6, 4.2, 4.1, 4, 3.9, 3.8, and 3.7 MPa, respectively.

800 °C in Air Atmosphere

Shi [32] investigated the tension–tension cyclic fatigue behavior of 2D woven SiC/SiC composite at 800 °C in air condition. When the fatigue peak stress is

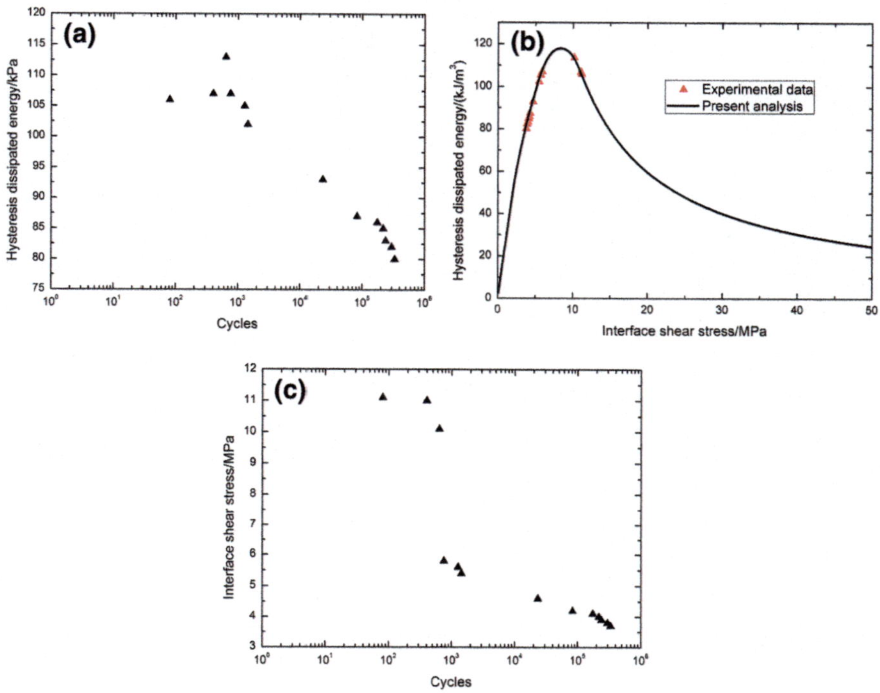


**Fig. 3.24** **a** Experimental fatigue stress–strain hysteresis loops for different applied cycle number; **b** the experimental and theoretical fatigue hysteresis dissipated energy versus the fiber/matrix interface shear stress curve; and **c** the estimated fiber/matrix interface shear stress versus applied cycle number curve of 2D woven SiC/SiC composite under the fatigue peak stress of $\sigma_{max}$ = 250 MPa at room temperature

$\sigma_{max} = 150$ MPa, the experimental fatigue hysteresis dissipated energy versus the applied cycle number curve is shown in Fig. 3.25a. The experimental fatigue hysteresis dissipated energy for the applied cycle number of $N = 5$, 10, 50, 1000, 15,000, 20,000, 25,000, 30,000, 32,000, 34,000, 36,000, and 36,500 are 5.5, 6.5, 8.3, 10.6, 12.6, 13.5, 17.3, 18.7, 19.6, 20.9, 23.6, and 26 kPa. The theoretical fatigue hysteresis dissipated energy versus the fiber/matrix interface shear stress is shown in Fig. 3.25b. With decreasing of the fiber/matrix interface shear stress, the fiber/matrix interface debonding ratio of $2l_d/l_c$ changes from $2l_d/l_c < 1$ to $2l_d/l_c = 1$, and the fiber/matrix interface sliding ratio of $y/l_d$ changes from $y/l_d = 1$ to $y/l_d < 1$, and then $y/l_d = 1$, corresponding to the fiber/matrix interface slip Case I, II, III and IV. The estimated fiber/matrix interface shear stress versus applied cycle number curve is shown in Fig. 3.25c. The estimated fiber/matrix interface shear stress for the applied cycle number of $N = 5$, 10, 50, 1000, 15,000, 20,000, 25,000, 30,000, 32,000, 34,000, 36,000, and 36,500 are 32.6, 27.6, 21.5, 16.8, 14.1, 13.2, 10.3, 9.5, 9, 8.5, 7.5, and 6.8 MPa, respectively.

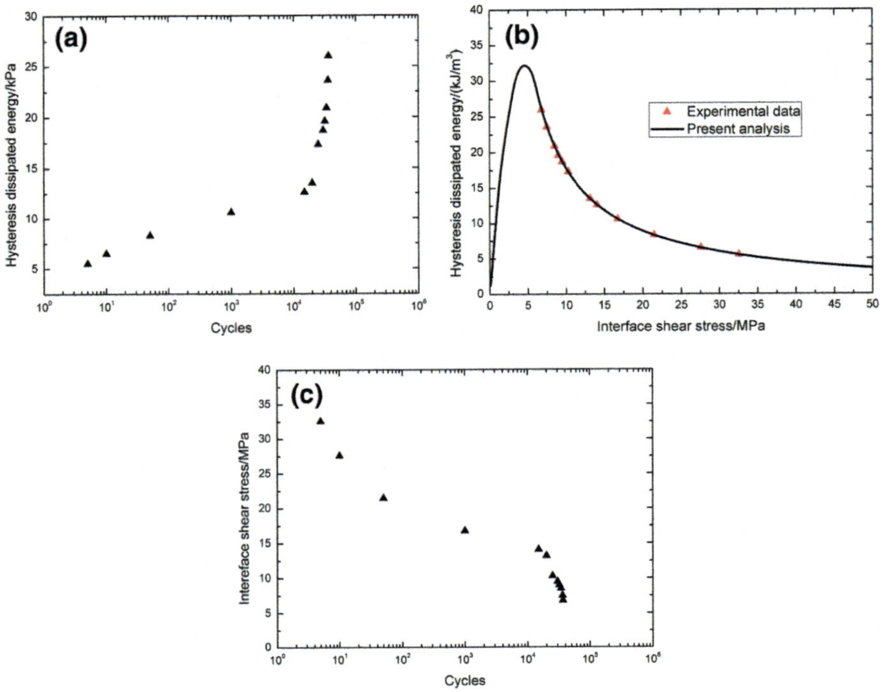


**Fig. 3.25** **a** Experimental fatigue hysteresis dissipated energy for different applied cycle number; **b** the experimental and theoretical fatigue hysteresis dissipated energy versus the fiber/matrix interface shear stress curve; and **c** the estimated fiber/matrix interface shear stress versus applied cycle number curve of 2D woven SiC/SiC composite under the fatigue peak stress of $\sigma_{max}$ = 150 MPa at 800 °C in air condition

600, 800, and 1000 °C in Inert Atmosphere

Reynaud [9] investigated the tension–tension cyclic fatigue behavior of 2D woven SiC/SiC composite under the fatigue peak stress of $\sigma_{max} = 130$ MPa at elevated temperatures of 600, 800 and 1000 °C in inert atmosphere. The theoretical fatigue hysteresis dissipated energy versus the fiber/matrix interface shear stress curve of 2D SiC/SiC composite under the fatigue peak stress of $\sigma_{max} = 130$ MPa is shown in Fig. 3.26a. With decreasing of the fiber/matrix interface shear stress, the fiber/matrix interface debonding ratio of $2l_d/l_c$ changes from $2l_d/l_c < 1$ to $2l_d/l_c = 1$, and the fiber/matrix interface sliding ratio of $y/l_d$ changes from $y/l_d = 1$ to $y/l_d < 1$, and then $y/l_d = 1$, corresponding to the fiber/matrix interface slip Case I, II, III, and IV. At an elevated temperature of 600 °C, the experimental fatigue hysteresis dissipated energy increases from 5.4 kPa at $N = 25$–9.4 kPa at $N = 333{,}507$, as shown in Fig. 3.26b. The estimated fiber/matrix interface shear stress versus applied cycle number curve is shown in Fig. 3.26c. The estimated fiber/matrix interface shear stress decreases from $\tau_i = 35$ MPa at the applied cycle number of $N = 1$ to

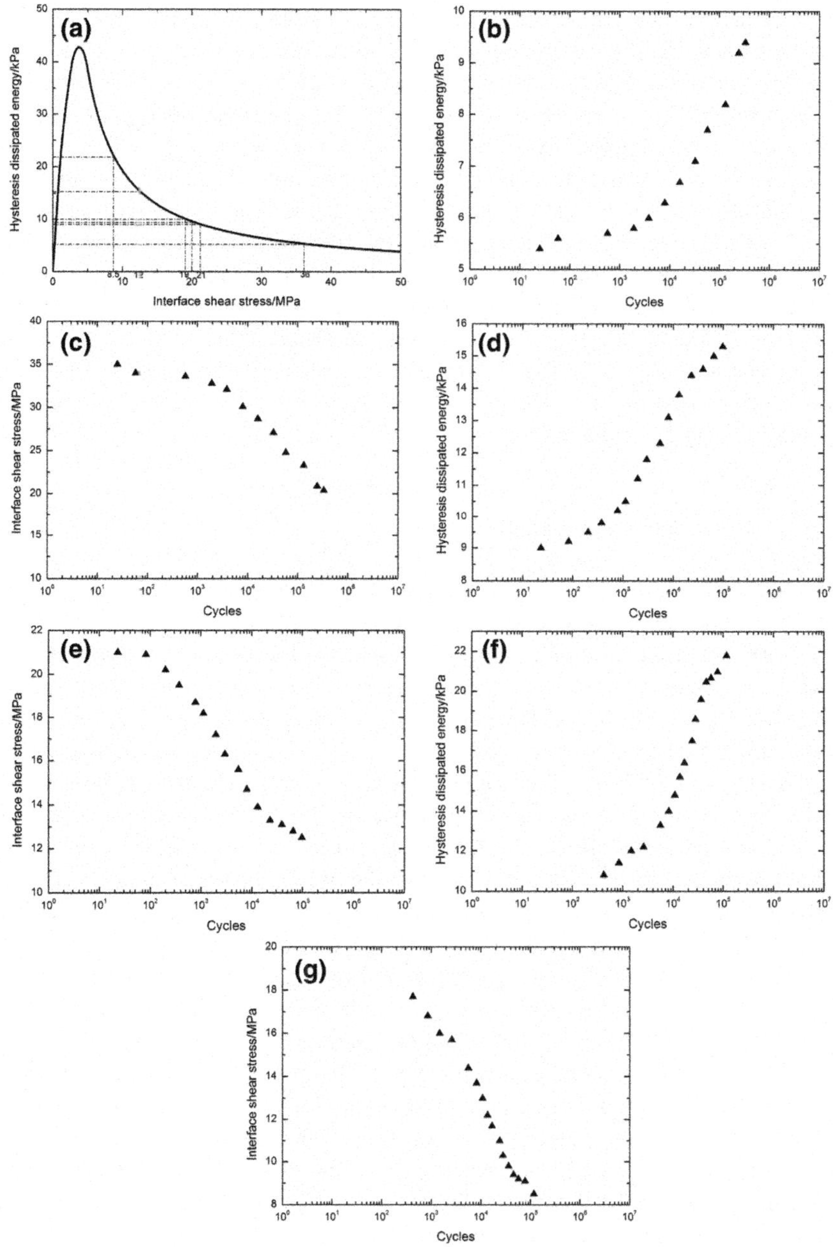


**Fig. 3.26** **a** Theoretical fatigue hysteresis dissipated energy versus the fiber/matrix interface shear stress curve; **b** the experimental fatigue hysteresis dissipated energy versus the cycle number curve at 600 °C; **c** the estimated fiber/matrix interface shear stress versus applied cycle number curve at 600 °C; **d** the experimental fatigue hysteresis dissipated energy versus the cycle number curve at 800 °C; **e** the estimated fiber/matrix interface shear stress versus applied cycle number curve at 800 °C; **f** the experimental fatigue hysteresis dissipated energy versus the cycle number curve at 1000 °C; and **g** the estimated fiber/matrix interface shear stress versus applied cycle number curve at 1000 °C

$\tau_i$ = 20.4 MPa at the applied cycle number of $N$ = 333,507. At an elevated temperature of 800 °C, the experimental fatigue hysteresis dissipated energy increases from 9.0 kPa at the applied cycle number of $N$ = 23–15.3 kPa at the applied cycle number of $N$ = 97,894, as shown in Fig. 3.26d. The estimated fiber/matrix interface shear stress versus applied cycle number curve is shown in Fig. 3.26e. The estimated fiber/matrix interface shear stress decreases from $\tau_i$ = 22 MPa at the applied cycle number of $N = 1$ to $\tau_i$ = 12.5 MPa at the applied cycle number of $N$ = 97,894. At an elevated temperature of 1000 °C, the experimental fatigue hysteresis dissipated energy increases from 10.8 kPa at the applied cycle number of $N$ = 425–21.8 kPa at the applied cycle number of $N$ = 117,055, as shown in Fig. 3.26f. The estimated fiber/matrix interface shear stress versus applied cycle number curve is shown in Fig. 3.26g. The estimated fiber/matrix interface shear stress decreases from $\tau_i$ = 18 MPa at the applied cycle number of $N = 1$ to $\tau_i$ = 8.5 MPa at the applied cycle number of $N$ = 117,055.

1000 °C in Air and Steam Conditions

Michael [33] investigated the tension–tension cyclic fatigue behavior of 2D woven SiC/SiC composite at 1000 °C in air and steam conditions.

At 1000 °C in air condition, when the fatigue peak stress is $\sigma_{max}$ = 80 MPa, the experimental fatigue hysteresis dissipated energy versus the applied cycle number curve is shown in Fig. 3.27a. The experimental fatigue hysteresis dissipated energy for the applied cycle number of $N$ = 2, 1000, 10,000, 20,000, and 30,000 are 4.6, 5.4, 5.8, 6.3, and 7 kPa. The theoretical fatigue hysteresis dissipated energy versus the fiber/matrix interface shear stress is shown in Fig. 3.27b. With decreasing fiber/matrix interface shear stress, the fiber/matrix interface debonding ratio of $2l_d/l_c$ changes from $2l_d/l_c < 1$ to $2l_d/l_c = 1$, and the fiber/matrix interface sliding ratio of $y/l_d$ changes from $y/l_d = 1$ to $y/l_d < 1$, and then $y/l_d = 1$, corresponding to the fiber/matrix interface slip Case I, II, III, and IV. The estimated fiber/matrix interface shear stress versus applied cycle number curve is shown in Fig. 3.27c. The estimated fiber/matrix interface shear stress for the applied cycle number of $N$ = 2, 1000, 10,000, 20,000, and 30,000 are 15, 13, 12, 11, and 10 MPa, respectively.

At 1000 °C in steam condition, when the fatigue peak stress is $\sigma_{max}$ = 100 MPa, the experimental fatigue hysteresis dissipated energy versus the applied cycle number curve is shown in Fig. 3.28a. The experimental fatigue hysteresis dissipated energy for the applied cycle number of $N$ = 2, 500, 3000, and 10,000 are 9, 10.4, 13.5, and 16.8 kPa. The theoretical fatigue hysteresis dissipated energy versus the fiber/matrix interface shear stress is shown in Fig. 3.28b. With decreasing fiber/matrix interface shear stress, the fiber/matrix interface debonding ratio of $2l_d/l_c$ changes from $2l_d/l_c < 1$ to $2l_d/l_c = 1$, and the fiber/matrix interface sliding ratio of $y/l_d$ changes from $y/l_d = 1$ to $y/l_d < 1$, and then $y/l_d = 1$, corresponding to the fiber/matrix interface slip Case I, II, III and IV. The estimated fiber/matrix interface shear stress

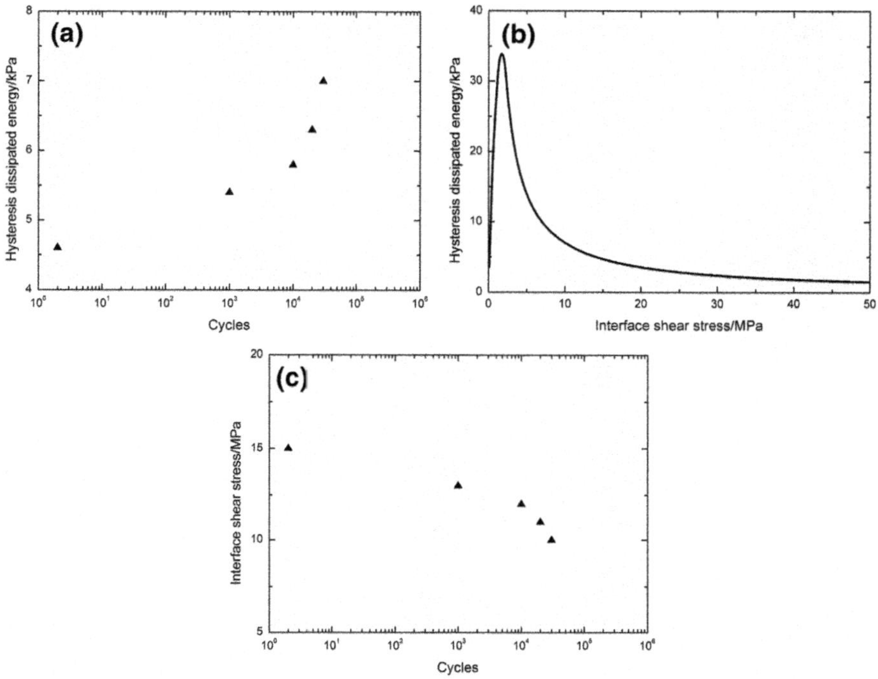


**Fig. 3.27** **a** Experimental fatigue hysteresis dissipated energy versus applied cycle number; **b** the theoretical fatigue hysteresis dissipated energy versus the fiber/matrix interface shear stress curve; and **c** the fiber/matrix interface shear stress versus the applied cycle number curve of 2D SiC/SiC composite under the fatigue peak stress of $\sigma_{max}$ = 80 MPa at 1000 °C in air condition

versus applied cycle number curve is shown in Fig. 3.28c. The estimated fiber/matrix interface shear stress for the applied cycle number of $N$ = 2, 500, 3000, and 10,000 are 15, 13, 10, and 8 MPa, respectively.

### 1100 °C in Air and Steam Condition

Groner [34] investigated the tension–tension cyclic fatigue behavior of 2D woven SiC/SiC composite at 1100 °C in air and steam conditions.

At 1100 °C in air condition, when the fatigue peak stress is $\sigma_{max}$ = 140 MPa, the experimental fatigue hysteresis dissipated energy versus the applied cycle number curve is shown in Fig. 3.29a. The experimental fatigue hysteresis dissipated energy for the applied cycle number of $N$ = 7429, 15,381 and 23,391 are 3.5, 8 and 10.8 kPa. The theoretical fatigue hysteresis dissipated energy versus the fiber/matrix interface shear stress is shown in Fig. 3.29b. With decreasing fiber/matrix interface shear stress, the fiber/matrix interface debonding ratio of $2l_d/l_c$ changes from $2l_d/l_c < 1$ to $2l_d/l_c = 1$, and the fiber/matrix interface sliding ratio of $y/l_d$ changes from $y/l_d = 1$ to $y/l_d < 1$, and then $y/l_d = 1$, corresponding to the fiber/

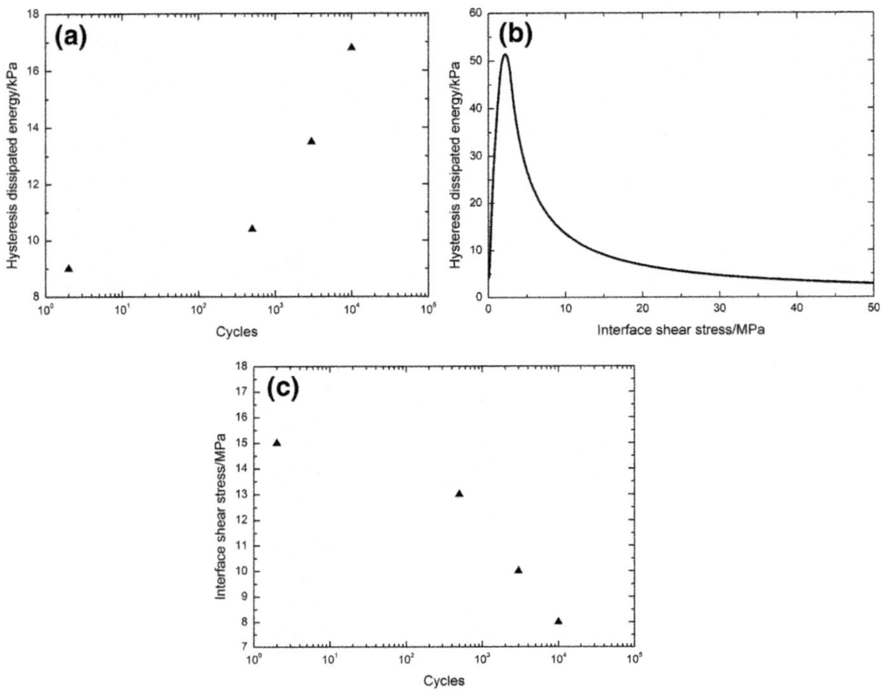


**Fig. 3.28** **a** Experimental fatigue hysteresis dissipated energy versus the applied cycle number curve; **b** the theoretical fatigue hysteresis dissipated energy versus the fiber/matrix interface shear stress curve; and **c** the fiber/matrix interface shear stress versus the applied cycle number curve of 2D woven SiC/SiC composite under the fatigue peak stress of $\sigma_{max}$ = 100 MPa at 1000 °C in steam condition

matrix interface slip Case I, II, III, and IV. The estimated fiber/matrix interface shear stress versus applied cycle number curve is shown in Fig. 3.29c. The estimated fiber/matrix interface shear stress for the applied cycle number of $N$ = 7429, 15,381 and 23,391 are 38.6, 16.6 and 12.3 MPa, respectively.

### 1200 °C in Air and Steam Conditions

Jacob [35] investigated the tension–tension cyclic fatigue behavior of 2D woven SiC/SiC composite at 1200 °C in air and steam conditions.

At 1200 °C in air condition, when the fatigue peak stress is $\sigma_{max}$ = 140 MPa, the experimental fatigue hysteresis dissipated energy versus the applied cycle number curve is shown in Fig. 3.30a. The experimental fatigue hysteresis dissipated energy for the applied cycle number of $N$ = 1000, 10,000 and 30,000 are 5.2, 22.4 and 25 kPa. The theoretical fatigue hysteresis dissipated energy versus the fiber/matrix interface shear stress is shown in Fig. 3.30b. With decreasing fiber/matrix

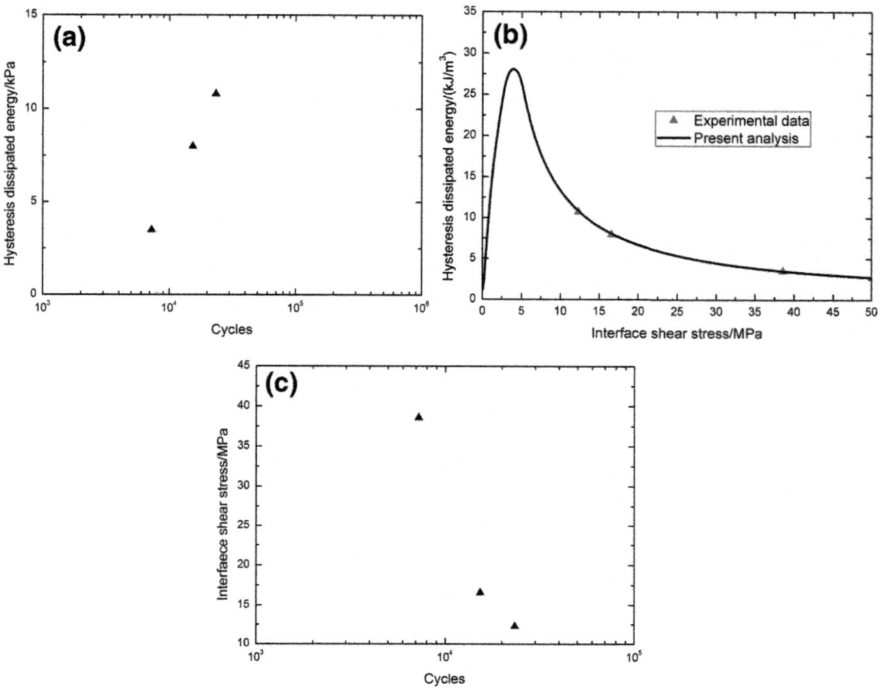


**Fig. 3.29** **a** Experimental fatigue hysteresis dissipated energy versus the applied cycle number curve; **b** the theoretical fatigue hysteresis dissipated energy versus the fiber/matrix interface shear stress curve; and **c** the estimated fiber/matrix interface shear stress versus applied cycle number curve of 2D woven SiC/SiC composite under the fatigue peak stress of $\sigma_{max}$ = 140 MPa at 1100 °C in air condition

interface shear stress, the fiber/matrix interface debonding ratio of $2l_d/l_c$ changes from $2l_d/l_c < 1$ to $2l_d/l_c = 1$, and the fiber/matrix interface sliding ratio of $y/l_d$ changes from $y/l_d = 1$ to $y/l_d < 1$, and then $y/l_d = 1$, corresponding to the fiber/matrix interface slip Case I, II, III, and IV. The estimated fiber/matrix interface shear stress versus applied cycle number curve is shown in Fig. 3.30c. The estimated fiber/matrix interface shear stress for the applied cycle number of $N$ = 1000, 10,000, and 30,000 are 15, 3.5, and 3 MPa, respectively.

At 1200 °C in steam condition, when the fatigue peak stress is $\sigma_{max}$ = 140 MPa, the experimental fatigue hysteresis dissipated energy versus the applied cycle number curve is shown in Fig. 3.31a. The experimental fatigue hysteresis dissipated energy for the applied cycle number of $N$ = 100, 1000, and 10,000 are 4.5, 19.3, and 24.6 kPa. The theoretical fatigue hysteresis dissipated energy versus the fiber/matrix interface shear stress is shown in Fig. 3.31b. With decreasing fiber/matrix interface shear stress, the fiber/matrix interface debonding ratio of $2l_d/l_c$ changes from $2l_d/l_c < 1$ to $2l_d/l_c = 1$, and the fiber/matrix interface sliding ratio of $y/l_d$ changes from $y/l_d = 1$ to $y/l_d < 1$, and then $y/l_d = 1$, corresponding to the fiber/

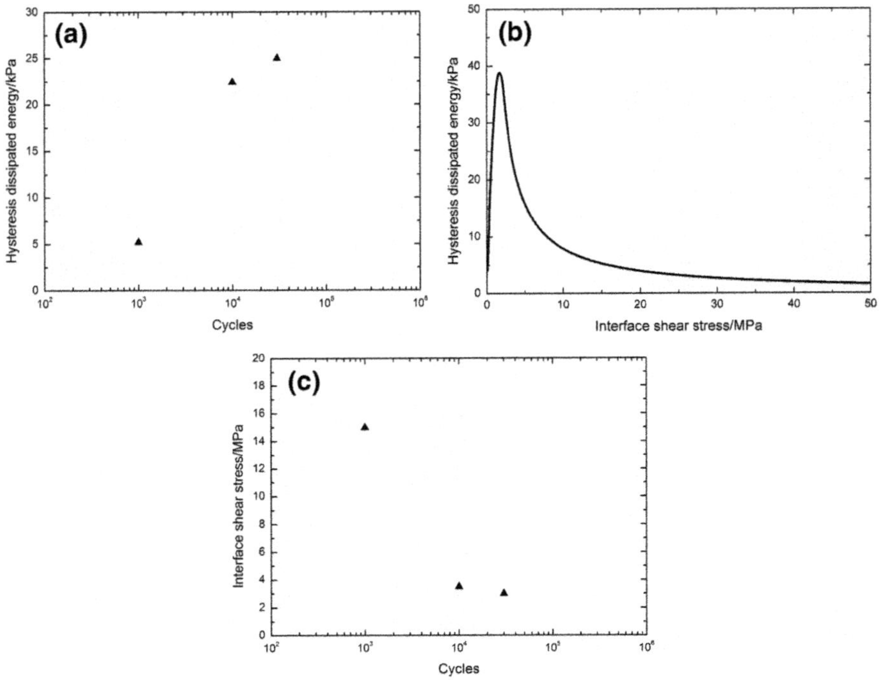


**Fig. 3.30** **a** Experimental fatigue hysteresis dissipated energy versus the applied cycle number curve; **b** the theoretical fatigue hysteresis dissipated energy versus the fiber/matrix interface shear stress curve; and **c** the estimated fiber/matrix interface shear stress curve of 2D SiC/SiC composite under the fatigue peak stress of $\sigma_{max}$ = 140 MPa at 1200 °C in air condition

matrix interface slip Case I, II, III, and IV. The estimated fiber/matrix interface shear stress versus applied cycle number curve is shown in Fig. 3.31c. The estimated fiber/matrix interface shear stress for the applied cycle number of $N$ = 100, 1000, and 10,000 are 17, 4, and 3.2 MPa, respectively.

### 1300 °C in Air Condition

Zhu et al. [36] investigated the tension–tension cyclic fatigue behavior of 2D woven SiC/SiC composite at 1300 °C in air condition.

When the fatigue peak stress is $\sigma_{max}$ = 120 MPa, the experimental fatigue hysteresis dissipated energy versus the applied cycle number curve is shown in Fig. 3.32a. The experimental fatigue hysteresis dissipated energy for the applied cycle number of $N$ = 100, 6000, 18,000, and 360,000 are 4, 7.2, 8.9, and 19 kPa. The theoretical fatigue hysteresis dissipated energy versus the fiber/matrix interface shear stress is shown in Fig. 3.32b. With decreasing fiber/matrix interface shear stress, the fiber/matrix interface debonding ratio of $2l_d/l_c$ changes from $2l_d/l_c < 1$ to $2l_d/l_c = 1$, and the fiber/matrix interface sliding ratio of $y/l_d$ changes from $y/l_d = 1$ to

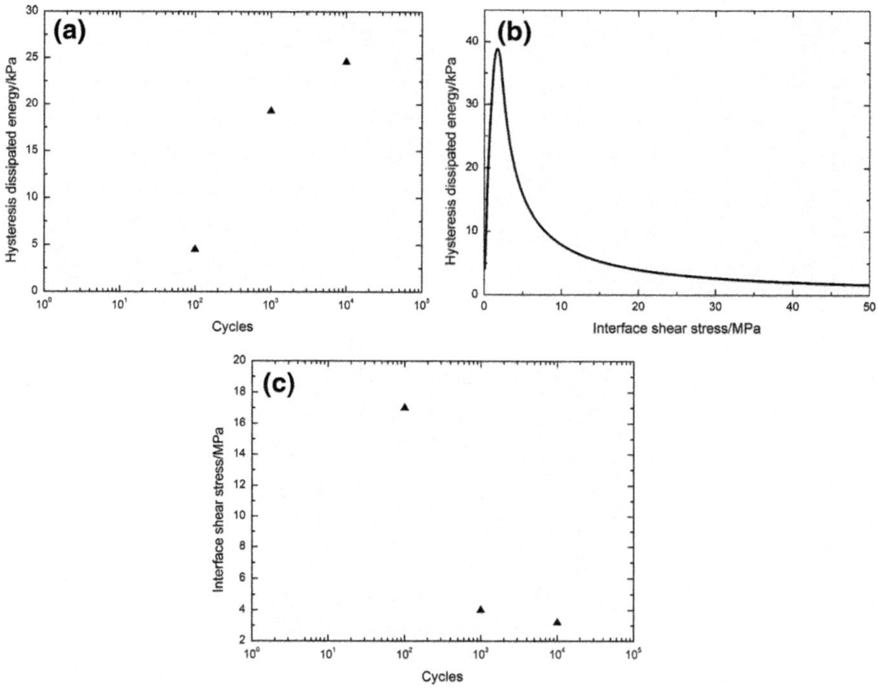


**Fig. 3.31** **a** Experimental fatigue hysteresis dissipated energy versus the applied cycle number curve; **b** the theoretical fatigue hysteresis dissipated energy versus the fiber/matrix interface shear stress curve; and **c** the estimated fiber/matrix interface shear stress curve of 2D SiC/SiC composite under the fatigue peak stress of $\sigma_{max}$ = 140 MPa at 1200 °C in steam condition

$y/l_d < 1$, and then $y/l_d = 1$, corresponding to the fiber/matrix interface slip Case I, II, III, and IV. The estimated fiber/matrix interface shear stress versus applied cycle number curve is shown in Fig. 3.32c. The estimated fiber/matrix interface shear stress for the applied cycle number of $N$ = 100, 6000, 18,000, and 360,000 are 18, 10, 8, and 3.7 MPa, respectively.

### 2.5D Woven C/SiC Composite

The experimental fatigue hysteresis dissipated energy versus the applied cycle number curves, the theoretical fatigue hysteresis dissipated energy versus the fiber/matrix interface shear stress curve, and the estimated fiber/matrix interface shear stress versus the applied cycle number curve of 2.5D woven C/SiC composite under different peak stresses at room temperature, 600 °C in inert condition, and 800 °C in air condition are analyzed.

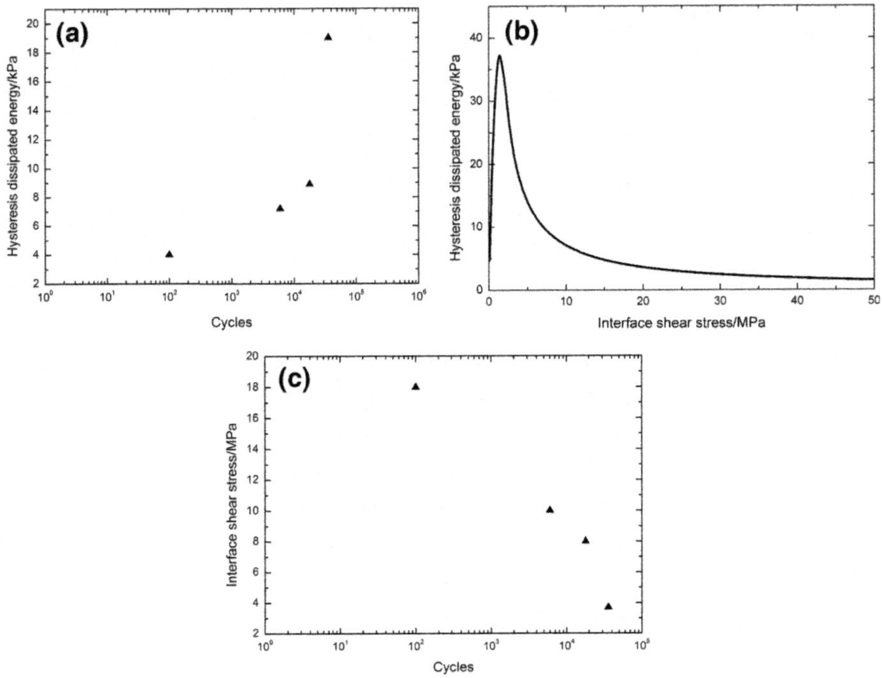


**Fig. 3.32** **a** Experimental fatigue hysteresis dissipated energy versus the applied cycle number curve; **b** the theoretical fatigue hysteresis dissipated energy versus the fiber/matrix interface shear stress curve; and **c** the estimated fiber/matrix interface shear stress curve of 2D SiC/SiC composite under the fatigue peak stress of $\sigma_{\max}$ = 120 MPa at 1300 °C in air condition

## Room Temperature

Yang [37] investigated the tension–tension cyclic fatigue behavior of 2.5D C/SiC composite at room temperature. When the fatigue peak stress is $\sigma_{\max}$ = 180 MPa, the experimental fatigue hysteresis dissipated energy versus the applied cycle number curve is shown in Fig. 3.33a. The experimental fatigue hysteresis dissipated energy for the applied cycle number of $N$ = 10, 1010, 4010, and 5210 are 7.8, 8.5, 9.4, and 12.8 kPa. The theoretical fatigue hysteresis dissipated energy versus the fiber/matrix interface shear stress is shown in Fig. 3.33b. With decreasing fiber/matrix interface shear stress, the fiber/matrix interface debonding ratio of $2l_d/l_c$ changes from $2l_d/l_c < 1$ to $2l_d/l_c = 1$, and the fiber/matrix interface sliding ratio of $y/l_d$ changes from $y/l_d = 1$ to $y/l_d < 1$, and then $y/l_d = 1$, corresponding to the fiber/matrix interface slip Case I, II, III, and IV. The estimated fiber/matrix interface shear stress versus applied cycle number curve is shown in Fig. 3.33c. The estimated fiber/matrix interface shear stress for the applied cycle number of $N$ = 10, 1010, 4010, and 5210 are 13.2, 12.3, 11.1, and 8.1 MPa, respectively.

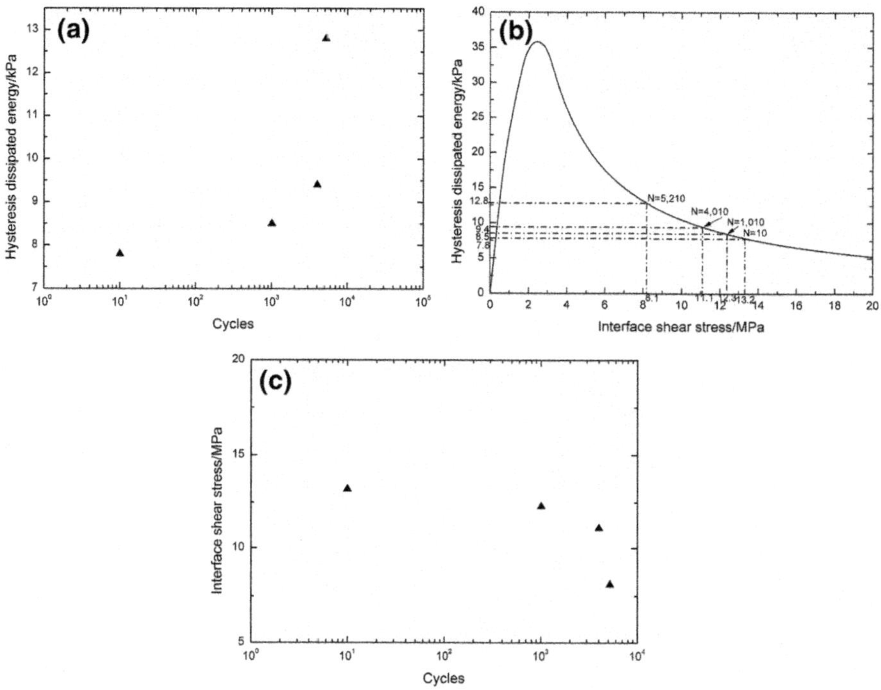


**Fig. 3.33** **a** Experimental fatigue hysteresis dissipated energy versus the applied cycle number curve; **b** the theoretical fatigue hysteresis dissipated energy versus the fiber/matrix interface shear stress curve; and **c** the estimated fiber/matrix interface shear stress curve of 2.5D C/SiC composite under the fatigue peak stress of $\sigma_{max}$ = 180 MPa at room temperature

### 800 °C in Air Condition

Yang [37] investigated the tension–tension cyclic fatigue behavior of 2.5D C/SiC composite at 800 °C in air condition. When the fatigue peak stress is $\sigma_{max}$ = 140 MPa, the experimental fatigue hysteresis dissipated energy versus the applied cycle number curve is shown in Fig. 3.34a. The experimental fatigue hysteresis dissipated energy for the applied cycle number of $N$ = 500, 15,000, 20,000, and 22,700 are 6.3, 7.2, 8.7, and 11.8 kPa. The theoretical fatigue hysteresis dissipated energy versus the fiber/matrix interface shear stress is shown in Fig. 3.34b. With decreasing fiber/matrix interface shear stress, the fiber/matrix interface debonding ratio of $2l_d/l_c$ changes from $2l_d/l_c < 1$ to $2l_d/l_c = 1$, and the fiber/matrix interface sliding ratio of $y/l_d$ changes from $y/l_d = 1$ to $y/l_d < 1$, and then $y/l_d = 1$, corresponding to the fiber/matrix interface slip Case I, II, III, and IV. The estimated fiber/matrix interface shear stress versus applied cycle number curve is shown in Fig. 3.34c. The estimated fiber/matrix interface shear stress for the applied cycle number of $N$ = 500, 15,000, 20,000, and 22,700 are 9.2, 8.2, 6.7, and 5 MPa, respectively.

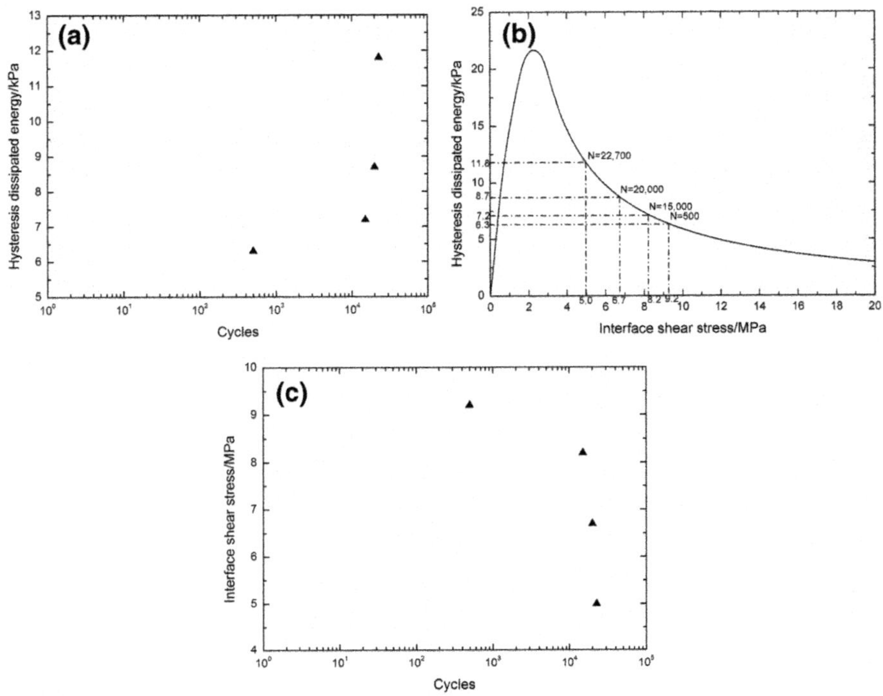


**Fig. 3.34** **a** Experimental fatigue hysteresis dissipated energy versus the applied cycle number curve; **b** the theoretical fatigue hysteresis dissipated energy versus the fiber/matrix interface shear stress curve; and **c** the estimated fiber/matrix interface shear stress curve of 2.5D C/SiC composite under the fatigue peak stress of $\sigma_{max}$ = 140 MPa at 800 °C in air condition

### 600 °C in Inert Condition

Dalmaz et al. [38] investigated the tension–tension cyclic fatigue behavior of 2.5D C/SiC composite at 600 °C in inert atmosphere. When the fatigue peak stress is $\sigma_{max}$ = 230 MPa, the experimental fatigue hysteresis dissipated energy versus the applied cycle number curve is shown in Fig. 3.35a. The experimental fatigue hysteresis dissipated energy for the applied cycle number of $N$ = 10, 10,000, and 100,000 are 33, 29, and 19 kPa. The theoretical fatigue hysteresis dissipated energy versus the fiber/matrix interface shear stress is shown in Fig. 3.35b. With decreasing fiber/matrix interface shear stress, the fiber/matrix interface debonding ratio of $2l_d/l_c$ changes from $2l_d/l_c < 1$ to $2l_d/l_c = 1$, and the fiber/matrix interface sliding ratio of $y/l_d$ changes from $y/l_d = 1$ to $y/l_d < 1$, and then $y/l_d = 1$, corresponding to the fiber/matrix interface slip Case I, II, III and IV. The estimated fiber/matrix interface shear stress versus applied cycle number curve is shown in Fig. 3.35c. The estimated fiber/matrix interface shear stress for the applied cycle number of $N$ = 10, 10,000, and 100,000 are 2.2, 1.9, and 1.2 MPa, respectively.

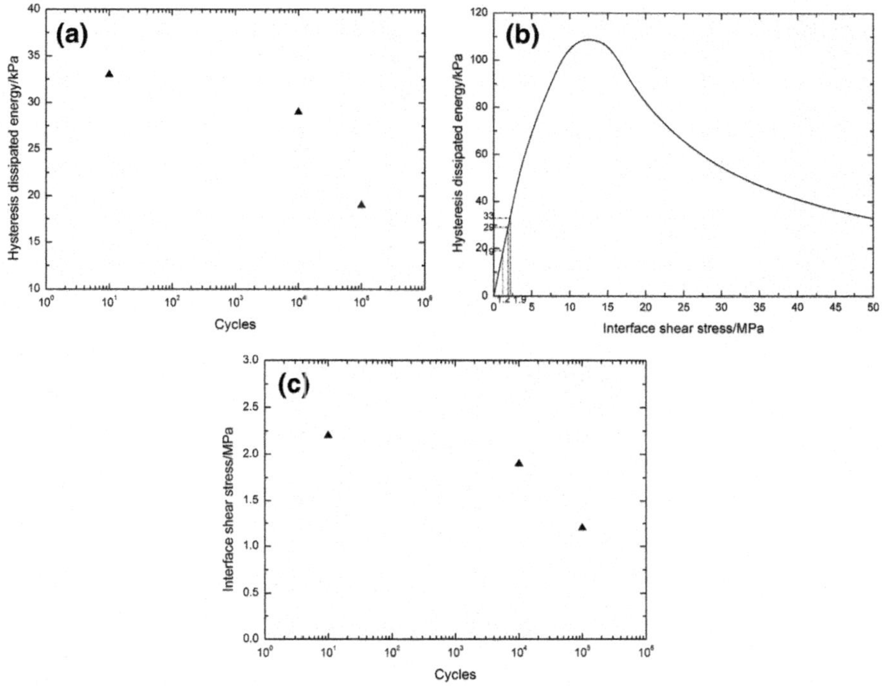


**Fig. 3.35** **a** Experimental fatigue hysteresis dissipated energy versus the applied cycle number curve; **b** the theoretical fatigue hysteresis dissipated energy versus the fiber/matrix interface shear stress curve; and **c** the estimated fiber/matrix interface shear stress curve of 2.5D C/SiC composite under the fatigue peak stress of $\sigma_{max}$ = 230 MPa at 600 °C in inert condition

#### 3.3.2.5 3D SiC/SiC Composite

Shi et al. [39] investigated the tension–tension cyclic fatigue behavior of 3D braided SiC/SiC composite at 1300 °C in air condition. When the fatigue peak stress is $\sigma_{max}$ = 100 MPa, the experimental fatigue hysteresis dissipated energy versus the applied cycle number curve is shown in Fig. 3.36a. The experimental fatigue hysteresis dissipated energy for the applied cycle number of $N$ = 10, 50, 100, 200, 300, and 400 are 7.4, 8.3, 9.9, 12, 14.2, and 25 kPa. The theoretical fatigue hysteresis dissipated energy versus the fiber/matrix interface shear stress is shown in Fig. 3.36b. With decreasing fiber/matrix interface shear stress, the fiber/matrix interface debonding ratio of $2l_d/l_c$ changes from $2l_d/l_c < 1$ to $2l_d/l_c = 1$, and the fiber/matrix interface sliding ratio of $y/l_d$ changes from $y/l_d = 1$ to $y/l_d < 1$, and then $y/l_d = 1$, corresponding to the fiber/matrix interface slip Case I, II, III, and IV. The estimated fiber/matrix interface shear stress versus applied cycle number curve is shown in Fig. 3.36c. The estimated fiber/matrix interface shear stress for the applied cycle number of $N$ = 10, 50, 100, 200, 300, and 400 are 11.6, 10.3, 8.6, 7.1, 6, and 2.5 MPa, respectively.

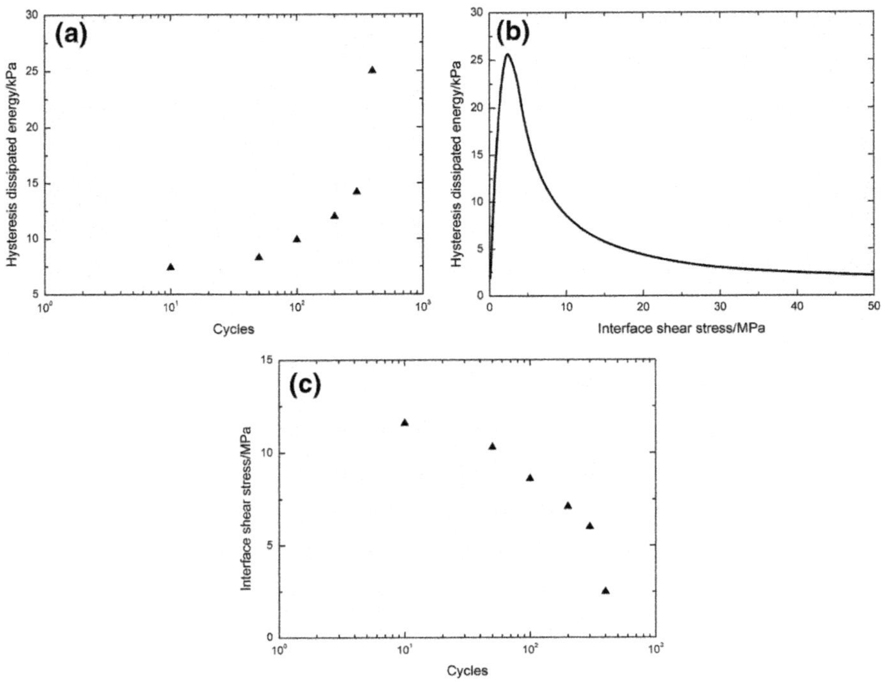


**Fig. 3.36** **a** Experimental fatigue hysteresis dissipated energy versus the applied cycle number curve; **b** the theoretical fatigue hysteresis dissipated energy versus the fiber/matrix interface shear stress curve; and **c** the estimated fiber/matrix interface shear stress curve of 3D braided SiC/SiC composite under the fatigue peak stress of $\sigma_{max}$ = 100 MPa at 1300 °C in air condition

## 3.4 Conclusions

In this chapter, the fiber/matrix interface shear stress of fiber-reinforced CMCs with different fiber preforms, i.e., unidirectional, 2D cross-ply and woven, 2.5D woven and 3D braided, has been estimated from the fatigue hysteresis dissipated energy at room and elevated temperatures. Under cyclic fatigue loading, the experimental fatigue hysteresis dissipated energy versus the applied cycles and the theoretical fatigue hysteresis dissipated energy versus the fiber/matrix interface shear stress have been analyzed. With decreasing of the fiber/matrix interface shear stress, the fatigue hysteresis dissipated energy increases to the peak value, and then decreases to zero, corresponding to the interface slip Case I, II, III, and IV. Using the experimental fatigue hysteresis dissipated energy, the fiber/matrix interface shear stress of unidirectional SiC/CAS, SiC/$Si_3N_4$ with the strong and weak fiber/matrix interface bonding, C/SiC composites at room temperature and 800 °C in air condition, cross-ply SiC/CAS and C/SiC composites at room temperature, 700, 800 and 850 °C in air condition, 2D C/SiC at room temperature, 550 °C in air and 1200 °C in vacuum conditions, 2D SiC/SiC at room temperature, 800 °C in air, 600, 800 and

1000 °C in inert, 1000, 1100 and 1200 °C in air and steam, 1300 °C in air conditions, 2.5D C/SiC at room temperature, 800 °C in air and 600 °C in inert conditions, and 3D braided SiC/SiC at 1300 °C in air conditions have been obtained.

## References


1. Rouby D, Reynaud P (1993) Fatigue behavior related to interface modification during load cycling in ceramic-matrix fibre composites. Compos Sci Technol 48(1–4):109–118. https://doi.org/10.1016/0266-3538(93)90126-2
2. Evans AG, Zok FW, McMeeking RM (1995) Fatigue of ceramic matrix composites. Acta Metall Mater 43(3):859–875. https://doi.org/10.1016/0956-7151(94)00304-Z
3. Rouby D, Louet N (2002) The frictional interface: a tribological approach of thermal misfit, surface roughness and sliding velocity effects. Compos A 33:1453–1459. https://doi.org/10.1016/S1359-835X(02)00145-8
4. Holmes JW, Cho CD (1992) Experimental observation of frictional heating in fiber-reinforced ceramics. J Am Ceram Soc 75(4):929–938. https://doi.org/10.1111/j.1151-2916.1992.tb04162.x
5. Kim J, Liaw PK (2005) Characterization of fatigue damage modes in nicalon/calcium aluminosilicate composites. J Eng Mater Technol 127:8–15. https://doi.org/10.1115/1.1836766
6. Liu CD, Cheng LF, Luan XG, Lin B, Zhou J (2008) Damage evolution and real-time non-destructive evaluation of 2D carbon-fiber/SiC-matrix composites under fatigue loading. Mater Lett 62:3922–3924. https://doi.org/10.1016/j.matlet.2008.04.063
7. Holmes JW, Sørensen BF (1995) High temperature mechanical behavior of ceramic matrix composites. In: Nair SV, Jakus K (eds) Butterworth-Hinemann, Boston MA, pp 261–326
8. Yang CP, Jiao GQ, Wang B, Du L (2009) Oxidation damages and a stiffness model for 2D-C/SiC composites. Acta Mater Compos Sin 26:175–181
9. Reynaud P (1996) Cyclic fatigue of ceramic-matrix composites at ambient and elevated temperatures. Compos Sci Technol 56(7):809–814. https://doi.org/10.1016/0266-3538(96)00025-5
10. Domergue JM, Vagaggini E, Evans AG (1995) Relationship between hysteresis measurements and the constituent properties of ceramic matrix composites: II, experimental studies on unidirectional materials. J Am Ceram Soc 78(10):2721–2731. https://doi.org/10.1111/j.1151-2916.1995.tb08047.x
11. Fantozzi G, Reynaud P (2009) Mechanical hysteresis in ceramic matrix composites. Mater Sci Eng A 521–522:18–23. https://doi.org/10.1016/j.msea.2008.09.128
12. Mall S, Engesser JM (2006) Effects of frequency on fatigue behavior of CVI C/SiC at elevated temperature. Compos Sci Technol 66:863–874. https://doi.org/10.1016/j.compscitech.2005.06.020
13. Moevus M, Reynaud P, R'Mili M, Godin N, Rouby D, Fantozzi G (2006) Static fatigue of a 2.5D SiC/[Si-B-C] composite at intermediate temperature under air. Adv Sci Technol 50:141–146. https://doi.org/10.4028/www.scientific.net/AST.50.141
14. Cho CD, Holmes JW, Barber JR (1991) Estimate of interfacial shear in ceramic composites from frictional heating measurements. J Am Ceram Soc 74(11):2802–2808. https://doi.org/10.1111/j.1151-2916.1991.tb06846.x
15. Vagaggini E, Domergue JM, Evans AG (1995) Relationships between hysteresis measurements and the constituent properties of ceramic matrix composites: I, Theory. J Am Ceram Soc 78(10):2709–2720. https://doi.org/10.1111/j.1151-2916.1995.tb08047.x

16. Solti JP, Robertson DD, Mall S (2000) Estimation of interfacial properties from hysteresis energy loss in unidirectional ceramic matrix composites. Adv Compos Mater 9(3):161–173. https://doi.org/10.1163/15685510051033322
17. Li LB, Song YD (2010) An approach to estimate interface shear stress of ceramic matrix composites from hysteresis loops. Appl Compos Mater 17:309–328. https://doi.org/10.1007/s10443-009-9122-6
18. Li LB, Reynaud P, Fantozzi G (2017) Tension-tension fatigue behavior of unidirectional SiC/$Si_3N_4$ composite with strong and weak interface bonding at room temperature. Ceram Int 43:8769–8777. https://doi.org/10.1016/j.ceramint.2017.03.211
19. Li LB (2013) Modeling hysteresis behavior of cross-ply C/SiC ceramic matrix composites. Compos B 53:36–45. https://doi.org/10.1016/j.compositesb.2013.04.029
20. Li LB (2013) Fatigue hysteresis behavior of cross-ply C/SiC ceramic matrix composites at room and elevated temperatures. Mater Sci Eng A 586:160–170. https://doi.org/10.1016/j.msea.2013.08.017
21. Li LB (2016) Comparisons of damage evolution between 2D C/SiC and SiC/SiC ceramic-matrix composites under tension-tension cyclic fatigue loading at room and elevated temepratures. Materials 9:844. https://doi.org/10.3390/ma9100844
22. Li LB (2017) Comparisons of interface shear stress degradation rate between C/SiC and SiC/SiC ceramic-matrix composites under cyclic fatigue loading at room and elevated temperatures. Compos Interfaces 24:171–202. https://doi.org/10.1080/09276440.2016.1196995
23. Li LB (2018) Synergistic effects of temperature, oxidation, and stress level on fatigue hysteresis behavior of cross-ply ceramic-matrix composites. J Aust Ceram Soc 54:11–22. https://doi.org/10.1007/s41779-017-0121-z
24. Li LB, Song YD, Sun YC (2013) Estimate interface shear stress of unidirectional C/SiC ceramic matrix composites from hysteresis loops. Appl Compos Mater 20:693–707. https://doi.org/10.1007/s10443-012-9297-0
25. Li LB (2014) Assessment of the interfacial properties from fatigue hysteresis loss energy in ceramic-matrix composites with different fiber preforms. Mater Sci Eng A 613:17–36. https://doi.org/10.1016/j.msea.2014.06.092
26. Li LB, Song YD, Sun YC (2014) Effect of matrix cracking on hysteresis behavior of cross-ply ceramic matrix composites. J Compos Mater 48:1505–1530. https://doi.org/10.1177/0021998313488149
27. Li LB, Song YD, Sun ZG (2009) Influence of interface deboning on the fatigue hysteresis loops of ceramic matrix composites. Chin J Solid Mech 30:8–14
28. Shuler SF, Holmes JW, Wu X, Roach D (1993) Influence of loading frequency on the room-temperature fatigue of a carbon-fiber/SiC-matrix composite. J Am Ceram Soc 76:2327–2336. https://doi.org/10.1111/j.1151-2916.1993.tb07772.x
29. Staehler JM, Mall S, Zawada LP (2003) Frequency dependence of high-cycle fatigue behavior of CVI C/SiC at room temperature. Compos Sci Technol 63:2121–2131. https://doi.org/10.1016/S0266-3538(03)00190-8
30. Li Y, Xiao P, Li Z, Zhou W, Liensdorf T, Freudenberg W, Langhof N, Krenkel W (2016) Tensile fatigue behavior of plain-weave reinforced $C_f$/C-SiC composites. Ceram Int 42:6850–6857. https://doi.org/10.1016/j.ceramint.2016.01.068
31. Rodrigues PA, Rosa LG, Steen M. (1995) Fatigue behavior of a ceramic matrix composite (CMC), 2D $C_{fiber}$/$SiC_{matrix}$. In: The 2nd International conference on high temperature ceramic matrix composites, Santa Barbara, CA, United States
32. Shi J (2001) Tensile fatigue and life prediction of a SiC/SiC composite. In: Proceeding of ASME Turbo Expo 2001, New Orleans, Louisiana
33. Michael K (2010) Fatigue behavior of a SiC/SiC composite at 1000°C in air and steam. Master thesis, Air Force Institute of Technology, Ohio, USA
34. Groner JD (1994) Characterization of fatigue behavior of 2D woven fabric reinforced ceramic matrix composite at elevated temperature. Master thesis, Air Force Institute of Technology, Ohio, USA

35. Jacob D (2010) Fatigue behavior of an advanced SiC/SiC composite with an oxidation inhibited matrix at 1200°C in air and in steam. Master thesis, Air Force Institute of Technology, Ohio, USA
36. Zhu SJ, Mizuno M, Nagano Y, Cao JW, Kagawa Y, Kaya H (1998) Creep and fatigue behavior in an enhanced SiC/SiC composite at high temperature. J Am Ceram Soc 81:2269–2277. https://doi.org/10.1111/j.1151-2916.1998.tb02621.x
37. Yang FS (2011) Research on fatigue behavior of 2.5d woven ceramic matrix composites. Master thesis, Nanjing University of Aeronautics and Astronautics, Nanjing, China
38. Dalmaz A, Reynaud P, Rouby D, Fantozzi G, Abbe F (1998) Mechanical behavior and damage development during cyclic fatigue at high-temperature of a 2.5D carbon/SiC composite. Compos Sci Technol 58:693–699. https://doi.org/10.1016/S0266-3538(97)00150-4
39. Shi DQ, Jing X, Yang XG (2015) Low cycle fatigue behavior of a 3D braided KD-I fiber reinforced ceramic matrix composite for coated and uncoated specimens at 1100 °C and 1300 °C. Mater Sci Eng A 631:38–44. https://doi.org/10.1016/j.msea.2015.01.078

# Chapter 4
# Fatigue Life Prediction of Ceramic-Matrix Composites

## 4.1 Introduction

Under cyclic fatigue loading, the damage mechanisms of interface wear and interface oxidation degraded the fiber/matrix interface shear stress and fibers strength. The fatigue hysteresis loops, fatigue hysteresis energy dissipation, fatigue hysteresis modulus can be used to monitor the damage evolution in fiber-reinforced CMCs. At room and elevated temperatures, the fatigue hysteresis dissipated energy and fatigue hysteresis modulus of C/SiC composites degrade with increasing applied cycles [1, 2], and the degradation rate and fatigue lifetime are influenced by the loading frequency and peak stress [3–9]. However, for the SiC/SiC composite, the evolution of the fatigue hysteresis dissipated energy and the fatigue lifetime depends on the peak stress, temperature, and testing environments [10–17]. The presence of steam at higher fatigue stresses caused noticeable degradation in fatigue life. Upon first loading to the fatigue peak stress, matrix cracking and fiber/matrix interface debonding occur, and fibers fractured in the interface debonded and bonded region. With increasing applied cycles, the degradation of the fiber/matrix interface shear stress reduces the load transfer capacity between fibers and the matrix [18], and the load carrying capacity of fibers [19]. At elevated temperature in oxidative environment, with increasing oxidation time, more fibers are broken as the interface shear stress and fibers strength degrade [20–24]. When the broken fibers fraction approaches to the critical value, the composites fatigue fracture [25].

In this chapter, the fatigue lifetime of fiber-reinforced CMCs at room and elevated temperatures is predicted for different testing conditions and fiber preforms. The relationships among the broken fibers fraction, applied cycle numbers, and the fatigue peak stress are established. At room temperature, the damage mechanisms of fiber/matrix interface wear degrade the fiber/matrix interface shear stress and fibers strength; and at elevated temperature in oxidative atmosphere, the damage mechanisms of fiber/matrix interface wear and oxidation degrade the fiber/matrix interface shear stress and fibers strength. The fibers broken fraction in



L. Li, *Damage, Fracture, and Fatigue of Ceramic-Matrix Composites*,
https://doi.org/10.1007/978-981-13-1783-5_4

different damage regions are determined, i.e., the interface wear region, interface oxidation region, interface debonded region, and interface bonded region. When the fibers broken fraction approaches to the critical value, the composites fatigue fracture. The fatigue limit stress and fatigue life *S–N* curve of fiber-reinforced CMCs with different fibers preform are predicted, i.e., the unidirectional C/SiC, SiC/CAS, SiC/1723 and SiC/$Si_3N_4$, cross-ply C/SiC, SiC/CAS, SiC/1723 and SiC/BMAS, 2D woven C/SiC and SiC/SiC, 2/5D woven C/SiC, and 3D braided C/SiC composites.

## 4.2 Fatigue Life Prediction Models

Under cyclic fatigue loading, the loading directions are along the fiber for unidirectional CMCs, 0° fiber ply for cross-ply and plain-weave 2D woven CMCs, warp yarn for 2.5D woven CMCs, and axial fibers at a small angle $\theta$ for 3D braided CMCs. An effective coefficient of the fiber volume content along the loading direction (ECFL) can be defined using the following equation.

$$\psi = \frac{V_{\mathrm{f\_axial}}}{V_{\mathrm{f}}} \tag{4.1}$$

where $V_{\mathrm{f}}$ and $V_{\mathrm{f\_axial}}$ denote the total fiber volume fraction in the composites and the effective fiber volume fraction in the cyclic loading direction. The values of parameter $\psi$ for the unidirectional, cross-ply, 2D woven, 2.5D woven, and 3D braided fiber-reinforced CMCs are 1.0, 0.5, 0.5, 0.75, and 0.93, respectively.

### 4.2.1 Life Prediction at Room Temperature

Under cyclic fatigue loading, the fiber/matrix interface shear stress and fibers strength decrease with increasing applied cycle number due to the fiber/matrix interface wear. Evans et al. [26] developed the fiber/matrix interface shear stress degradation model.

$$\tau_i(N) = \tau_{\mathrm{io}} + \left[1 - \exp\left(-\omega N^{\lambda}\right)\right]\left(\tau_{\mathrm{imin}} - \tau_{\mathrm{io}}\right) \tag{4.2}$$

where $\tau_{\mathrm{io}}$ denotes the initial interface shear stress; $\tau_{\mathrm{imin}}$ denotes the steady-state interface shear stress under cyclic loading; $N$ denotes the cycle number; $\tau_i(N)$ denotes the interface shear stress at the $N$th cycle; and $\omega$ and $\lambda$ are empirical constants.

The fibers strength degradation can be described using the following equation [27].

$$\sigma_{\text{o}}(N) = \sigma_{\text{o}}[1 - p_1(\log N)^{p_2}] \tag{4.3}$$

where $p_1$ and $p_2$ are empirical parameters.

The two-parameter Weibull model is adopted to describe the fibers strength distribution. The fibers fracture probability $P_f$ can be described using the following equation [28].

$$P_f = 1 - \exp\left(-\int_{L_0} \frac{1}{l_0}\left[\frac{\sigma_{\text{f}}(x)}{\sigma_0}\right]^{m_f} \mathrm{d}x\right) \tag{4.4}$$

The stress carried by intact and fracture fibers can be described using the following equation.

$$\frac{\sigma}{V_{\text{f}}} = T\left[1 - P_f\right] + \langle T_{\text{b}} \rangle P_f \tag{4.5}$$

where $<T_{\text{b}}>$ denotes the average stress carried by broken fibers.

$$\begin{aligned}\langle T_{\text{b}} \rangle = & \frac{T}{P_f}\left(\frac{\sigma_{\text{c}}}{T}\right)^{m+1}\left(\frac{\sigma_{\text{o}}(N)}{\sigma_{\text{o}}}\right)^m \frac{\tau_i(N)}{\tau_i}\left\{1 - \exp\left[-\left(\frac{T}{\sigma_{\text{c}}}\right)^{m+1}\left(\frac{\sigma_{\text{o}}}{\sigma_{\text{o}}(N)}\right)^m \frac{\tau_i}{\tau_i(N)}\right]\right\} \\ & - \frac{T}{P_f}\exp\left\{-\left(\frac{T}{\sigma_{\text{c}}}\right)^{m+1}\left(\frac{\sigma_{\text{o}}}{\sigma_{\text{o}}(N)}\right)^m \frac{\tau_i}{\tau_i(N)}\right\}\end{aligned} \tag{4.6}$$

and

$$P_f = 1 - \exp\left\{-\left(\frac{T}{\sigma_{\text{c}}}\right)^{m+1}\left(\frac{\sigma_{\text{o}}}{\sigma_{\text{o}}(N)}\right)^m \frac{\tau_i}{\tau_i(N)}\right\} \tag{4.7}$$

Substituting Eqs. (4.6) and (4.7) into Eq. (4.5), it leads to the following equation.

$$\frac{\sigma}{V_{\text{f}}} = T\left(\frac{\sigma_{\text{c}}}{T}\right)^{m+1}\left(\frac{\sigma_{\text{o}}(N)}{\sigma_{\text{o}}}\right)^m \frac{\tau_i(N)}{\tau_i}\left\{1 - \exp\left[-\left(\frac{T}{\sigma_{\text{c}}}\right)^{m+1}\left(\frac{\sigma_{\text{o}}}{\sigma_{\text{o}}(N)}\right)^m \frac{\tau_i}{\tau_i(N)}\right]\right\} \tag{4.8}$$

Using Eqs. (4.2), (4.3), and (4.8), the intact fibers stress $T$ at the matrix cracking plane can be determined for different peak stresses. Substituting Eqs. (4.2), (4.3),

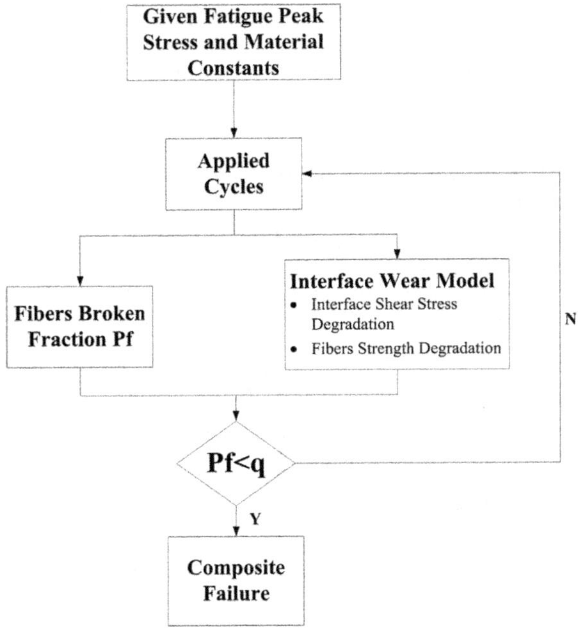


**Fig. 4.1** Flowchart for fatigue life prediction at room temperature

and $T$ into the Eq. (4.7), the fibers broken fraction can be determined. The flowchart of the fatigue life prediction at room temperature is given in Fig. 4.1.

### *4.2.2 Life Prediction at Elevated Temperature*

At elevated temperature, matrix cracking serves as the avenues for the ingress of oxidation gas into the composite [29–31]. The fibers oxidation is assumed to be controlled by diffusion of oxygen gas in the present analysis, as shown in Fig. 4.2. The fibers strength would be time-dependent and controlled by surface defects resulting from the oxidation [32]. Assuming the fibers fracture toughness remains to be constant and that the fibers strength $\sigma_0$ is related to the mean oxidized layer thickness, the fibers time dependence strength can be described using the following equations [33].

$$\sigma_0(t) = \sigma_0, t \leq \frac{1}{k}\left(\frac{K_{\mathrm{IC}}}{Y\sigma_0}\right)^4 \tag{4.9}$$

$$\sigma_0(t) = \frac{K_{\mathrm{IC}}}{Y\sqrt[4]{kt}}, \quad t > \frac{1}{k}\left(\frac{K_{\mathrm{IC}}}{Y\sigma_0}\right)^4 \tag{4.10}$$

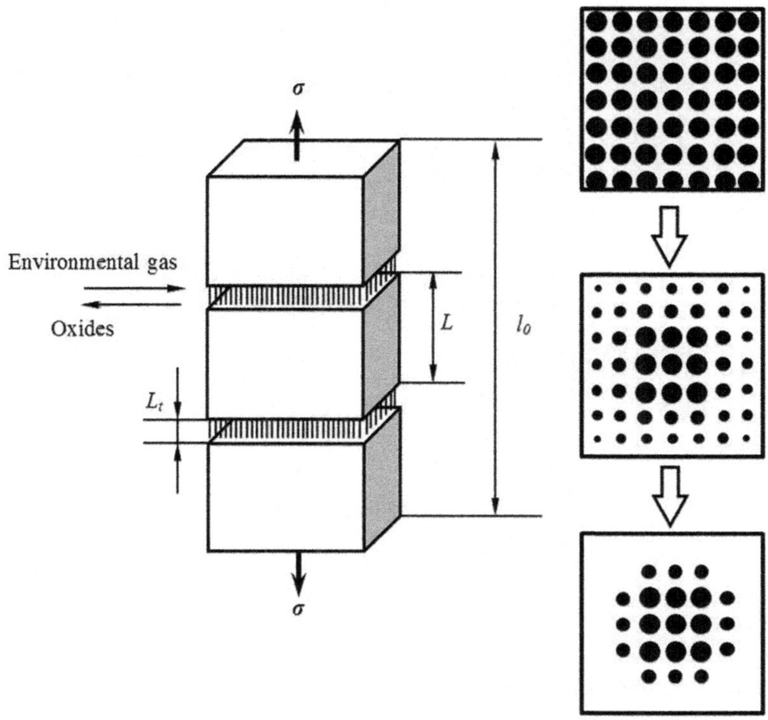


**Fig. 4.2** Schematic of fiber oxidation in multiple cracked CMCs

where $K_{IC}$ denotes the critical stress intensity factor; $Y$ is a geometric parameter; and $k$ is the parabolic rate constant.

The fiber/matrix interface oxidation length of $l_t$ can be described using the following equation [34].

$$l_t = \varphi_1 \left(1 - e^{-\frac{\varphi_2 t}{b}}\right) \tag{4.11}$$

where $\varphi_1$ and $\varphi_2$ are fitting parameters dependent on temperature, and $b$ is the delay factor.

The fracture probabilities of oxidized fibers in the oxidation region, unoxidized fibers in the oxidation region, fibers in the interface debonded region, and interface bonded region of $P_{fa}$, $P_{fb}$, $P_{fc}$, and $P_{fd}$ can be described using the following equations.

$$P_{fa} = 1 - \exp\left\{-2\frac{l_t}{l_0}\left[\frac{T}{\sigma_0(t)}\right]^{m_f}\right\} \tag{4.12}$$

$$P_{\mathrm{fb}} = 1 - \exp\left\{-2\frac{l_{\mathrm{t}}}{l_0}\left(\frac{T}{\sigma_0}\right)^{m_{\mathrm{f}}}\right\} \tag{4.13}$$

$$P_{\mathrm{fc}} = 1 - \exp\left\{-\frac{r_{\mathrm{f}}T^{m_{\mathrm{f}}+1}}{l_0(\sigma_0(N))^{m_{\mathrm{f}}}\tau_i(N)(m_{\mathrm{f}}+1)}\left[1 - \left(1 - \frac{l_{\mathrm{d}}(N)}{l_{\mathrm{f}}(N)}\right)^{m_{\mathrm{f}}+1}\right]\right\} \tag{4.14}$$

$$P_{\mathrm{fd}} = 1 - \exp\left\{-\frac{2r_{\mathrm{f}}T^{m_{\mathrm{f}}}}{\rho l_0(\sigma_0(N))^{m_{\mathrm{f}}}(m_{\mathrm{f}}+1)\left(1 - \frac{\sigma_{\mathrm{fo}}}{T} - \frac{l_{\mathrm{d}}(N)}{l_{\mathrm{s}}(N)}\right)} \times \left[\left(1 - \frac{l_{\mathrm{d}}(N)}{l_{\mathrm{f}}(N)}\right.\right.\right.$$
$$\left.\left.\left. - \left(1 - \frac{\sigma_{\mathrm{fo}}}{T} - \frac{l_{\mathrm{d}}(N)}{l_{\mathrm{f}}(N)}\right)\frac{\rho l_{\mathrm{d}}(N)}{r_{\mathrm{f}}}\right)^{m_{\mathrm{f}}+1} - \left(1 - \frac{l_{\mathrm{d}}(N)}{l_{\mathrm{f}}(N)} - \left(1 - \frac{\sigma_{\mathrm{fo}}}{T} - \frac{l_{\mathrm{d}}(N)}{l_{\mathrm{f}}(N)}\right)\frac{\rho l_{\mathrm{c}}}{2r_{\mathrm{f}}}\right)^{m_{\mathrm{f}}+1}\right]\right\} \tag{4.15}$$

where $l_{\mathrm{f}}$ denotes the slip length over which the fiber stress would decay to zero if not interrupted by the far-field equilibrium stresses.

$$l_{\mathrm{f}}(N) = \frac{r_{\mathrm{f}}T}{2\tau_i(N)} \tag{4.16}$$

The GLS assumption is used to determine the stress carried by intact and fracture fibers [28].

$$\frac{\sigma}{V_{\mathrm{f}}} = \left[1 - P_f\left(1 + \frac{2l_{\mathrm{f}}}{l_{\mathrm{c}}}\right)\right]T + P_r\frac{2l_{\mathrm{f}}}{l_{\mathrm{c}}}\langle T_{\mathrm{b}}\rangle \tag{4.17}$$

where

$$P_f = \varphi[\eta P_{\mathrm{fa}} + (1-\eta)P_{\mathrm{fb}}] + P_{\mathrm{fc}} + P_{\mathrm{fd}} \tag{4.18}$$

$$P_r = P_{\mathrm{fc}} + P_{\mathrm{fd}} \tag{4.19}$$

where $\eta$ denotes the oxidation fibers fraction in the oxidized region, and $\varphi$ denotes the fraction of oxidation in the multiple matrix cracks.

$$\varphi = \frac{l_{\mathrm{sat}}}{l_{\mathrm{f}} - 2l_{\mathrm{t}}} \tag{4.20}$$

The average stress carried by broken fibers can be described using the following equation.

$$
\begin{aligned}
\langle T_\mathrm{b}\rangle &= \int_0^{l_\mathrm{f}} T_\mathrm{b}(x)f(x)\mathrm{d}x \\
&= \frac{T}{P_r}\left(\frac{\sigma_\mathrm{c}}{T}\right)^{m_\mathrm{f}+1}\left(\frac{\sigma_\mathrm{o}(N)}{\sigma_\mathrm{o}}\right)^{m_\mathrm{f}}\frac{\tau_i(N)}{\tau_i}\left\{1-\exp\left[-\left(\frac{T}{\sigma_\mathrm{c}}\right)^{m_\mathrm{f}+1}\left(\frac{\sigma_\mathrm{o}}{\sigma_\mathrm{o}(N)}\right)^{m_\mathrm{f}}\frac{\tau_i}{\tau_i(N)}\right]\right\} \\
&\quad -\frac{T}{P_r}\exp\left\{-\left(\frac{T}{\sigma_\mathrm{c}}\right)^{m_\mathrm{f}+1}\left(\frac{\sigma_\mathrm{o}}{\sigma_\mathrm{o}(N)}\right)^{m_\mathrm{f}}\frac{\tau_i}{\tau_i(N)}\right\}
\end{aligned}
\tag{4.21}
$$

Substituting Eqs. (4.18), (4.19), and (4.21) into Eq. (4.17), the intact fibers stress $T$ at the matrix crack plane can be determined for different applied cycle number and fatigue peak stresses. Substituting Eqs. (4.2), (4.3), (4.9), (4.10), and (4.11) and $T$ into Eqs. (4.18) and (4.19), the fibers broken fraction can be determined. The flowchart of the fatigue life prediction at elevated temperature is given in Fig. 4.3.

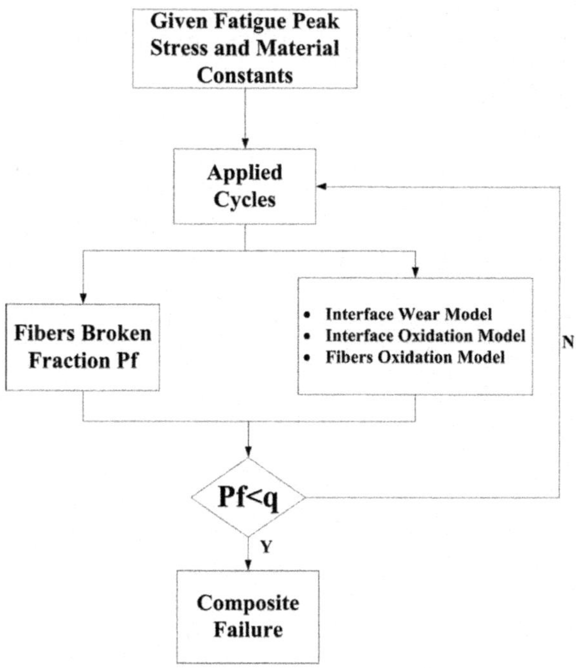


**Fig. 4.3** Flowchart for fatigue life prediction at elevated temperature

## 4.3 Experimental Comparisons

### 4.3.1 Unidirectional Ceramic-Matrix Composites

#### 4.3.1.1 C/SiC Composite at Room Temperature and 800 °C in Air Condition

The experimental and theoretical fatigue life *S–N* curves and the broken fibers fraction versus the applied cycle number curves of unidirectional C/SiC composite at room temperature are shown in Fig. 4.4. At room temperature, the monotonic tensile strength of C/SiC composite is 270 MPa, and the fatigue limit stress is 88% of the tensile strength. When the fatigue peak stresses are 96 and 88% of the tensile strength, the composite experienced 93,424 and 1,000,000 applied cycles, respectively.

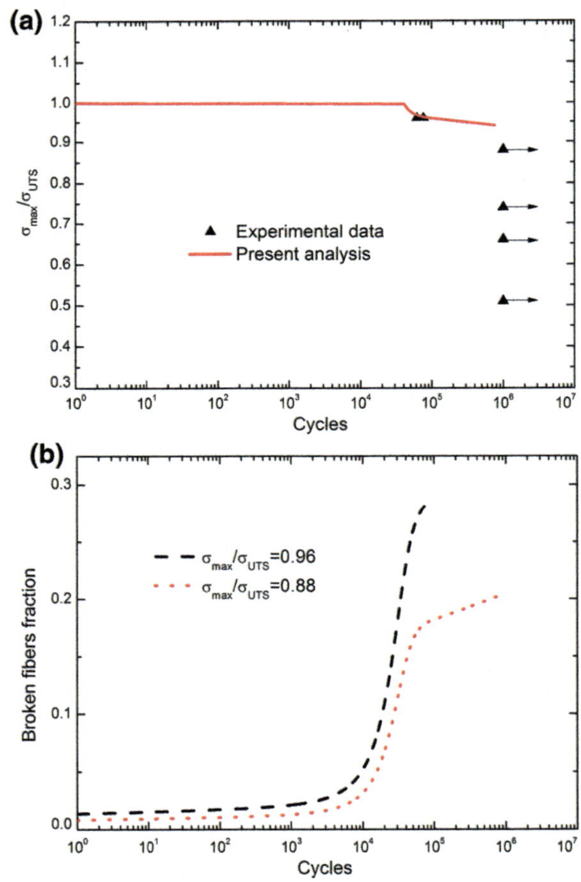


**Fig. 4.4** **a** Experimental and theoretical fatigue life *S–N* curves and **b** the broken fibers fraction versus the applied cycle number curves at the fatigue peak stresses of $0.96\sigma_{UTS}$ and $0.88\sigma_{UTS}$ of unidirectional C/SiC composite at room temperature

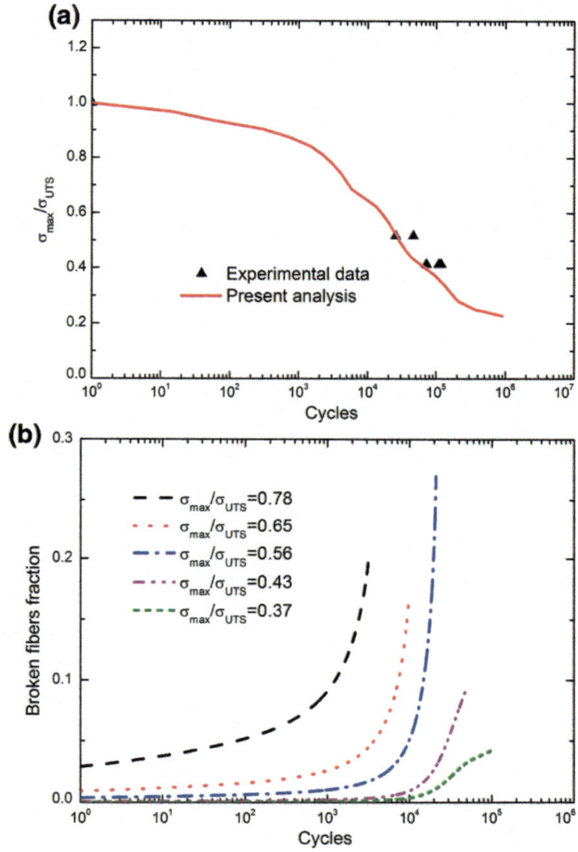


**Fig. 4.5** **a** Experimental and theoretical fatigue life *S–N* curves and **b** the broken fibers fraction versus the applied cycle number curves at the fatigue peak stresses of $0.78\sigma_{UTS}$, $0.65\sigma_{UTS}$, $0.56\sigma_{UTS}$, $0.43\sigma_{UTS}$, and $0.37\sigma_{UTS}$ of unidirectional C/SiC composite at 800 °C in air condition

The experimental and theoretical fatigue life *S–N* curves and the broken fibers fraction versus the applied cycle number curves of unidirectional C/SiC composite at 800 °C in air condition are shown in Fig. 4.5. At 800 °C in air condition, the monotonic tensile strength of C/SiC composite is 320 MPa, and the fatigue limit stress is 22% of the tensile strength. When the fatigue peak stresses are 78, 65, 56, 43, and 37% of the tensile strength, the composite experienced 3133, 9957, 21,245, 48,578, and 102,952 applied cycles, respectively.

#### 4.3.1.2 SiC/CAS Composite at Room Temperature

The experimental and theoretical fatigue life *S–N* curves and the broken fibers fraction versus the applied cycle number curves of unidirectional SiC/CAS composite at room temperature are shown in Fig. 4.6. At room temperature, the monotonic tensile strength of C/SiC composite is 450 MPa, and the fatigue limit stress is 61% of the tensile strength. When the fatigue peak stresses are 93, 84, 75,

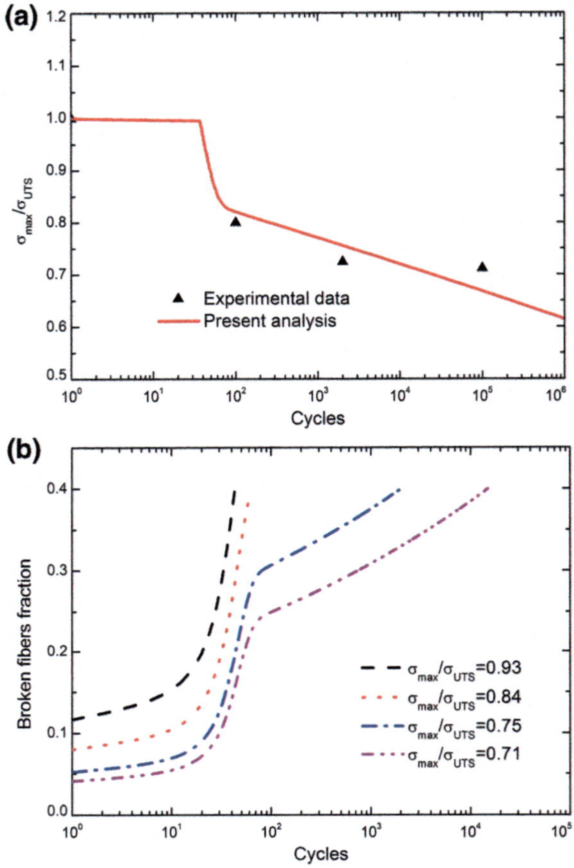


**Fig. 4.6** **a** Experimental and theoretical fatigue life *S–N* curves and **b** the broken fibers fraction versus the applied cycle number curves at the fatigue peak stresses of $0.93\sigma_{UTS}$, $0.84\sigma_{UTS}$, $0.75\sigma_{UTS}$, and $0.71\sigma_{UTS}$ of unidirectional SiC/CAS composite at room temperature

and 71% of the tensile strength, the composite experienced 43, 62, 2005, and 14,778 applied cycles, respectively.

#### 4.3.1.3 SiC/1723 Composite at Room Temperature

The experimental and theoretical fatigue life *S–N* curves and the broken fibers fraction versus the applied cycle number curves of unidirectional SiC/1723 composite at room temperature are shown in Fig. 4.7. At room temperature, the monotonic tensile strength of C/SiC composite is 650 MPa, and the fatigue limit stress is 70% of the tensile strength. When the fatigue peak stresses are 98, 92, 85, 77, and 74% of the tensile strength, the composite experienced 2872, 15,026, 70,823, 305,366, and 535,575 applied cycles, respectively.

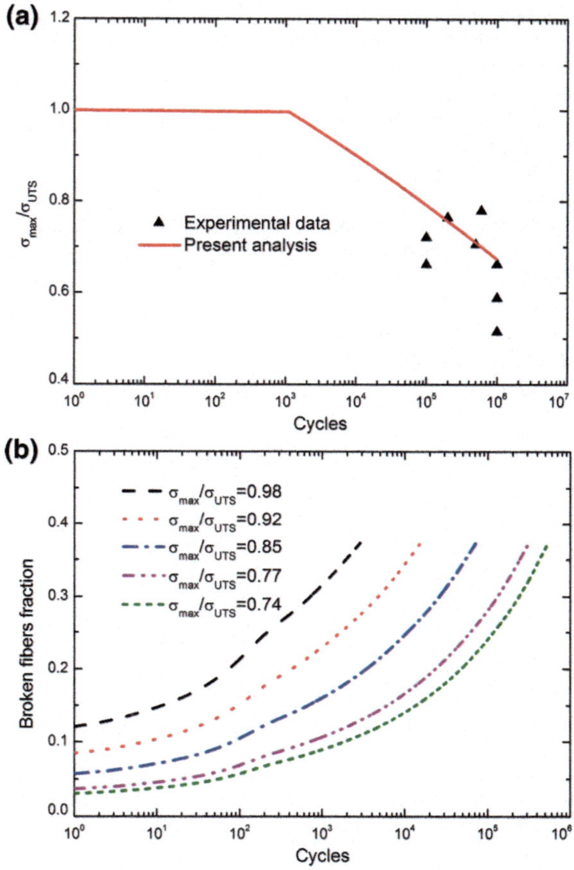


**Fig. 4.7** **a** Experimental and theoretical fatigue life *S–N* curves and **b** the broken fibers fraction versus the applied cycle number curves at the fatigue peak stresses of $0.98\sigma_{UTS}$, $0.92\sigma_{UTS}$, $0.85\sigma_{UTS}$, $0.77\sigma_{UTS}$, and $0.74\sigma_{UTS}$ of unidirectional SiC/1723 composite at room temperature

#### 4.3.1.4 SiC/$Si_3N_4$ Composite

Strong Interface Bonding

The experimental and theoretical fatigue life *S–N* curves and the broken fibers fraction versus the applied cycle number curves of unidirectional SiC/$Si_3N_4$ composite with the strong fiber/matrix interface bonding at room temperature are shown in Fig. 4.8. The fatigue limit stress approaches to 75% of the tensile strength. When the fatigue peak stresses are 97, 90, and 84% of the tensile strength, the composite experienced 110, 2124, and 33,142 applied cycles, respectively.

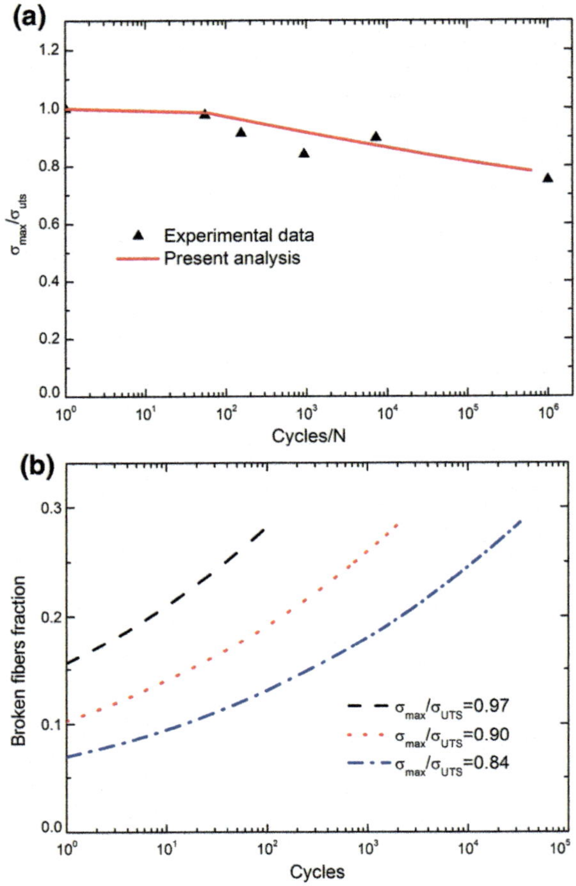


**Fig. 4.8** **a** Experimental and theoretical fatigue life *S–N* curves and **b** the broken fibers fraction versus the applied cycle number curves at the fatigue peak stresses of $0.97\sigma_{UTS}$, $0.9\sigma_{UTS}$, and $0.84\sigma_{UTS}$ of unidirectional $SiC/Si_3N_4$ composite with fiber/matrix strong interface bonding at room temperature. Reprinted with permission from Ref. [24]. Copyright 2017 Elsevier Ltd and Techna Group S.r.l

Weak Interface Bonding

The experimental and theoretical fatigue life *S–N* curves and the broken fibers fraction versus the applied cycle number curves of unidirectional $SiC/Si_3N_4$ composite with the weak fiber/matrix interface bonding at room temperature are shown in Fig. 4.9. The fatigue limit stress approaches to 58% of the tensile strength. When the fatigue peak stresses are 85, 80, and 73% of the tensile strength, the composite experienced 53, 262, and 2764 applied cycles, respectively.

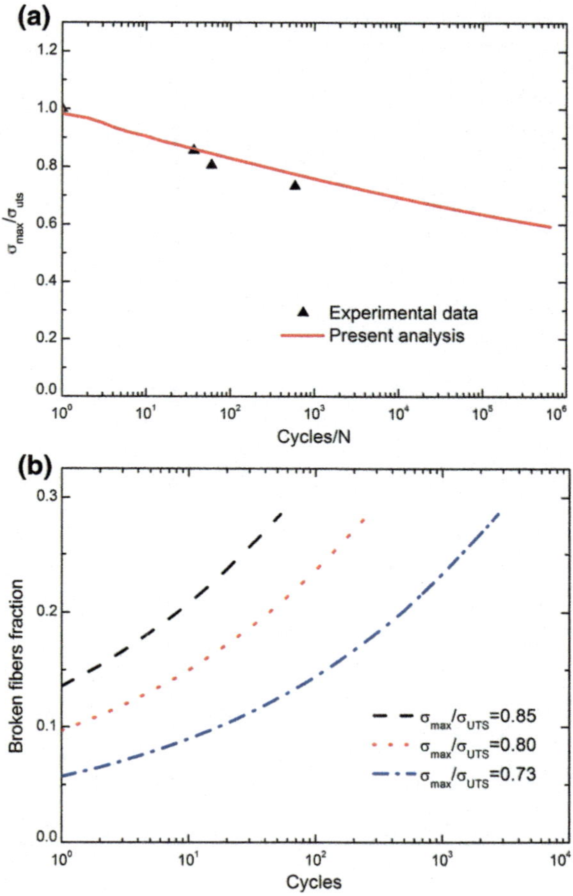


**Fig. 4.9** **a** Experimental and theoretical fatigue life *S–N* curves and **b** the broken fibers fraction versus the applied cycle number curves at the fatigue peak stresses of $0.85\sigma_{UTS}$, $0.8\sigma_{UTS}$, and $0.73\sigma_{UTS}$ of unidirectional $SiC/Si_3N_4$ composite with fiber/matrix weak interface bonding at room temperature. Reprinted with permission from Ref. [24]. Copyright 2017 Elsevier Ltd and Techna Group S.r.l

## *4.3.2 Cross-Ply and 2D Woven Ceramic-Matrix Composites*

### 4.3.2.1 Cross-Ply C/SiC at Room Temperature and 800 °C in Air

The experimental and theoretical fatigue life *S–N* curves and the broken fibers fraction versus the applied cycle number curves of cross-ply C/SiC composite at room temperature are shown in Fig. 4.10. At room temperature, the monotonic tensile strength of cross-ply C/SiC composite is 124 MPa, and the fatigue limit stress approaches to 85% of the tensile strength. When the fatigue peak stresses are 90 and 85% of the tensile strength, the composite experienced 389,371 and 1,000,000 cycles, respectively.

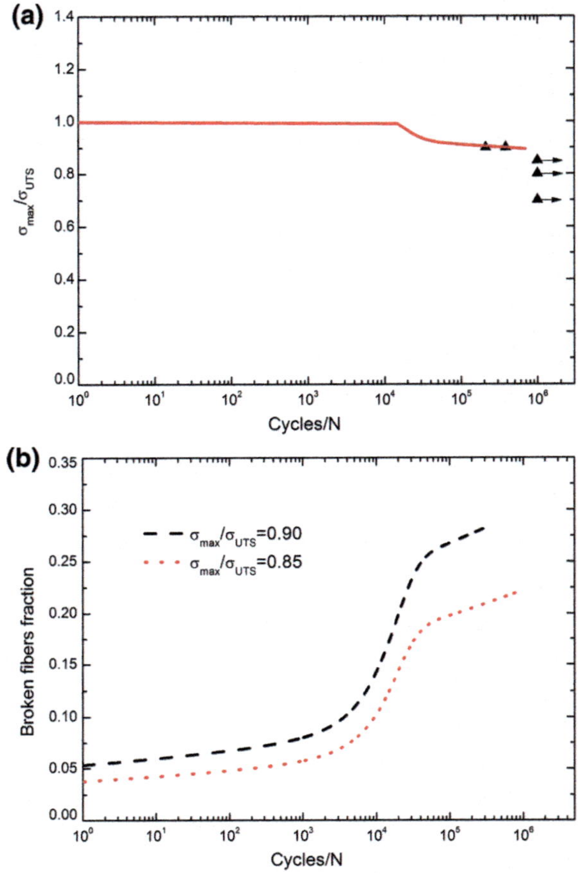


**Fig. 4.10** **a** Experimental and theoretical fatigue life *S–N* curves and **b** the broken fibers fraction versus the applied cycle number curves at the fatigue peak stresses of $0.9\sigma_{UTS}$ and $0.85\sigma_{UTS}$ of cross-ply C/SiC composite at room temperature

The experimental and theoretical fatigue life *S–N* curves and the broken fibers fraction versus the applied cycle number curves of cross-ply C/SiC composite at 800 °C in air condition are shown in Fig. 4.11. At 800 °C in air condition, the monotonic tensile strength of C/SiC composite is 150 MPa, and the fatigue limit stress approaches to 26% of the tensile strength. When the fatigue peak stresses are 70, 65, and 60% of the tensile strength, the composite experienced 1474, 2486, and 3643 cycles, respectively.

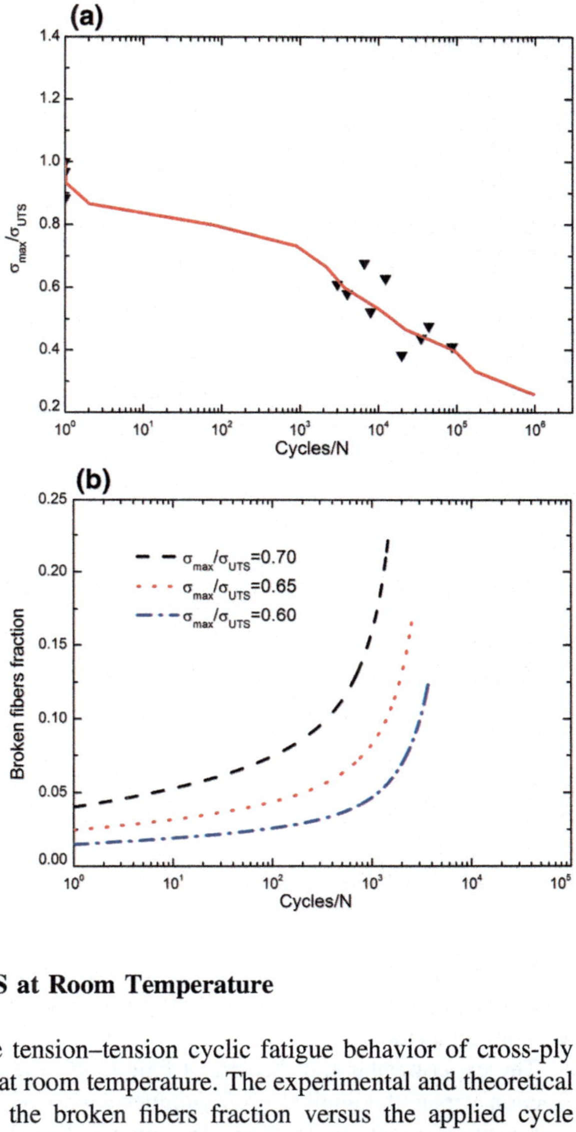


**Fig. 4.11** **a** Experimental and theoretical fatigue life *S–N* curves and **b** the broken fibers fraction versus the applied cycle number curves at the fatigue peak stresses of $0.7\sigma_{UTS}$, $0.65\sigma_{UTS}$, and $0.6\sigma_{UTS}$ of cross-ply C/SiC composite at 800 °C in air condition

#### 4.3.2.2 Cross-Ply SiC/CAS at Room Temperature

Opalski [35] investigated the tension–tension cyclic fatigue behavior of cross-ply $[0/90]_{2s}$ SiC/CAS composite at room temperature. The experimental and theoretical fatigue life *S–N* curves and the broken fibers fraction versus the applied cycle number curves of cross-ply $[0/90]_{2s}$ SiC/CAS composite at room temperature are shown in Fig. 4.12. At room temperature, the monotonic tensile strength of $[0/90]_{2s}$ SiC/CAS composite is 275 MPa, and the fatigue limit stress approaches to 47% of the tensile strength. When the fatigue peak stresses are 87, 72, and 65% of the tensile strength, the composite experienced 290, 3442, and 19,477 cycles, respectively.

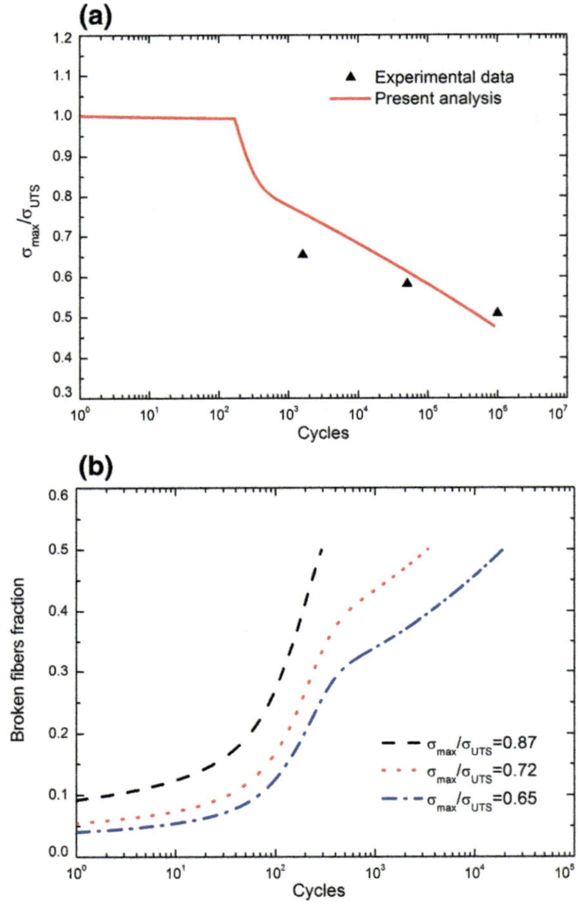


**Fig. 4.12** **a** Experimental and theoretical fatigue life *S–N* curves and **b** the broken fibers fraction versus the applied cycle number curves at the fatigue peak stresses of $0.87\sigma_{UTS}$, $0.72\sigma_{UTS}$, and $0.65\sigma_{UTS}$ of cross-ply $[0/90]_{2s}$ SiC/CAS composite at room temperature

Karandikar [36] investigated the tension–tension cyclic fatigue behavior of cross-ply $[0_3/90/0_3]$, $[0_3/90_3/0_3]$, and [0/90/0/90/0/90/0/90/0] SiC/CAS composite at room temperature.

The experimental and theoretical fatigue life *S–N* curves and the broken fibers fraction versus the applied cycle number curves of cross-ply $[0_3/90/0_3]$ SiC/CAS composite at room temperature are shown in Fig. 4.13. At room temperature, the monotonic tensile strength of $[0_3/90/0_3]$ SiC/CAS composite is 336 MPa, and the fatigue limit stress is 47.3% of the tensile strength. When the fatigue peak stresses are 95, 83, and 68% of the tensile strength, the composite experienced 29, 234, and 9631 applied cycles, respectively.

The experimental and theoretical fatigue life *S–N* curves and the broken fibers fraction versus the applied cycle number curves of cross-ply $[0_3/90_3/0_3]$ SiC/CAS

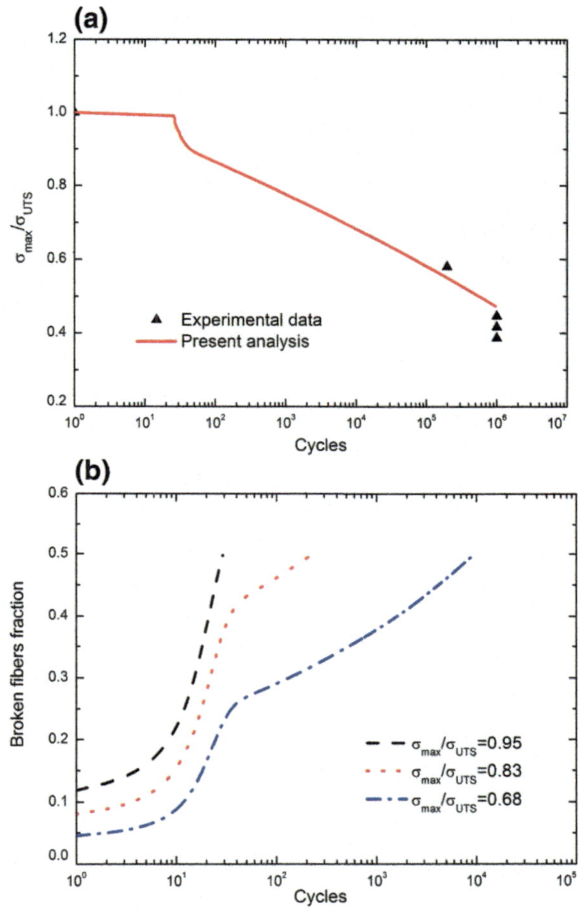


**Fig. 4.13** **a** Experimental and theoretical fatigue life *S–N* curves and **b** the broken fibers fraction versus the applied cycle number curves at the fatigue peak stresses of $0.95\sigma_{UTS}$, $0.83\sigma_{UTS}$, and $0.68\sigma_{UTS}$ of cross-ply $[0_3/90/0_3]$ SiC/CAS composite at room temperature

composite at room temperature are shown in Fig. 4.14. At room temperature, the monotonic tensile strength of $[0_3/90_3/0_3]$ SiC/CAS composite is 261 MPa, and the fatigue limit stress approaches to 45% of the tensile strength. When the fatigue peak stresses are 88, 76, and 69% of the tensile strength, the composite experienced 182, 1310, and 8538 applied cycles, respectively.

The experimental and theoretical fatigue life *S–N* curves and the broken fibers fraction versus the applied cycle number curves of cross-ply [0/90/0/90/0/90/0/90/0] SiC/CAS composite at room temperature are shown in Fig. 4.15. At room temperature, the monotonic tensile strength of [0/90/0/90/0/90/0/90/0] SiC/CAS composite is 235 MPa, and the fatigue limit stress approaches to 55% of the tensile

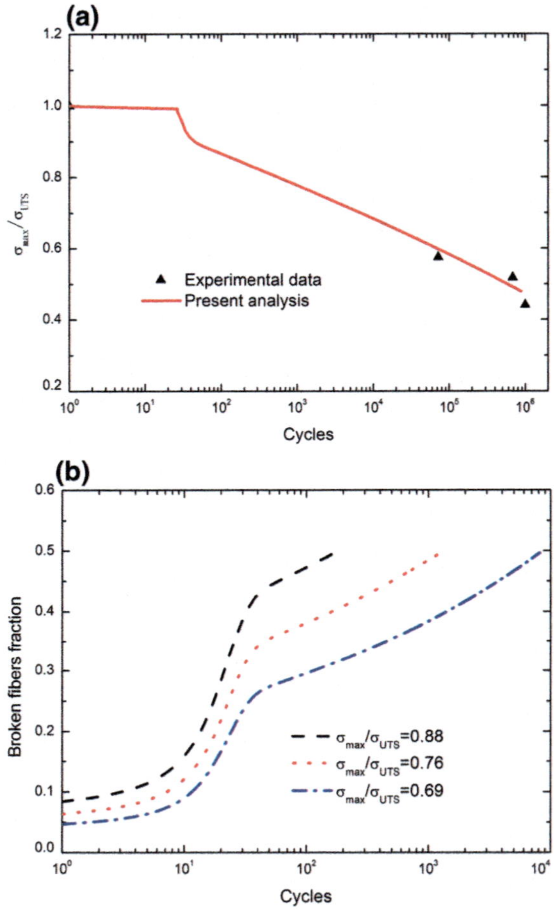


**Fig. 4.14** **a** Experimental and theoretical fatigue life *S–N* curves; and **b** the broken fibers fraction versus the applied cycle number curves at the fatigue peak stresses of $0.88\sigma_{UTS}$, $0.76\sigma_{UTS}$ and $0.69\sigma_{UTS}$ of cross-ply $[0_3/90_3/0_3]$ SiC/CAS composite at room temperature

strength. When the fatigue peak stresses are 94, 76, and 68% of the tensile strength, the composite experienced 41, 1193, and 21,717 applied cycles, respectively.

Kim and Liaw [37] investigated the tension–tension cyclic fatigue behavior of cross-ply $[0/90]_{4s}$ SiC/CAS composite at room temperature. The experimental and theoretical fatigue life *S–N* curves and the broken fibers fraction versus the applied cycle number curves of cross-ply $[0/90]_{4s}$ SiC/CAS composite at room temperature are shown in Fig. 4.16. At room temperature, the monotonic tensile strength of $[0/90]_{4s}$ SiC/CAS composite is 180 MPa, and the fatigue limit stress approaches to 60% of the tensile strength. When the fatigue peak stresses are 95, 83, and 68% of the tensile strength, the composite experienced 29, 234, and 9631 applied cycles, respectively.

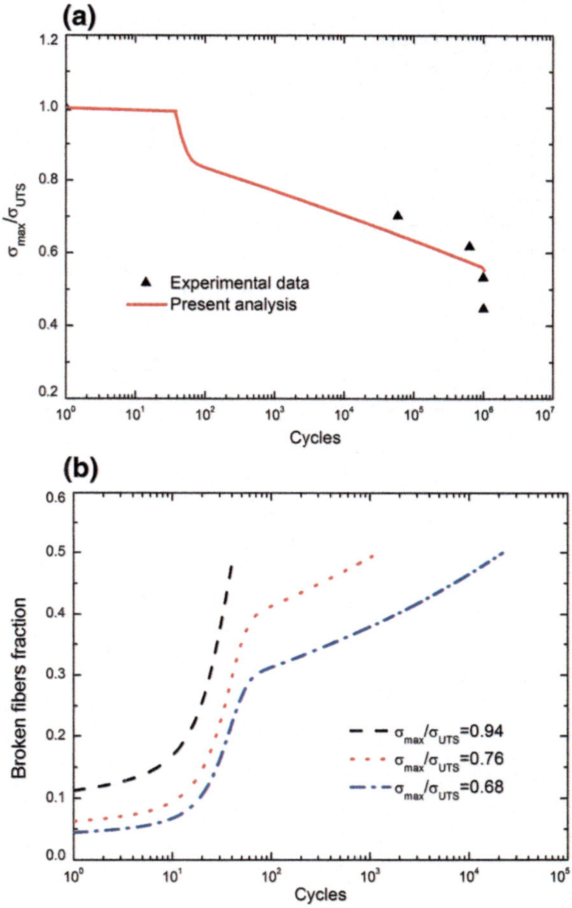


**Fig. 4.15** **a** Experimental and theoretical fatigue life *S–N* curves and **b** the broken fibers fraction versus the applied cycle number curves at the fatigue peak stresses of $0.94\sigma_{UTS}$, $0.76\sigma_{UTS}$, and $0.68\sigma_{UTS}$ of cross-ply [0/90/0/90/0/90/0/90/0] SiC/CAS composite at room temperature

### 4.3.2.3 Cross-Ply SiC/1723 at Room Temperature

Zawada et al. [38] investigated the tension–tension cyclic fatigue behavior of cross-ply $[0/90]_{3s}$ SiC/1723 composite at room temperature. The experimental and theoretical fatigue life *S–N* curves and the broken fibers fraction versus the applied cycle number curves of cross-ply $[0/90]_{3s}$ SiC/1723 composite at room temperature are shown in Fig. 4.17. At room temperature, the monotonic tensile strength of $[0/90]_{3s}$ SiC/1723 composite is 284 MPa, and the fatigue limit stress approaches to 65% of the tensile strength. When the fatigue peak stresses are 98, 91, and 84% of the tensile strength, the composite experienced 1053, 5619, and 26,380 applied cycles, respectively.

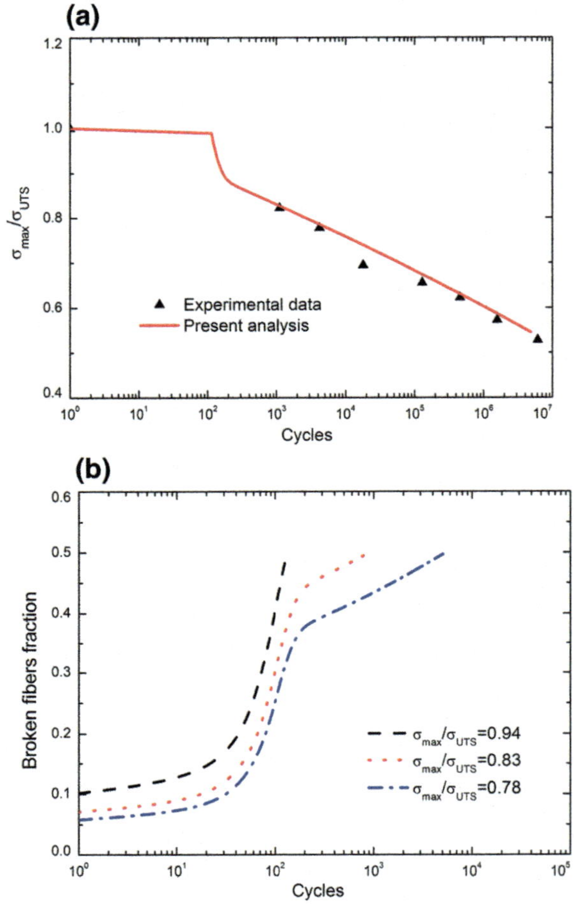


**Fig. 4.16** **a** Experimental and theoretical fatigue life *S–N* curves and **b** the broken fibers fraction versus the applied cycle number curves at the fatigue peak stresses of $0.94\sigma_{UTS}$, $0.83\sigma_{UTS}$, and $0.78\sigma_{UTS}$ of cross-ply $[0/90]_{4s}$ SiC/CAS composite at room temperature

#### 4.3.2.4 Cross-Ply SiC/BMAS at Room Temperature

Vanwijgenhoven et al. [39] investigated the tension–tension cyclic fatigue behavior of cross-ply $[0/90]_{3s}$ SiC/BMAS composite at room temperature. The experimental and theoretical fatigue life *S–N* curves and the broken fibers fraction versus the applied cycle number curves of cross-ply $[0/90]_{3s}$ SiC/BMAS composite at room temperature are shown in Fig. 4.18. At room temperature, the monotonic tensile strength of $[0/90]_{3s}$ SiC/BMAS composite is 408 MPa, and the fatigue limit stress approaches to 65% of the tensile strength. When the fatigue peak stresses are 93, 83, and 73% of the tensile strength, the composite experienced 132, 240, and 2826 applied cycles, respectively.

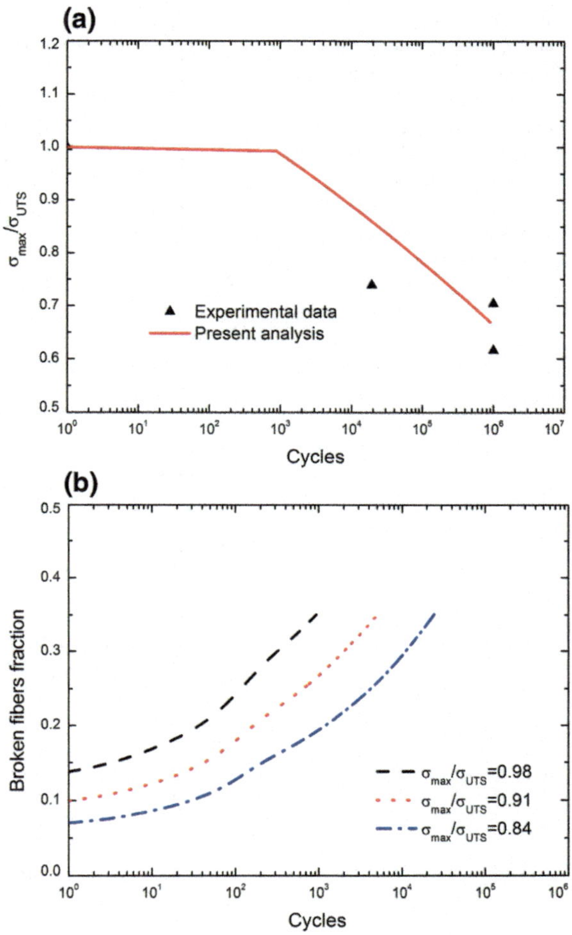


**Fig. 4.17** **a** Experimental and theoretical fatigue life *S–N* curves and **b** the broken fibers fraction versus the applied cycle number curves at the fatigue peak stresses of $0.98\sigma_{UTS}$, $0.91\sigma_{UTS}$, and $0.84\sigma_{UTS}$ of cross-ply $[0/90]_{3s}$ SiC/1723 composite at room temperature

#### 4.3.2.5 2D Woven C/SiC at Room Temperature and 550 °C in Air Condition

The experimental and theoretical fatigue life *S–N* curves and the broken fibers fraction versus the applied cycle number curves of 2D woven C/SiC composite at room temperature are shown in Fig. 4.19. At room temperature, the monotonic tensile strength of 2D woven C/SiC composite is 420 MPa, and the fatigue limit stress approaches to 83% of the tensile strength. When the fatigue peak stresses are 96, 91, 89, 86, 83 and 80% of the tensile strength, the composite experienced 2009, 28,237, 93,029, 550,740, 1,000,000, and 1,000,000 cycles, respectively.

The experimental and theoretical fatigue life *S–N* curves and the broken fibers fraction versus the applied cycle number curves of 2D woven C/SiC composite at 550 °C in air condition are shown in Fig. 4.20. At 550 °C in air condition, the monotonic tensile strength of 2D woven C/SiC composite is 487 MPa, and the

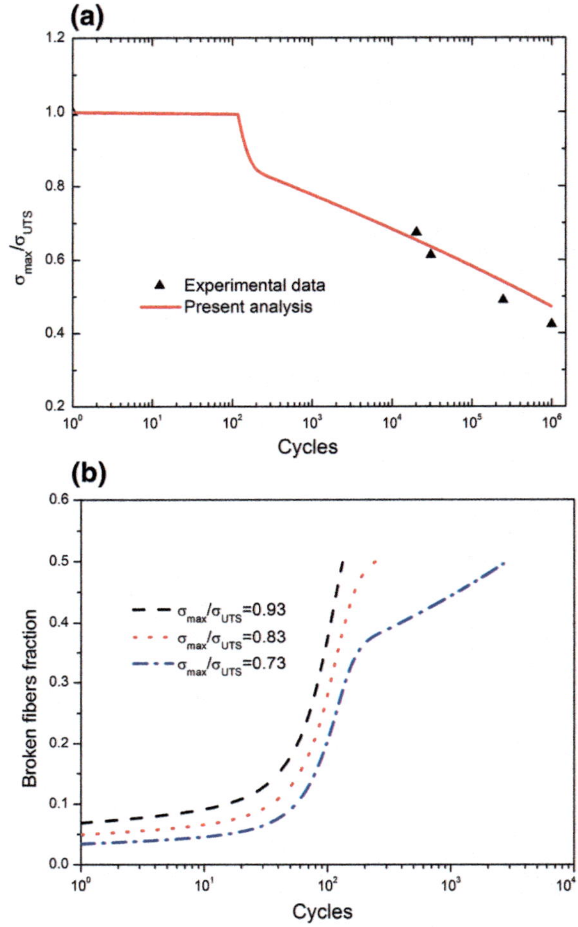


**Fig. 4.18** **a** Experimental and theoretical fatigue life *S–N* curves and **b** the broken fibers fraction versus the applied cycle number curves at the fatigue peak stresses of $0.93\sigma_{UTS}$, $0.83\sigma_{UTS}$, and $0.73\sigma_{UTS}$ of cross-ply $[0/90]_{3s}$ SiC/BMAS composite at room temperature

fatigue limit stress approaches to 30% of the tensile strength. When the fatigue peak stresses are 72, 56, 36, and 22% of the tensile strength, the composite experienced 3514, 24,840, 183,651, and 1,000,000 cycles, respectively.

#### 4.3.2.6 2D Woven SiC/SiC Composites

Room Temperature

Rouby and Reynaud [40] investigated the tension–tension cyclic fatigue behavior of 2D woven SiC/SiC composite at room temperature. The experimental and theoretical fatigue life *S–N* curves and the broken fibers fraction versus the applied cycle

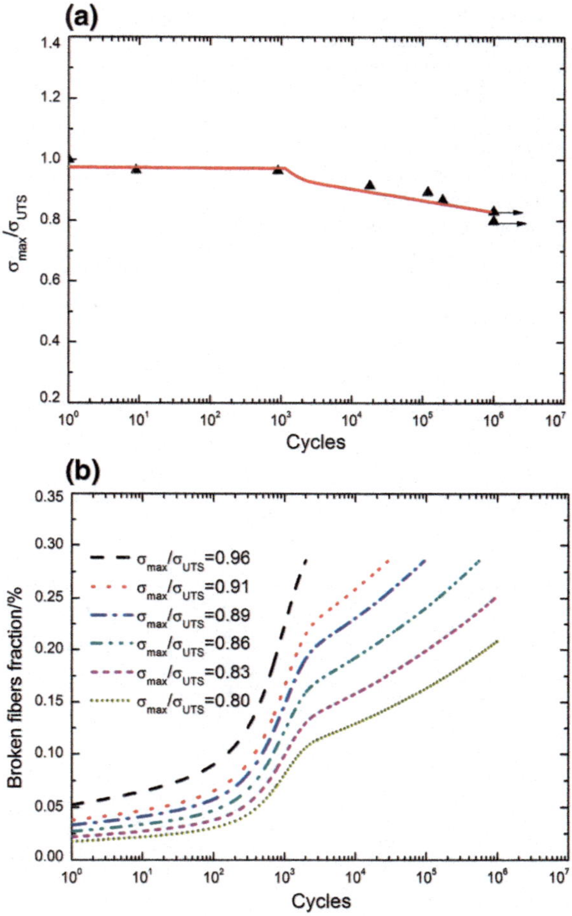


**Fig. 4.19** **a** Experimental and theoretical fatigue life *S–N* curves and **b** the broken fibers fraction versus the applied cycle number curves at the fatigue peak stresses of $0.96\sigma_{UTS}$, $0.91\sigma_{UTS}$, $0.89\sigma_{UTS}$, $0.86\sigma_{UTS}$, $0.83\sigma_{UTS}$, and $0.8\sigma_{UTS}$ of 2D woven C/SiC composite at room temperature

number curves of 2D woven SiC/SiC composite at room temperature are shown in Fig. 4.21. At room temperature, the monotonic tensile strength of 2D woven SiC/SiC composite is 170 MPa, and the fatigue limit stress approaches to 75% of the tensile strength. When the fatigue peak stresses are 95, 90, and 85% of the tensile strength, the composite experienced 150, 1532, and 22,443 cycles, respectively.

Mizuno et al. [12] investigated the tension–tension cyclic fatigue behavior of 2D woven SiC/SiC composite at room temperature. The experimental and theoretical fatigue life *S–N* curves and the broken fibers fraction versus the applied cycle number curves of 2D woven SiC/SiC composite at room temperature are shown in

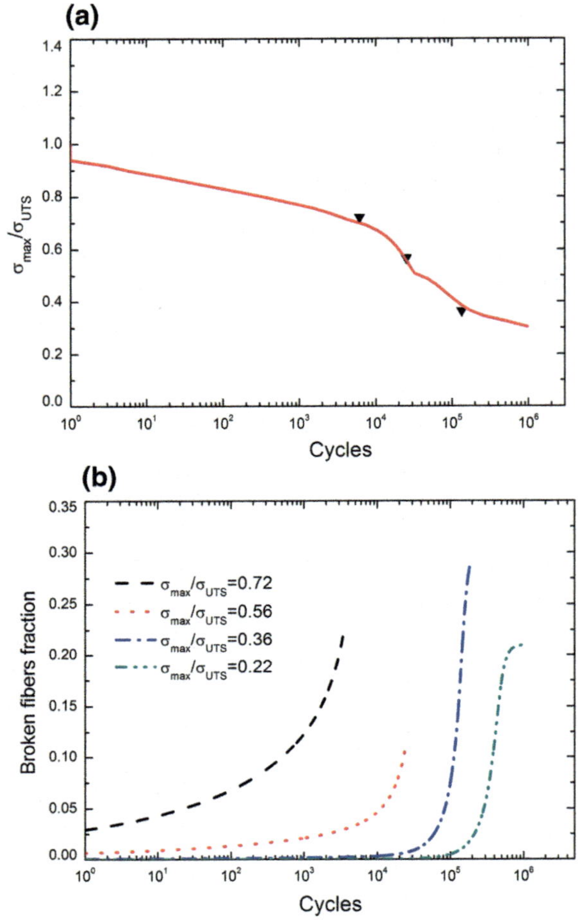


**Fig. 4.20** **a** Experimental and theoretical fatigue life *S–N* curves and **b** the broken fibers fraction versus the applied cycle number curves at the fatigue peak stresses of $0.72\sigma_{UTS}$, $0.56\sigma_{UTS}$, $0.36\sigma_{UTS}$, and $0.22\sigma_{UTS}$ of 2D woven C/SiC composite at 550 °C in air condition

Fig. 4.22. At room temperature, the monotonic tensile strength of 2D woven SiC/SiC composite is 209 MPa, and the fatigue limit stress approaches to 85% of the tensile strength. When the fatigue peak stresses are 95, 90, and 85% of the tensile strength, the composite experienced 216, 542, and 1,000,000 applied cycles, respectively.

750 °C in Moisture Condition

Mall [11] investigated the tension–tension cyclic fatigue behavior of 2D woven SiC/SiC composite under the 0% (dry) and 60% moisture content conditions at 750 °C. The experimental and theoretical fatigue life *S–N* curves and the broken fibers fraction versus the applied cycle number curves of 2D woven SiC/SiC composite at 750 °C under the 0% (dry) moisture content condition are shown in

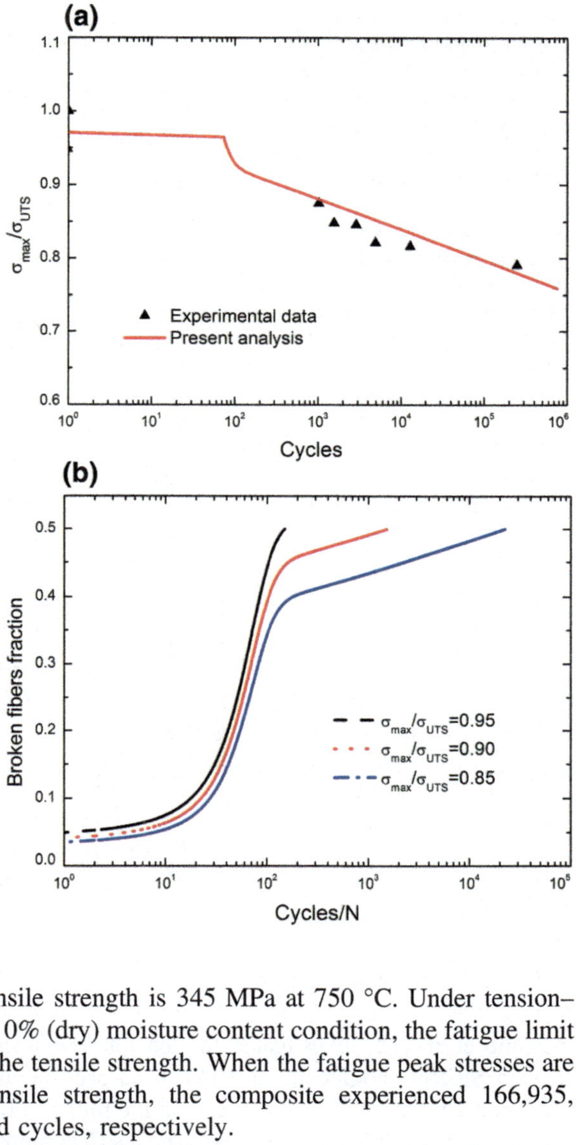


**Fig. 4.21** **a** Experimental and theoretical fatigue life *S–N* curves and **b** the broken fibers fraction versus the applied cycle number curves at the fatigue peak stresses of $0.95\sigma_{UTS}$, $0.9\sigma_{UTS}$, and $0.85\sigma_{UTS}$ of 2D woven SiC/SiC composite at room temperature. Reprinted with permission from Ref. [22]. Copyright 2016, Springer Science Business Media New York

Fig. 4.23. The monotonic tensile strength is 345 MPa at 750 °C. Under tension–tension fatigue loading in the 0% (dry) moisture content condition, the fatigue limit stress approaches to 67% of the tensile strength. When the fatigue peak stresses are 90, 80, and 70% of the tensile strength, the composite experienced 166,935, 421,262, and 871,439 applied cycles, respectively.

The experimental and theoretical fatigue life *S–N* curves and the broken fibers fraction versus the applied cycle number curves of 2D woven SiC/SiC composite at 750 °C under the 60% moisture content condition are shown in Fig. 4.24. Under tension–tension fatigue loading in the 60% moisture content condition, the fatigue limit stress approaches to 49% of the tensile strength. When the fatigue peak stresses are 90, 80, and 70% of the tensile strength, the composite experienced 23,665, 61,028, and 116,039 applied cycles, respectively.

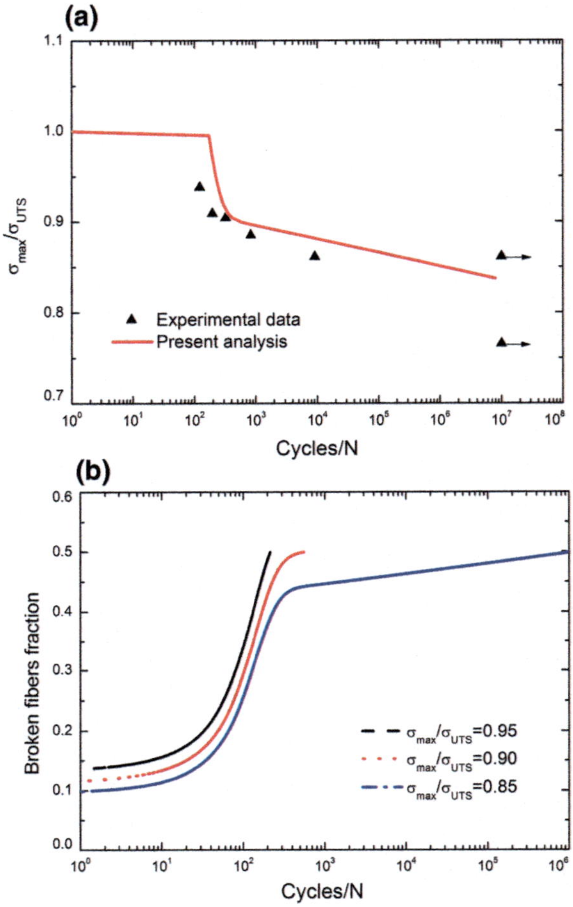


**Fig. 4.22** **a** Experimental and theoretical fatigue life *S–N* curves and **b** the broken fibers fraction versus the applied cycle number curves at the fatigue peak stresses of $0.95\sigma_{UTS}$, $0.9\sigma_{UTS}$, and $0.85\sigma_{UTS}$ of 2D woven SiC/SiC composite at room temperature. Reprinted with permission from Ref. [22]. Copyright 2016, Springer Science Business Media New York

1000 °C in Argon Condition

Mizuno et al. [12] investigated the tension–tension cyclic fatigue behavior of 2D woven SiC/SiC composite at 1000 °C in argon condition. The experimental and theoretical fatigue life *S–N* curves and the broken fibers fraction versus the applied cycle number curves of 2D woven SiC/SiC composite at 1000 °C in argon condition are shown in Fig. 4.25. The monotonic tensile strength is 251 MPa at 1000 °C, and the fatigue limit stress approaches to 30% of the tensile strength. When the fatigue peak stresses are 80, 70, and 60% of the tensile strength, the composite experienced 141, 1148, and 7726 applied cycles, respectively.

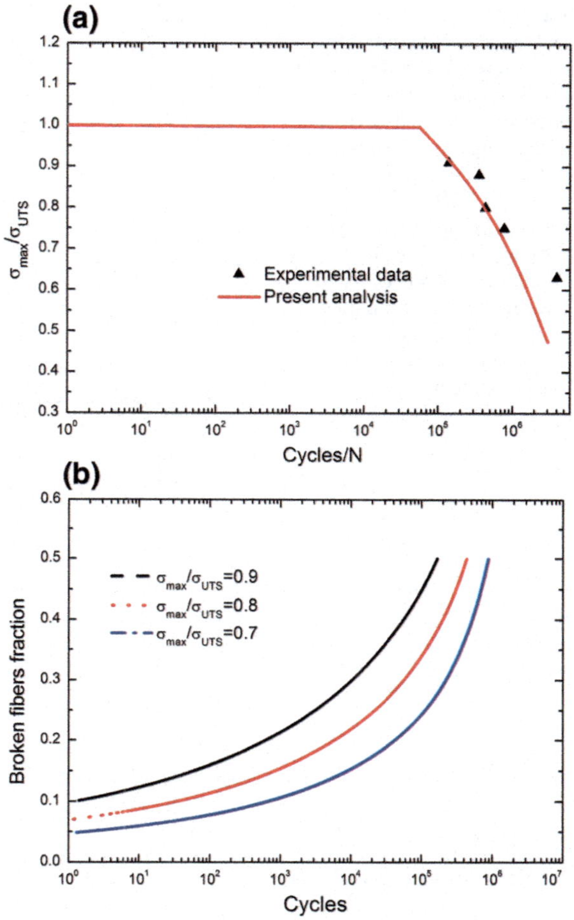


**Fig. 4.23** **a** Experimental and theoretical fatigue life *S–N* curves and **b** the broken fibers fraction versus the applied cycle number curves at the fatigue peak stresses of $0.9\sigma_{UTS}$, $0.8\sigma_{UTS}$, and $0.7\sigma_{UTS}$ of 2D woven SiC/SiC composite under the 0% (dry) moisture content conditions at 750 °C. Reprinted with permission from Ref. [22]. Copyright 2016, Springer Science Business Media New York

1000 °C in Air and Steam Condition

Michael [13] investigated the tension–tension cyclic fatigue behavior of 2D woven SiC/SiC composite at 1000 °C in air and steam conditions. The monotonic tensile strength is 114 MPa at 1000 °C. The experimental and theoretical fatigue life *S–N* curves and the broken fibers fraction versus the applied cycle number curves of 2D woven SiC/SiC composite at 1000 °C in air condition are shown in Fig. 4.26. The fatigue limit stress approaches to 20% of the tensile strength. When the fatigue peak stresses are 90, 80, and 70% of the tensile strength, the composite experienced 70,606, 103,781, and 146,108 applied cycles, respectively.

The experimental and theoretical fatigue life *S–N* curves and the broken fibers fraction versus the applied cycle number curves of 2D woven SiC/SiC composite at

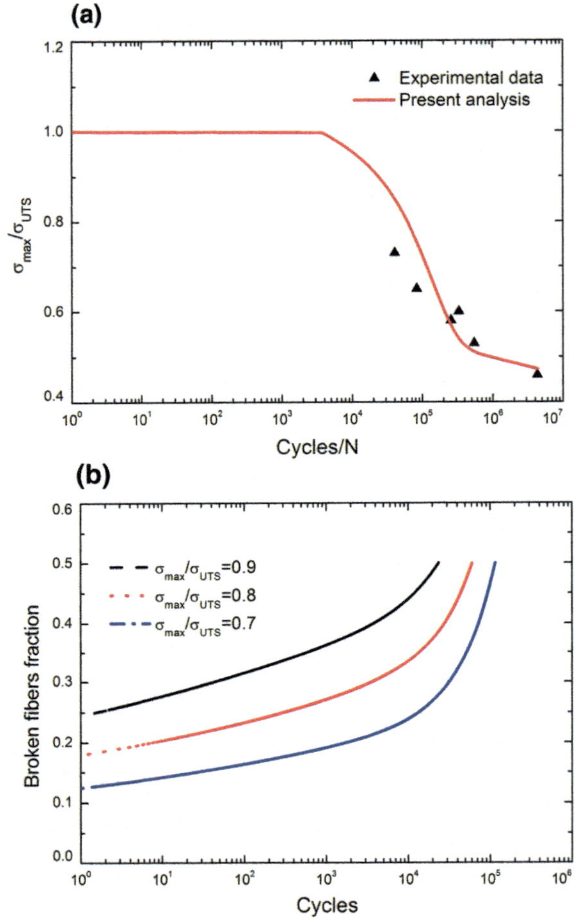


**Fig. 4.24** **a** Experimental and theoretical fatigue life *S–N* curves; and **b** the broken fibers fraction versus the applied cycle number curves at the fatigue peak stresses of $0.9\sigma_{UTS}$, $0.8\sigma_{UTS}$, and $0.7\sigma_{UTS}$ of 2D woven SiC/SiC composite under the 60% moisture content conditions at 750 °C. Reprinted with permission from Ref. [22]. Copyright 2016, Springer Science Business Media New York

1000 °C in steam condition are shown in Fig. 4.27. The fatigue limit stress approaches to 12% of the tensile strength. When the fatigue peak stresses are 95, 90, and 85% of the tensile strength, the composite experienced 36,780, 49,997, and 62,946 applied cycles, respectively.

### 1200 °C in Air and Steam Condition

Jacob [14] investigated the tension–tension cyclic fatigue behavior of 2D SiC/SiC composite at 1200 °C in air and steam conditions. The monotonic tensile strength is 306 MPa. The experimental and theoretical fatigue life *S–N* curves and the broken fibers fraction versus the applied cycle number curves of 2D woven SiC/SiC composite at 1200 °C in air condition are shown in Fig. 4.28. The fatigue limit stress approaches to 18% of the tensile strength. When the fatigue peak stresses are

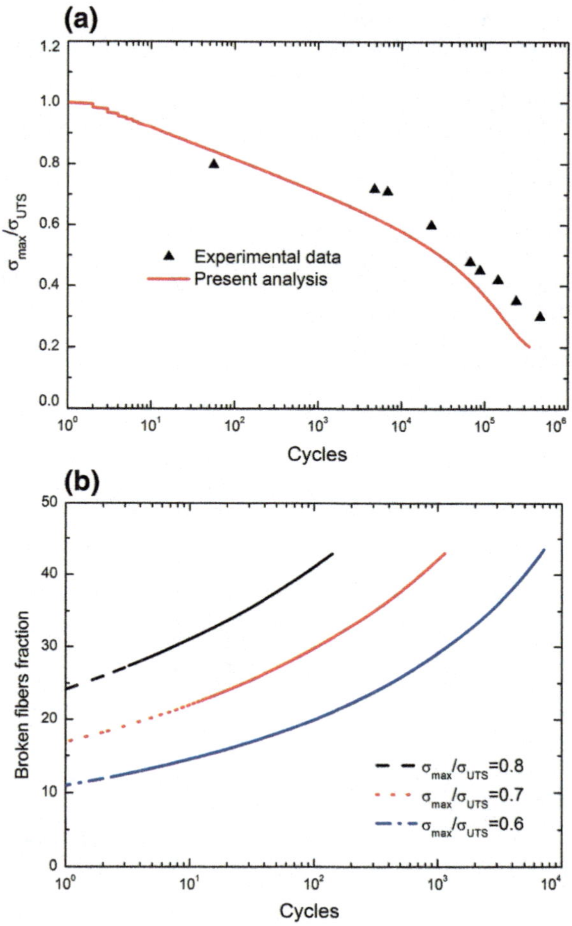


**Fig. 4.25** **a** Experimental and theoretical fatigue life *S–N* curves and **b** the broken fibers fraction versus the applied cycle number curves at the fatigue peak stresses of $0.8\sigma_{UTS}$, $0.7\sigma_{UTS}$, and $0.6\sigma_{UTS}$ of 2D woven SiC/SiC composite at 1000 °C in argon condition. Reprinted with permission from Ref. [22]. Copyright 2016, Springer Science Business Media New York

90, 80, and 70% of the tensile strength, the composite experienced 458, 1457, and 2918 applied cycles, respectively.

The experimental and theoretical fatigue life *S–N* curves and the broken fibers fraction versus the applied cycle number curves of 2D woven SiC/SiC composite at 1200 °C in steam condition are shown in Fig. 4.29. The fatigue limit stress approaches to 10% of the tensile strength. When the fatigue peak stresses are 80, 70, and 60% of the tensile strength, the composite experienced 123, 417, and 884 applied cycles, respectively.

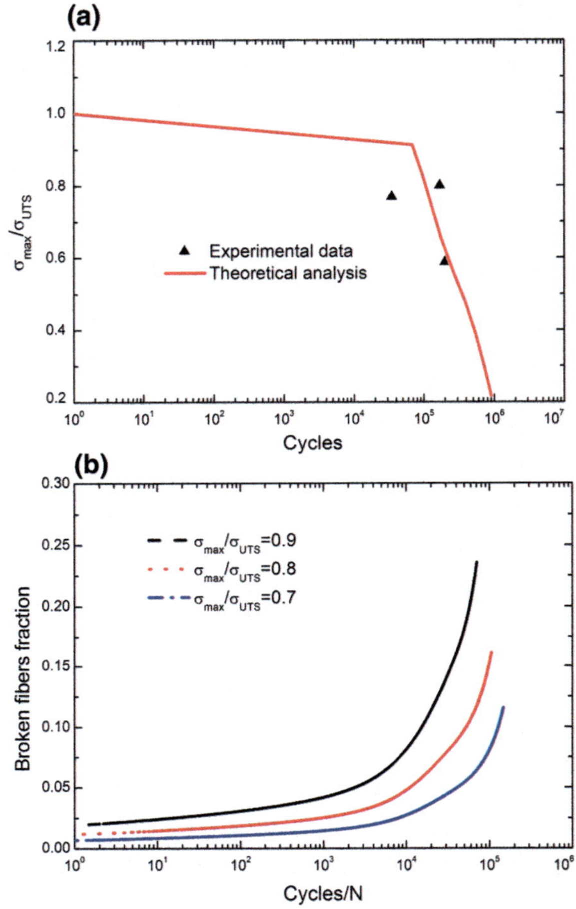


**Fig. 4.26** **a** Experimental and theoretical fatigue life *S–N* curves and **b** the broken fibers fraction versus the applied cycle number curves at the fatigue peak stresses of $0.9\sigma_{UTS}$, $0.8\sigma_{UTS}$, and $0.7\sigma_{UTS}$ of 2D woven SiC/SiC composite at 1000°C in air condition. Reprinted with permission from Ref. [22]. Copyright 2016, Springer Science Business Media New York

1300 °C in Air and Steam Conditions

Ruggles-Wrenn and Sharma [16] investigated the cyclic tension–tension fatigue behavior of 2D SiC/SiC composite at 1300 °C in air and steam conditions. The monotonic tensile strength is 241 MPa. The experimental and theoretical fatigue life *S–N* curves and the broken fibers fraction versus the applied cycle number curves of 2D woven SiC/SiC composite at 1300 °C in air condition are shown in Fig. 4.30. The fatigue limit stress approaches to 42% of the tensile strength. When the fatigue peak stresses are 90, 80, and 70% of the tensile strength, the composite experienced 18,102, 33,552, and 56,802 applied cycles, respectively.

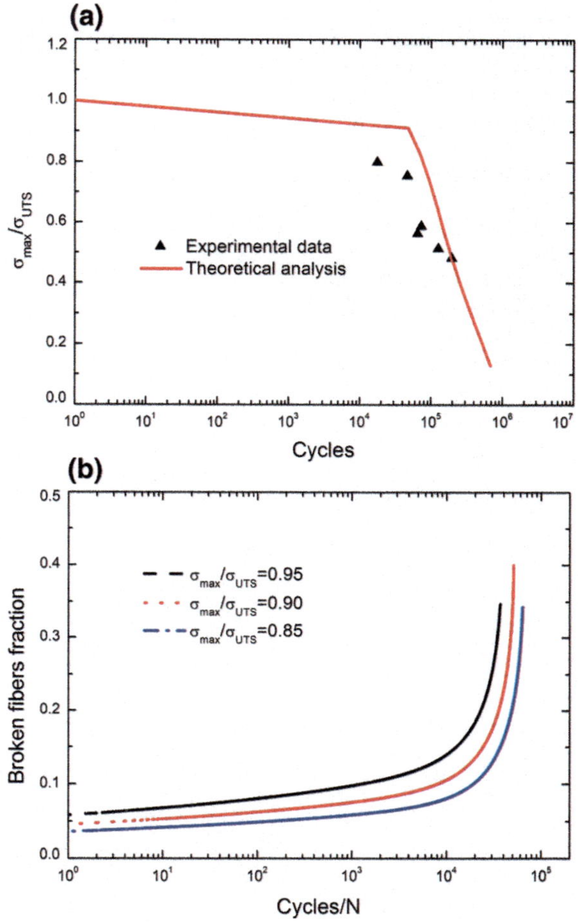


**Fig. 4.27** **a** Experimental and theoretical fatigue life *S–N* curves and **b** the broken fibers fraction versus the applied cycle number curves at the fatigue peak stresses of $0.95\sigma_{UTS}$, $0.90\sigma_{UTS}$, and $0.85\sigma_{UTS}$ of 2D woven SiC/SiC composite at 1000 °C in steam condition. Reprinted with permission from Ref. [22]. Copyright 2016, Springer Science Business Media New York

The experimental and theoretical fatigue life *S–N* curves and the broken fibers fraction versus the applied cycle number curves of 2D woven SiC/SiC composite at 1300 °C in steam condition are shown in Fig. 4.31. The fatigue limit stress approaches to 40% of the tensile strength. When the fatigue peak stresses are 90, 80, and 70% of the tensile strength, the composite experienced 493, 13,646, and 53,023 applied cycles, respectively.

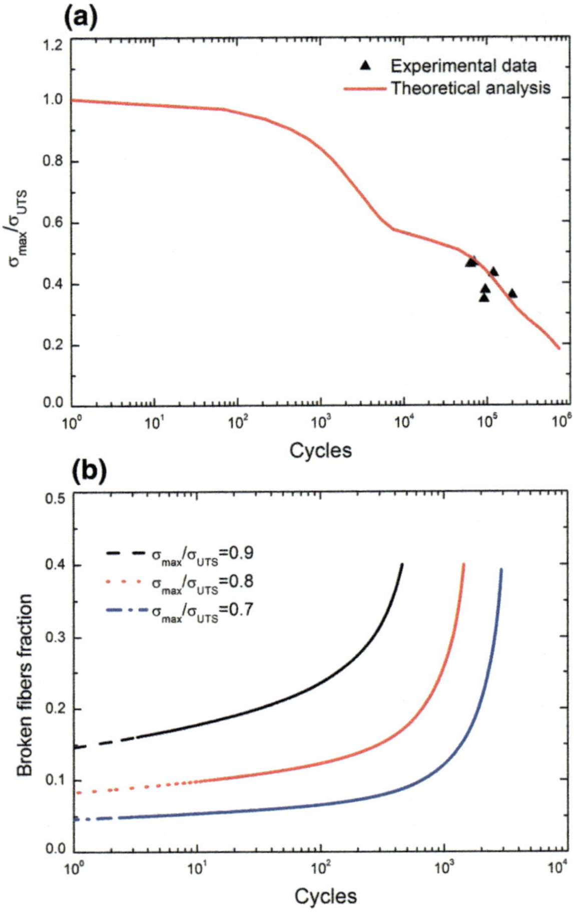


**Fig. 4.28** **a** Experimental and theoretical fatigue life *S–N* curves and **b** the broken fibers fraction versus the applied cycle number curves at the fatigue peak stresses of $0.9\sigma_{UTS}$, $0.8\sigma_{UTS}$, and $0.7\sigma_{UTS}$ of 2D woven SiC/SiC composite at 1200 °C in air condition. Reprinted with permission from Ref. [22]. Copyright 2016, Springer Science Business Media New York

#### 4.3.2.7 2.5D Woven C/SiC Composite at Room Temperature and 800 °C in Air Condition

The experimental and theoretical fatigue life *S–N* curves and the broken fibers fraction versus the applied cycle number curves of 2.5D woven C/SiC composite at room temperature are shown in Fig. 4.32. The monotonic tensile strength of 2.5D C/SiC composite is 212 MPa at room temperature. The fatigue limit stress

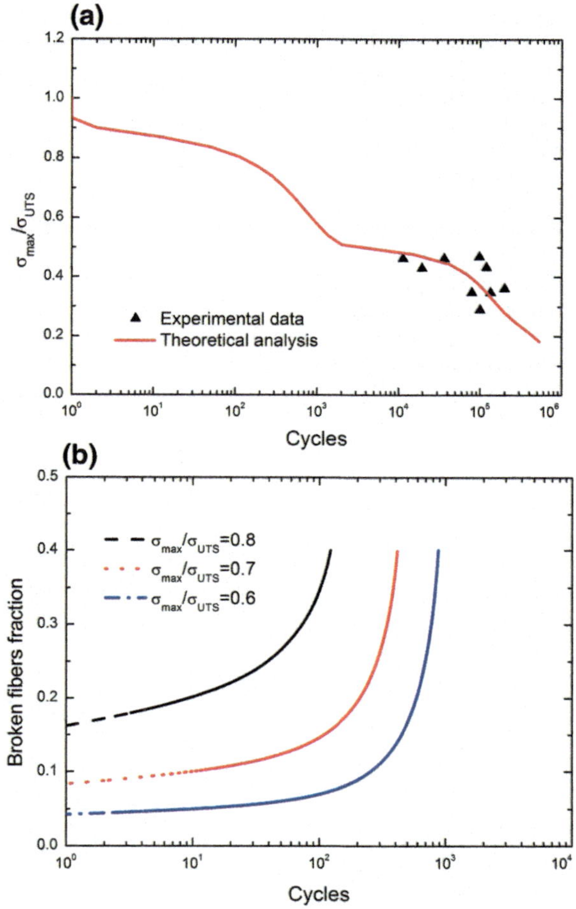


**Fig. 4.29** **a** Experimental and theoretical fatigue life *S–N* curves and **b** the broken fibers fraction versus the applied cycle number curves at the fatigue peak stresses of $0.8\sigma_{UTS}$, $0.7\sigma_{UTS}$, and $0.6\sigma_{UTS}$ of 2D woven SiC/SiC composite at 1200 °C in steam condition. Reprinted with permission from Ref. [22]. Copyright 2016, Springer Science Business Media New York

approaches to 85% of the tensile strength. When the fatigue peak stresses are 99, 96, 94, 89, and 85% of the tensile strength, the composite experienced 6182, 9084, 12,173, 77,354, and 1,000,000 cycles, respectively.

The experimental and theoretical fatigue life *S–N* curves and the broken fibers fraction versus the applied cycle number curves of 2.5D woven C/SiC composite at 800 °C in air condition are shown in Fig. 4.33. The monotonic tensile strength of

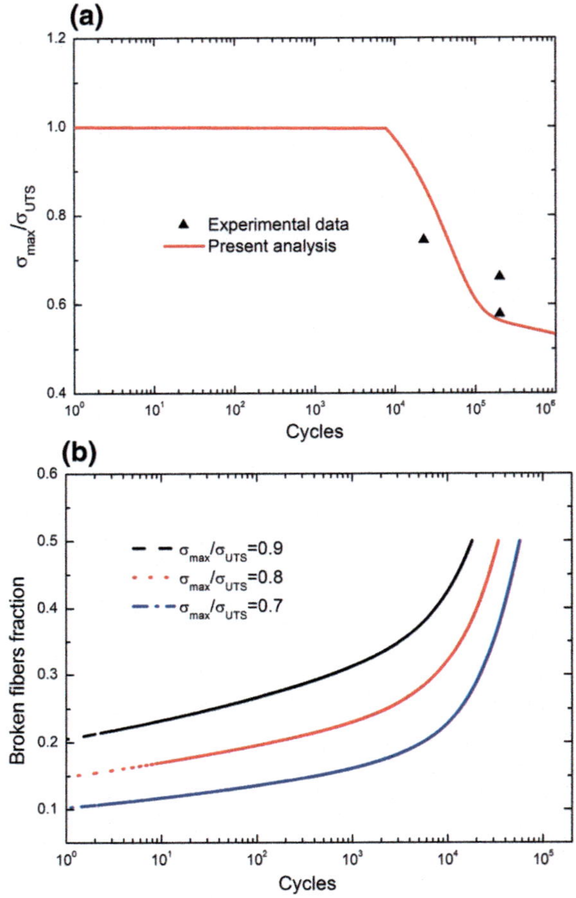


**Fig. 4.30** **a** Experimental and theoretical fatigue life *S–N* curves and **b** the broken fibers fraction versus the applied cycle number curves at the fatigue peak stresses of $0.9\sigma_{UTS}$, $0.8\sigma_{UTS}$, and $0.7\sigma_{UTS}$ of 2D woven SiC/SiC composite at 1300 °C in air condition. Reprinted with permission from Ref. [22]. Copyright 2016, Springer Science Business Media New York

2.5D C/SiC composite is 280 MPa at 800 °C in air condition. The fatigue limit stress approaches to 28% of the tensile strength. When the fatigue peak stresses are 80, 70, 60, and 50% of the tensile strength, the composite experienced 736, 3525, 8314, and 37,308 cycles, respectively.

#### 4.3.2.8 3D Braided C/SiC Composite

The experimental and theoretical fatigue life *S–N* curves and the broken fibers fraction versus the applied cycle number curves of 3D braided C/SiC composite at room temperature are shown in Fig. 4.34. The monotonic tensile strength of 3D

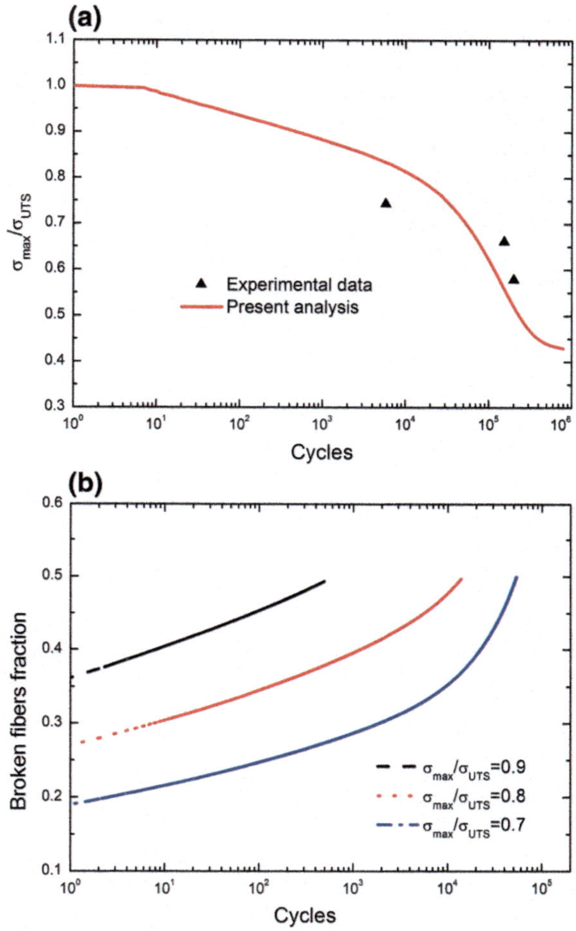


**Fig. 4.31** **a** Experimental and theoretical fatigue life *S–N* curves and **b** the broken fibers fraction versus the applied cycle number curves at the fatigue peak stresses of $0.9\sigma_{UTS}$, $0.8\sigma_{UTS}$, and $0.7\sigma_{UTS}$ of 2D woven SiC/SiC composite at 1300 °C in steam condition. Reprinted with permission from Ref. [22]. Copyright 2016, Springer Science Business Media New York

braided C/SiC composite is 276 MPa at room temperature. The fatigue limit stress approaches to 86% of the tensile strength. When the fatigue peak stresses are 94, 90, 89, 87, and 83% of the tensile strength, the composite experienced 349, 17,233, 45,914, 323,787, and 1,000,000 cycles, respectively.

The experimental and theoretical fatigue life *S–N* curves and the broken fibers fraction versus the applied cycle number curves of 3D braided C/SiC composite at

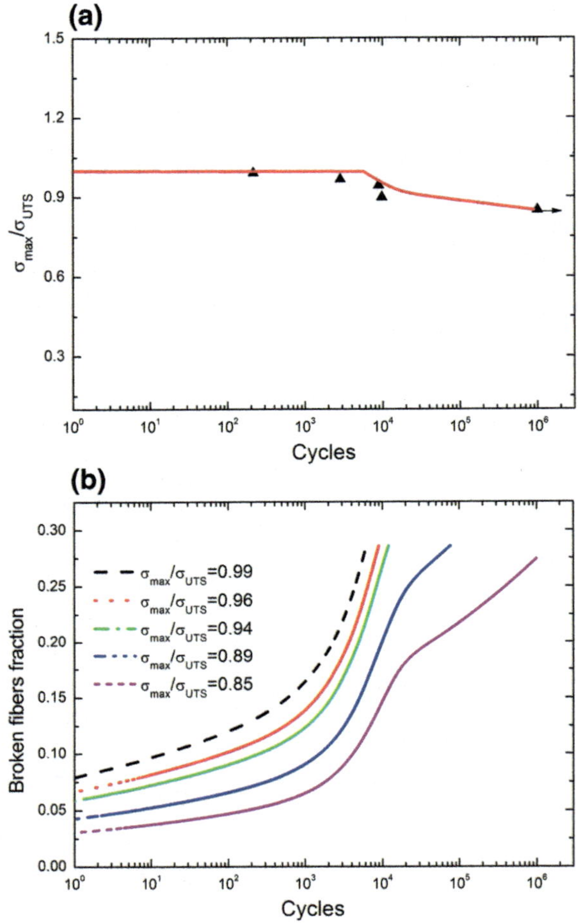


**Fig. 4.32** **a** Experimental and theoretical fatigue life *S–N* curves and **b** the broken fibers fraction versus the applied cycle number curves at the fatigue peak stresses of $0.99\sigma_{UTS}$, $0.96\sigma_{UTS}$, $0.94\sigma_{UTS}$, $0.89\sigma_{UTS}$, and $0.85\sigma_{UTS}$ of 2.5D woven C/SiC composite at room temperature

1100 °C in $10^{-4}$ Pa vacuum condition are shown in Fig. 4.35. The monotonic tensile strength of 3D braided C/SiC composite is 360 MPa at 1100 °C. The fatigue limit stress approaches to 95% of the tensile strength. When the fatigue peak stresses are 97 and 88% of the tensile strength, the composite experienced 171,281 and 1,000,000 cycles, respectively.

The experimental and theoretical fatigue life *S–N* curves and the broken fibers fraction versus the applied cycle number curves of 3D braided C/SiC composite at

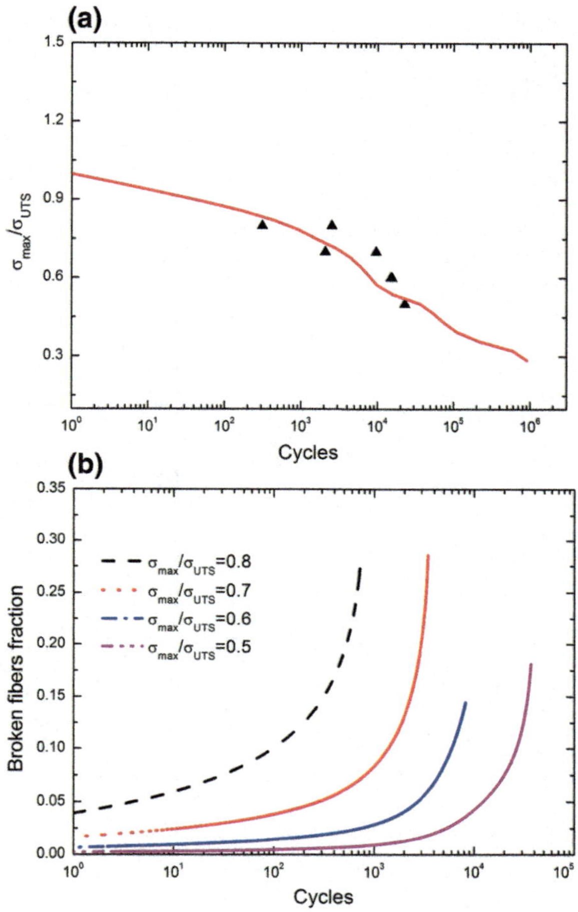


**Fig. 4.33** **a** Experimental and theoretical fatigue life *S–N* curves and **b** the broken fibers fraction versus the applied cycle number curves at the fatigue peak stresses of $0.8\sigma_{UTS}$, $0.7\sigma_{UTS}$, $0.6\sigma_{UTS}$, and $0.5\sigma_{UTS}$ of 2.5D woven C/SiC composite at 800 °C in air condition

1300 °C in $10^{-4}$ Pa vacuum condition are shown in Fig. 4.36. The monotonic tensile strength of 3D braided C/SiC composite is 304 MPa at 1300 °C. The fatigue limit stress approaches to 93% of the tensile strength. When the fatigue peak stresses are 99, 98, 93, and 83% of the tensile strength, the composite experienced 14,008, 17,740, 870,731, and 1,000,000 cycles, respectively.

The experimental and theoretical fatigue life *S–N* curves and the broken fibers fraction versus the applied cycle number curves of 3D braided C/SiC composite at

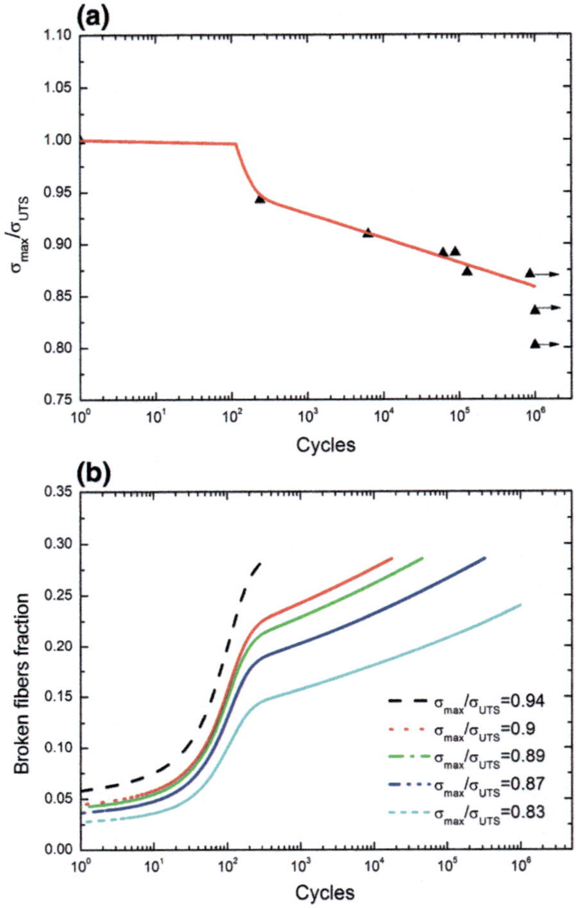


**Fig. 4.34** **a** The experimental and theoretical fatigue life *S–N* curves; and **b** the broken fibers fraction versus the applied cycle number curves at the fatigue peak stresses of $0.94\sigma_{UTS}$, $0.9\sigma_{UTS}$, $0.89\sigma_{UTS}$, $0.87\sigma_{UTS}$ and $0.83\sigma_{UTS}$ of 3D braided C/SiC composite at room temperature

1500 °C in $10^{-4}$ Pa vacuum condition are shown in Fig. 4.37. The monotonic tensile strength of 3D braided C/SiC composite is 261 MPa at 1500 °C. The fatigue limit stress approaches to 90% of the tensile strength. When the fatigue peak stresses are 98, 96, 95, 92, and 90% of the tensile strength, the composite experienced 2493, 11,799, 25,608, 259,213, and 1,000,000 cycles, respectively.

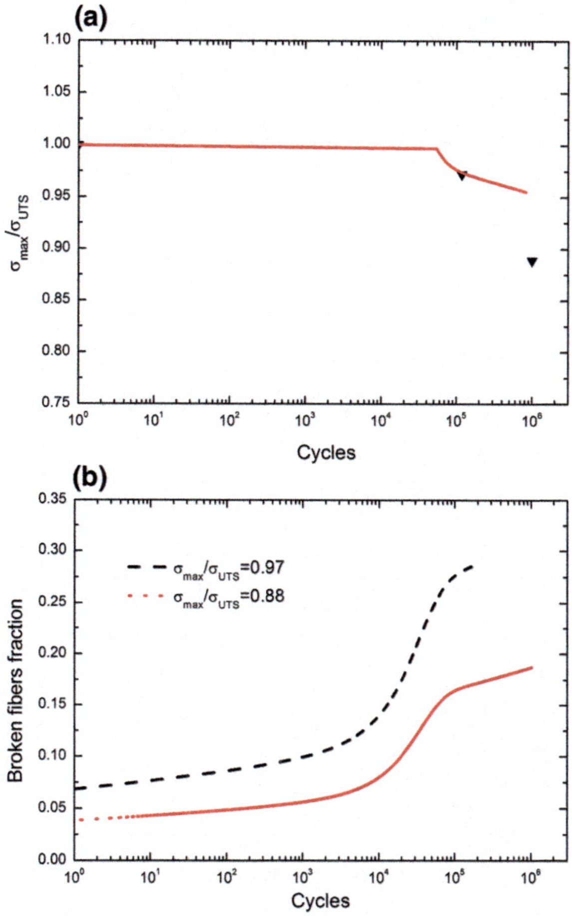


**Fig. 4.35** **a** Experimental and theoretical fatigue life *S*–*N* curves and **b** the broken fibers fraction versus the applied cycle number curves at the fatigue peak stresses of $0.97\sigma_{UTS}$ and $0.88\sigma_{UTS}$ of 3D braided C/SiC composite at 1100 °C in vacuum condition

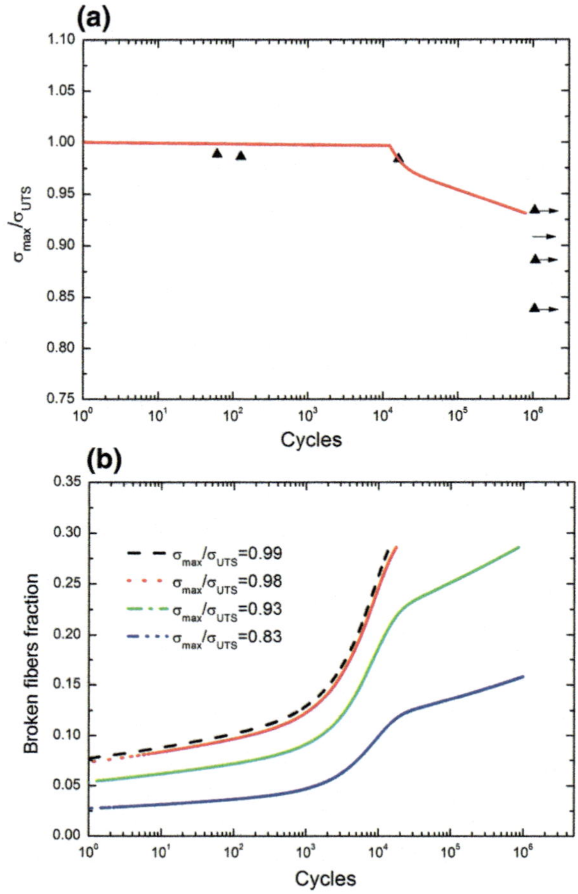


**Fig. 4.36** **a** Experimental and theoretical fatigue life *S–N* curves and **b** the broken fibers fraction versus the applied cycle number curves at the fatigue peak stresses of $0.97\sigma_{UTS}$ and $0.88\sigma_{UTS}$ of 3D braided C/SiC composite at 1300 °C in vacuum condition

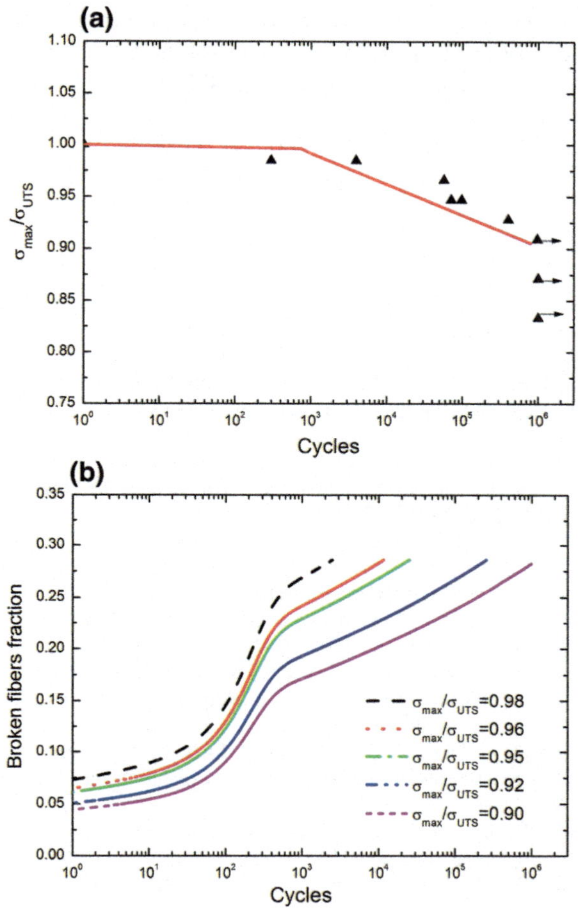


**Fig. 4.37** **a** Experimental and theoretical fatigue life *S–N* curves and **b** the broken fibers fraction versus the applied cycle number curves at the fatigue peak stresses of $0.98\sigma_{UTS}$, $0.96\sigma_{UTS}$, $0.95\sigma_{UTS}$, $0.92\sigma_{UTS}$, and $0.9\sigma_{UTS}$ of 3D braided C/SiC composite at 1500 °C in vacuum condition

## 4.4 Conclusions

In this chapter, the fatigue lifetime of fiber-reinforced CMCs at room and elevated temperatures has been predicted for different testing conditions and fiber preforms. The relationships among the broken fibers fraction, applied cycle numbers, and the fatigue peak stress have been established. The fibers broken fraction in different damage regions have been determined, i.e., the interface wear region, interface oxidation region, interface debonded region, and interface bonded region. When the fibers broken fraction approaches to the critical value, the composites fatigue fracture. The fatigue limit stress and fatigue life *S–N* curve of fiber-reinforced

CMCs with different fibers preform have been predicted, i.e., the unidirectional C/SiC, SiC/CAS, SiC/1723 and SiC/$Si_3N_4$, cross-ply C/SiC, SiC/CAS, SiC/1723 and SiC/BMAS, 2D woven C/SiC and SiC/SiC, 2/5D woven C/SiC, and 3D braided C/SiC composites.

## References


1. Li LB (2010) Fatigue damage models and life prediction of long-fiber-reinforced ceramic matrix composites. Ph.D. thesis, Nanjing University of Aeronautics and Astronautics, Nanjing, China
2. Li LB (2013) Fatigue hysteresis behavior of cross-ply C/SiC ceramic matrix composites at room and elevated temperatures. Mater Sci Eng A 586:160–170. https://doi.org/10.1016/j.msea.2013.08.017
3. Shuler SF, Holmes JW, Wu X (1993) Influence of loading frequency on the room-temperature fatigue of a carbon-fiber/SiC-matrix composite. J Am Ceram Soc 76:2327–2336. https://doi.org/10.1111/j.1151-2916.1993.tb07772.x
4. Lu MP, Tong XY, Ren SH, Yao LJ (2014) Fatigue behavior and life prediction of 2D C/SiC under room temperature. Adv Aeronaut Sci Eng 5:104–108
5. Min JB, Xue D, Shi Y (2014) Micromechanics modeling for fatigue damage analysis designed for fabric reinforced ceramic matrix composites. Compos Struct 111:213–223. https://doi.org/10.1016/j.compstruct.2013.12.025
6. Mall S, Engesser JM (2006) Effects of frequency on fatigue behavior of CVI C/SiC at elevated temperature. Compos Sci Technol 66:863–874. https://doi.org/10.1016/j.compscitech.2005.06.020
7. Cheng QY, Tong XY, Zheng X, Zhou J, Yao LJ, Li B (2010) Experimental investigation on the fatigue characteristics about high temperature of plain-woven C/SiC composite. J Mech Strength 32:819–823
8. Du SM, Qiao SR, Ji GC, Han D (2002) Tension-tension fatigue behavior of 3D-C/SiC composite at room temperature and 1300 °C. Mater Eng 9:22–25
9. Du SM, Qiao SR (2011) Tension-tension fatigue behavior of 3D-C/SiC composite at 1500 °C. Mater Eng 5:34–37
10. Reynaud P (1996) Cyclic fatigue of ceramic-matrix composites at ambient and elevated temperatures. Compos Sci Technol 56:809–814. https://doi.org/10.1016/0266-3538(96)00025-5
11. Mall S (2005) Effects of moisture on fatigue behavior of SiC/SiC composite at elevated temperature. Mater Sci Eng A 412:165–170. https://doi.org/10.1016/j.msea.2005.08.040
12. Mizuno M, Zhu SJ, Nagano Y, Sakaida Y, Kagawa Y, Watanabe M (1996) Cyclic-fatigue behavior of SiC/SiC composites at room and high temperatures. J Am Ceram Soc 79:3065–3018. https://doi.org/10.1111/j.1151-2916.1996.tb08078.x
13. Michael WK (2010) Fatigue behavior of a SiC/SiC composite at 1000 °C in air and steam. AFIT/GAE/ENY/10-D01
14. Jacob D (2010) Fatigue behavior of an advanced SiC/SiC composite with an oxidation inhibited matrix at 1200 °C in air and in steam. AFIT/GEA/ENY/10-M07
15. Zhu SJ, Mizuno M, Nagano Y, Cao JW, Kagawa Y, Kaya H (1998) Creep and fatigue behavior in an enhanced SiC/SiC composite at high temperature. J Am Ceram Soc 81:2269–2277. https://doi.org/10.1111/j.1151-2916.1998.tb02621.x
16. Ruggles-Wrenn MB, Sharma V (2011) Effects of steam environment on fatigue behavior of two SiC/[SiC + $Si_3N_4$] ceramic composites at 1300 °C. Appl Compos Mater 18(5):385–396. https://doi.org/10.1007/s10443-010-9163-x

17. Shi DQ, Jing X, Yang XG (2015) Low cycle fatigue behavior of a 3D braided KD-I fiber reinforced ceramic matrix composite for coated and uncoated specimens at 1100 °C and 1300 °C. Mater Sci Eng A 631:38–44. https://doi.org/10.1016/j.msea.2015.01.078
18. Li LB (2016) Fatigue hysteresis of carbon fiber-reinforced ceramic-matrix composites at room and elevated temperatures. Appl Compos Mater 23:1–27. https://doi.org/10.1007/s10443-015-9448-1
19. Li LB (2014) Modeling fatigue hysteresis behavior of unidirectional C/SiC ceramic-matrix composite. Compos B 66:466–474. https://doi.org/10.1016/j.compositesb.2014.06.014
20. Li LB (2016) Fatigue life prediction of carbon fiber-reinforced ceramic-matrix composites at room and elevated temperatures. Part I: experimental analysis. Appl Compos Mater 23:101–117. https://doi.org/10.1007/s10443-015-9446-3
21. Li LB (2016) Effects of temperature, oxidation and fiber preforms on fatigue life of carbon fiber-reinforced ceramic-matrix composites. Appl Compos Mater 23:799–819. https://doi.org/10.1007/s10443-016-9486-3
22. Li LB (2016) Comparisons of fatigue life between C/SiC and SiC/SiC ceramic-matrix composites at room and elevated temperatures. Appl Compos Mater 23:913–952. https://doi.org/10.1007/s10443-016-9492-5
23. Li LB (2016) Fatigue life prediction of fiber-reinforced ceramic-matrix composites with different fiber preforms at room and elevated temperatures. Materials 9:207
24. Li LB, Reynaud P, Fantozzi G (2017) Tension-tension fatigue behavior of unidirectional SiC/$Si_3N_4$ composite with strong and weak interface bonding at room temperature. Ceram Int 43:8769–8777. https://doi.org/10.1016/j.ceramint.2017.03.211
25. Li LB (2015) Fatigue life prediction of carbon fiber-reinforced ceramic-matrix composites at room and elevated temperatures. Part II: experimental comparisons. Appl Compos Mater 22:961–972. https://doi.org/10.1007/s10443-015-9445-4
26. Evans AG, Zok FW, McMeeking RM (1995) Fatigue of ceramic matrix composites. Acta Metall Mater 43(3):859–875. https://doi.org/10.1016/0956-7151(94)00304-Z
27. Lee SS, Stinchcomb WW (1994) Damage mechanisms of cross-ply Nicalon/CAS-II laminate under cyclic tension. Ceram Eng Sci Proc 15(4):40–48
28. Curtin WA, Ahn BK, Takeda N (1998) Modeling brittle and tough stress-strain behavior in unidirectional ceramic matrix composites. Acta Mater 46(10):3409–3420. https://doi.org/10.1016/S1359-6454(98)00041-X
29. Lamouroux F, Camus G, Thebault J (1994) Kinetics and mechanisms of oxidation of 2D woven C/SiC composites: I, experimental approach. J Am Ceram Soc 77(8):2049–2057. https://doi.org/10.1111/j.1151-2916.1994.tb07096.x
30. Halbig MC, McGuffin-Cawley JD, Eckel AJ, Brewer DN (2008) Oxidation kinetics and stress effects for the oxidation of continuous carbon fibers within a microcracked C/SiC ceramic matrix composite. J Am Ceram Soc 91(2):519–526. https://doi.org/10.1111/j.1551-2916.2007.02170.x
31. Filipuzzi L, Naslain R (1994) Oxidation mechanisms and kinetics of 1D-SiC/C/SiC composite materials: II, Modelling. J Am Ceram Soc 77(2):467–480. https://doi.org/10.1111/j.1151-2916.1994.tb07016.x
32. Naslain R (2004) Design, preparation and properties of non-oxide CMCs for application in engines and nuclear reactors: an overview. Compos Sci Technol 64(2):155–170. https://doi.org/10.1016/S0266-3538(03)00230-6
33. Lara-Curzio E (1999) Analysis of oxidation-assisted stress-rupture of continuous fiber-reinforced ceramic matrix composites at intermediate temperatures. Compos A 30(4):549–554. https://doi.org/10.1016/S1359-835X(98)00148-1
34. Casas L, Martinez-Esnaola JM (2003) Modelling the effect of oxidation on the creep behavior of fiber-reinforced ceramic matrix composites. Acta Mater 51(13):3745–3757. https://doi.org/10.1016/S1359-6454(03)00189-7
35. Opalski FA (1992) Fatigue behavior of a cross-ply ceramic matrix composite under tension-tension and tension-compression loading. M.S. thesis, AFIT/GAE/ENY/92D, Air Force Institute of Technology, Wright-Patterson AFB, Ohio

36. Karandikar PG (1992) Evolution of damage and mechanical response in ceramic matrix composites under monotonic and fatigue loading: experimental characterization and modeling. University of Delaware, PhD
37. Kim J, Liaw PK (2005) Characterization of fatigue damage modes in Nicalon/Calcium Aluminosilicate composites. J Eng Mater Technol 127(1):8–15. https://doi.org/10.1115/1.1836766
38. Zawada LP, Butkus LM, Hartman GA (1991) Tensile and fatigue behavior of silicon carbide fiber-reinforced aluminosilicate glass. J Am Ceram Soc 74(11):2851–2858. https://doi.org/10.1111/j.1151-2916.1991.tb06854.x
39. Vanwijgenhoven E, Weavers M, Biest O (1999) Influence of the laminate lay-up on the fatigue behavior of SiC-fiber/BMAS-matrix composites. Compos A 30(5):623–635. https://doi.org/10.1016/S1359-835X(98)00177-8
40. Rouby D, Reynaud P (1993) Fatigue behavior related to interface modification during load cycling in ceramic-matrix fiber composites. Compos Sci Technol 48(1–4):109–118. https://doi.org/10.1016/0266-3538(93)90126-2

Printed by Printforce, the Netherlands